AF508661

Latest Advances in Nanoplasmonics and Use of New Tools for Plasmonic Characterization

Latest Advances in Nanoplasmonics and Use of New Tools for Plasmonic Characterization

Editor

Grégory Barbillon

MDPI • Basel • Beijing • Wuhan • Barcelona • Belgrade • Manchester • Tokyo • Cluj • Tianjin

Editor
Grégory Barbillon
EPF-Ecole d'Ingénieurs (Graduate School of Engineering)
France

Editorial Office
MDPI
St. Alban-Anlage 66
4052 Basel, Switzerland

This is a reprint of articles from the Special Issue published online in the open access journal *Photonics* (ISSN 2304-6732) (available at: https://www.mdpi.com/journal/photonics/special_issues/Nanoplasmonics).

For citation purposes, cite each article independently as indicated on the article page online and as indicated below:

LastName, A.A.; LastName, B.B.; LastName, C.C. Article Title. *Journal Name* **Year**, *Volume Number*, Page Range.

ISBN 978-3-0365-3703-0 (Hbk)
ISBN 978-3-0365-3704-7 (PDF)

Contents

About the Editor

Grégory Barbillon completed his Ph.D. in Physics (2007) with greatest distinction at the University of Technology of Troyes (France). He then obtained his Habilitation (HDR) in Physics (2013) at the University of Paris Sud (Orsay, France). He has been a Professor of Physics at the Graduate School of Engineering "EPF-Ecole d'Ingénieurs" (Cachan, France) since his appointment in September 2017. His research interests focus on Plasmonics, Nano-Optics, Non-Linear Optics, Biosensing, Optical Sensing, Condensed Matter Physics, Nanophotonics, Nanotechnology, Nanomaterials, Surface Enhanced Spectroscopies, Sum Frequency Generation Spectroscopy, Materials Chemistry, Physical Chemistry, and Fluorescence.

Editorial

Latest Advances in Nanoplasmonics and Use of New Tools for Plasmonic Characterization

Grégory Barbillon

EPF-Ecole d'Ingénieurs, 55 Avenue du Président Wilson, 94230 Cachan, France; gregory.barbillon@epf.fr

Nanoplasmonics is a research topic that takes advantage of the light coupling to electrons in metals, and can break the diffraction limit for light confinement into subwavelength zones allowing strong field enhancements [1–4]. In the past two decades, a very significant explosion of this research topic and its applications has occurred. The applications cover a great number of fields such as plasmonic devices [5–8], plasmonic biosensing [9–14], plasmonic photocatalysis [15–19], plasmonic photovoltaics [20–23], surface-enhanced Raman scattering (SERS) [24–29] and its derivatives as the photo-induced enhanced Raman spectroscopy [30–35], SERS effect induced by high pressure [36] and the shell-isolated nanoparticle-enhanced Raman spectroscopy [37–41], other surface-enhanced spectroscopies, such as sum frequency generation (SFG) [42,43] and second harmonic generation (SHG) [44,45]. Thus, this Special Issue is focused on recent advances and insights in the research topic of nanoplasmonics and its applications.

This Special Issue is composed of nine research articles, and three review articles. The first part of the Issue is devoted to the surface plasmon resonance (SPR) spectroscopy [46,47]. Daniyal et al. have shown the use of the SPR spectroscopy for the optical characterization of a thin film based on nanocrystalline cellulose (NCC) [46]. Andam et al. have also used this SPR spectroscopy for determining the optical properties of ultra-thin films of azo-dye-doped polymers [47]. The second part is dedicated to plasmonic devices [48,49]. Firstly, Zhang et al. have reported on a plasmonic narrowband filter based on an equilateral triangle-shaped cavity and a metal–insulator–metal waveguide [48]. Lastly, Adibzadeh et al. have investigated the performances of plasmonic InP nanowire array solar cells [49]. In the third part, Gonçalves et al. demonstrated surface plasmon and Fano resonances in titanium carbide nanoparticles in the spectral range from visible to infrared [50]. In the fourth part, the plasmonic sensing is addressed [51–53]. At first, Cardoso et al. have reported on a second-order dispersion sensor based on multi-plasmonic resonances in D-shaped photonic crystal fibers [51]. Next, Ramdzan et al. have demonstrated a plasmonic sensing of mercury ions in an aqueous medium by using as a sensitive layer, a thin film composed of NCC and poly(3,4-ethylenethiophene) (PEDOT) which is deposited on a gold plasmonic film [52]. To finish this part, Barchiesi et al. have reported on the performance of plasmonic sensors based on copper/copper oxide films [53]. In the following part, the addressed topics are focused on surface-enhanced spectroscopies [54–56]. At first, Humbert et al. have highlighted a plasmonic coupling with the vibrations of the thiophenol molecule by using two-colour sum-frequency generation spectroscopy with an enhancement factor of the intensity around two orders of magnitude from blue to green–yellow due to the presence of a significant number of hotspots between Au nanosphere aggregates [54]. Yang et al. present in a review paper the applications of the SERS effect to agriculture and food safety [55]. Barbillon introduced a review paper on the recent applications of the shell-isolated nanoparticle-enhanced Raman spectroscopy [56]. To conclude this Special Issue on the latest advances in nanoplasmonics, Barbillon exhibited a short review paper on nanoplasmonics in high pressure environments [57].

Citation: Barbillon, G. Latest Advances in Nanoplasmonics and Use of New Tools for Plasmonic Characterization. *Photonics* **2022**, *9*, 112. https://doi.org/10.3390/photonics9020112

Received: 11 February 2022
Accepted: 14 February 2022
Published: 17 February 2022

Publisher's Note: MDPI stays neutral with regard to jurisdictional claims in published maps and institutional affiliations.

To realize this Special Issue entitled "Latest Advances in Nanoplasmonics and Use of New Tools for Plasmonic Characterization", we have obtained various contributions from authors of the high standard around the world. I want to thank all these authors as well as the whole editorial office of the journal "Photonics" for their great support and help in the management process of a great number of tasks associated to manuscript submissions. Finally, I expect that you will find this special issue dedicated to nanoplasmonics and their applications useful and attractive, which is aimed to the students or researchers who are or wish to be interested in this topic.

Funding: This research received no external funding.

Conflicts of Interest: The author declares no conflict of interest.

References

1. Shahbazyan, T.V.; Stockman, M.I. *Plasmonics: Theory and Applications*; Springer: Dordrecht, The Netherlands, 2013; pp. 1–577.
2. Maier, S.A. *Plasmonics: Fundamentals and Applications*; Springer: New York, NY, USA, 2007; pp. 3–220.
3. Barbillon, G.; Ivanov, A.; Sarychev, A.K. Applications of Symmetry Breaking in Plasmonics. *Symmetry* **2020**, *12*, 896. [CrossRef]
4. Maccaferri, N.; Barbillon, G.; Koya, A.N.; Lu, G.; Acuna, G.P.; Garoli, D. Recent advances in plasmonic nanocavities for single-molecule spectrocopy. *Nanoscale Adv.* **2021**, *3*, 633–642. [CrossRef]
5. Salamin, Y.; Ma, P.; Baeuerle, B.; Emboras, A.; Fedoryshyn, Y.; Heni, W.; Cheng, B.; Josten, A.; Leuthold, J. 100 GHz Plasmonic Photodetector. *ACS Photonics* **2018**, *5*, 3291–3297. [CrossRef]
6. Thomaschewski, M.; Yang, Y.Q.; Bozhevolnyi, S.I. Ultra-compact branchless plasmonic interferometers. *Nanoscale* **2018**, *10*, 16178–16183. [CrossRef] [PubMed]
7. Ayata, M.; Fedoryshyn, Y.; Heni, W.; Baeuerle, B.; Josten, A.; Zahner, M.; Koch, U.; Salamin, Y.; Hoessbacher, C.; Haffner, C.; et al. High-speed plasmonic modulator in a single metal layer. *Science* **2017**, *358*, 630–632. [CrossRef] [PubMed]
8. Haffner, C.; Heni, W.; Fedoryshyn, Y.; Niegemann, J.; Melikyan, A.; Elder, D.L.; Baeuerle, B.; Salamin, Y.; Josten, A.; Koch, U.; et al. All-plasmonic Mach-Zehnder modulator enabling optical high-speed communication at the microscale. *Nat. Photonics* **2015**, *9*, 525–528. [CrossRef]
9. Barbillon, G.; Bijeon, J.-L.; Lérondel, G.; Plain, J.; Royer, P. Detection of chemical molecules with integrated plasmonic glass nanotips. *Surf. Sci.* **2008**, *602*, L119–L122. [CrossRef]
10. Dhawan, A.; Duval, A.; Nakkach, M.; Barbillon, G.; Moreau, J.; Canva, M.; Vo-Dinh, T. Deep UV nano-microstructuring of substrates for surface plasmon resonance imaging. *Nanotechnology* **2011**, *22*, 165301. [CrossRef]
11. Pichon, B.P.; Barbillon, G.; Marie, P.; Pauly, M.; Begin-Colin, S. Iron oxide magnetic nanoparticles used as probing agents to study the nanostructure of mixed self-assembled monolayers. *Nanoscale* **2011**, *3*, 4696–4705. [CrossRef] [PubMed]
12. Dolci, M.; Bryche, J.-F.; Leuvrey, C.; Zafeiratos, S.; Gree, S.; Begin-Colin, S.; Barbillon, G.; Pichon, B.P. Robust clicked assembly based on iron oxide nanoparticles for a new type of SPR biosensor. *J. Mater. Chem. C* **2018**, *6*, 9102–9110. [CrossRef]
13. Dolci, M.; Bryche, J.-F.; Moreau, J.; Leuvrey, C.; Begin-Colin, S.; Barbillon, G.; Pichon, B.P. Investigation of the structure of iron oxide nanoparticle assemblies in order to optimize the sensitivity of surface plasmon resonance-based sensors. *Appl. Surf. Sci.* **2020**, *527*, 146773. [CrossRef]
14. Sarychev, A.K.; Ivanov, A.; Lagarkov, A.; Barbillon, G. Light Concentration by Metal-Dilectric Micro-Resonators for SERS Sensing. *Materials* **2019**, *12*, 103. [CrossRef] [PubMed]
15. Verma, R.; Belgamwar, R.; Polshettiwar, V. Plasmonic Photocatalysis for CO_2 Conversion to Chemicals and Fuels. *ACS Materials Lett.* **2021**, *3*, 574–598. [CrossRef]
16. Warkentin, C.L.; Yu, Z.; Sarkar, A.; Frontiera, R.R. Decoding Chemical and Physical Processes Driving Plasmonic Photocatalysis Using Surface-Enhanced Raman Spectroscopies. *Acc. Chem. Res.* **2021**, *54*, 2457–2466. [CrossRef] [PubMed]
17. Li, J.; Miao, P.; Zhang, Y.; Wu, J.; Zhang, B.; Du, Y.; Han, X.; Sun, J.; Xu, P. Recent Advances in Plasmonic Nanostructures for Enhanced Photocatalysis and Electrocatalysis. *Adv. Mater.* **2021**, *33*, 2000086. [CrossRef]
18. Koya, A.N.; Zhu, X.C.; Ohannesian, N.; Yanik, A.A.; Alabastri, A.; Zaccaria, R.P.; Krahne, R.; Shih, W.C.; Garoli, D. Nanoporous Metals: From Plasmonic Properties to Applications in Enhanced Spectroscopy and Photocatalyis. *ACS Nano* **2021**, *15*, 6038–6060. [CrossRef]
19. Mascaretti, L.; Naldoni, A. Hot electron and thermal effects in plasmonic photocatalysis. *J. Appl. Phys.* **2020**, *128*, 041101. [CrossRef]
20. Shao, W.J.; Liang, Z.Q.; Guan, T.F.; Chen, J.M.; Wang, Z.F.; Wu, H.H.; Zheng, J.Z.; Abdulhalim, I.; Jiang, L. One-step integration of a multiple-morphology gold nanoparticle array on a TiO_2 film via a facile sonochemical method for highly efficient organic photovoltaics. *J. Mater. Chem. A* **2018**, *6*, 8419–8429. [CrossRef]
21. Vangelidis, I.; Theodosi, A.; Beliatis, M.J.; Gandhi, K.K.; Laskarakis, A.; Patsalas, P.; Logothetidis, S.; Silva, S.R.P.; Lidorikis, E. Plasmonic Organic Photovoltaics: Unraveling Plasmonic Enhancement for Realistic Cell Geometries. *ACS Photonics* **2018**, *5*, 1440–1452. [CrossRef]

22. Li, M.Z.; Guler, U.; Li, Y.A.; Rea, A.; Tanyi, E.K.; Kim, Y.; Noginov, M.A.; Song, Y.L.; Boltasseva, A.; Shalaev, V.M.; et al. Plasmonic Biomimetic Nanocomposite with Spontaneous Subwavelength Structuring as Broadband Absorbers. *ACS Energy Lett.* **2018**, *3*, 1578–1583. [CrossRef]

23. Chen, X.; Fang, J.; Zhang, X.D.; Zhao, Y.; Gu, M. Optical/Electrical Integrated Design of Core-Shell Aluminum-Based Plasmonic Nanostructures for Record-Breaking Efficiency Enhancements in Photovoltaic Devices. *ACS Photonics* **2017**, *4*, 2102–2110. [CrossRef]

24. Huang, J.A.; Mousavi, M.Z.; Zhao, Y.Q.; Hubarevich, A.; Omeis, F.; Giovannini, G.; Schutte, M.; Garoli, D.; De Angelis, F. SERS discrimination of single DNA bases in single oligonucleotides by electro-plasmonic trapping. *Nat. Commun.* **2019**, *10*, 5321. [CrossRef]

25. Graniel, O.; Iatsunskyi, I.; Coy, E.; Humbert, C.; Barbillon, G.; Michel, T.; Maurin, D.; Balme, S.; Miele, P.; Bechelany, M. Au-covered hollow urchin-like ZnO nanostructures for surface-enhanced Raman scattering sensing. *J. Mater. Chem. C* **2019**, *7*, 15066–15073. [CrossRef]

26. Hubarevich, A.; Huang, J.-A.; Giovanni, G.; Schirato, A.; Zhao, Y.; Maccaferri, N.; De Angelis, F.; Alabastri, A.; Garoli, D. λ-DNA through Porous Materials–Surface-Enhanced Raman Scattering in a Single Plasmonic Nanopore. *J. Phys. Chem. C* **2020**, *124*, 22663–22670. [CrossRef]

27. Castro-Grijalba, A.; Montes-Garcia, V.; Cordero-Ferradas, M.J.; Coronado, E.; Perez-Juste, J.; Pastoriza-Santos, I. SERS-Based Molecularly Imprinted Plasmonic Sensor for Highly Sensitive PAH Detection. *ACS Sens.* **2020**, *5*, 693–702. [CrossRef]

28. Barbillon, G.; Ivanov, A.; Sarychev, A.K. SERS Amplification in Au/Si Asymmetric Dimer Array Coupled to Efficient Adsorption of Thiophenol Molecules. *Nanomaterials* **2021**, *11*, 1521. [CrossRef]

29. Barbillon, G.; Graniel, O.; Bechelany, M. Assembled Au/ZnO Nano-Urchins for SERS Sensing of the Pesticide Thiram. *Nanomaterials* **2021**, *11*, 2174. [CrossRef]

30. Ben-Jaber, S.; Peveler, W.J.; Quesada-Cabrera, R.; Cortés, E.; Sotelo-Vazquez, C.; Abdul-Karim, N.; Maier, S.A.; Parkin, I.P. Photo-induced enhanced Raman spectroscopy for universal ultra-trace detection of explosives, pollutants and biomolecules. *Nat. Commun.* **2016**, *7*, 12189. [CrossRef]

31. Almohammed, S.; Zhang, F.; Rodriguez, B.J.; Rice, J.H. Photo-induced surface-enhanced Raman spectroscopy from a diphenylalanine peptide nanotube-metal nanoparticle template. *Sci. Rep.* **2018**, *8*, 3880. [CrossRef]

32. Zhang, M.; Chen, T.; Liu, Y.; Zhu, J.; Liu, J.; Wu, Y. Three-Dimensional TiO_2–Ag Nanopore Arrays for Powerful Photoinduced Enhanced Raman Spectroscopy (PIERS) and Versatile Detection of Toxic Organics. *ChemNanoMat* **2019**, *5*, 55–60. [CrossRef]

33. Barbillon, G.; Noblet, T.; Humbert, C. Highly crystalline ZnO film decorated with gold nanospheres for PIERS chemical sensing. *Phys. Chem. Chem. Phys.* **2020**, *22*, 21000–21004. [CrossRef] [PubMed]

34. Barbillon, G. Oxygen Vacancy Dynamics in Highly Crystalline Zinc Oxide Film Investigated by PIERS Effect. *Materials* **2021**, *14*, 4423. [CrossRef] [PubMed]

35. Zhao, J.; Wang, Z.; Lan, J.; Khan, I.; Ye, X.; Wan, J.; Fei, Y.; Huang, S.; Li, S.; Kang, J. Recent advances and perspectives in photo-induced enhanced Raman spectroscopy. *Nanoscale* **2021**, *13*, 8707–8721. [CrossRef] [PubMed]

36. Sun, H.H.; Yao, M.G.; Song, Y.P.; Zhu, L.Y.; Dong, J.J.; Liu, R.; Li, P.; Zhao, B.; Liu, B.B. Pressure-induced SERS enhancement in a MoS_2/Au/R6G system by a two-step charge transfer process. *Nanoscale* **2019**, *11*, 21493–21501. [CrossRef] [PubMed]

37. Forato, F.; Talebzadeh, S.; Rousseau, N.; Mevellec, J.-Y.; Bujoli, B.; Knight, D.A.; Queffélec, C.; Humbert, B. Functionalized core–shell Ag@TiO_2 nanoparticles for enhanced Raman spectroscopy: a sensitive detection method for Cu(II) ions. *Phys. Chem. Chem. Phys.* **2019**, *21*, 3066–3072. [CrossRef]

38. Li, C.-Y.; Le, J.-B.; Wang, Y.-H.; Chen, S.; Yang, Z.-L.; Li, J.-F.; Cheng, J.; Tian, Z.-Q. In situ probing electrified interfacial water structures at atomically flat surfaces. *Nat. Mater.* **2019**, *18*, 697–701. [CrossRef] [PubMed]

39. Zhang, S.-P.; Lin, J.-S.; Lin, R.-K.; Radjenovic, P.M.; Yang, W.-M.; Xu, J.; Dong, J.-C.; Yang, Z.-L.; Hang, W.; Tian, Z.-Q.; et al. *In situ* Raman study of the photoinduced behavior of dye molecules on TiO_2(hkl) single crystal surfaces. *Chem. Sci.* **2020**, *11*, 6431–6435. [CrossRef] [PubMed]

40. Guan, S.; Attard, G.A.; Wain, A.J. Observation of Substituent Effects in the Electrochemical Adsorption and Hydrogenation of Alkynes on Pt{hkl} Using SHINERS. *ACS Catal.* **2020**, *10*, 10999–11010. [CrossRef]

41. Saeed, K.H.; Forster, M.; Li, J.-F.; Hardwick, L.J.; Cowan, A.J. Water oxidation intermediates on iridium oxide electrodes probed by *in situ* electrochemical SHINERS. *Chem. Commun.* **2020**, *56*, 1129–1132. [CrossRef] [PubMed]

42. Dalstein, L.; Humbert, C.; Ben Haddada, M.; Boujday, S.; Barbillon, G.; Busson, B. The Prevailing Role of Hotspots in Plasmon-Enhanced Sum-Frequency Generation Spectroscopy. *J. Phys. Chem. Lett.* **2019**, *10*, 7706–7711. [CrossRef]

43. Dalstein, L.; Ben Haddada, M.; Barbillon, G.; Humbert, C.; Tadjeddine, A.; Boujday, S.; Busson, B. Revealing the Interplay between Adsorbed Molecular Layers and Gold Nanoparticles by Linear and Nonlinear Optical Properties. *J. Phys. Chem. C* **2015**, *115*, 17146–17155. [CrossRef]

44. Shi, J.; Liang, W.-Y.; Raja, S.; Sang, Y.; Zhang, X.-Q.; Chen, C.-A.; Wang, Y.; Yang, X.; Lee, Y.-H.; Ahn, H.; et al. Plasmonic Enhancement and Manipulation of Optical Nonlinearity in Monolayer Tungsten Disulfide. *Laser Photonics Rev.* **2018**, *12*, 1800188. [CrossRef]

45. Tsai, W.-Y.; Chung, T.L.; Hsiao, H.-H.; Chen, J.-W.; Lin, R.J.; Wu, P.C.; Sun, G.; Wang, C.-M.; Misawa, H.; Tsai, D.P. Second Harmonic Light Manipulation with Vertical Split Ring Resonators. *Adv. Mater.* **2019**, *31*, 1806479. [CrossRef] [PubMed]

46. Daniyal, W.M.E.M.M.; Fen, Y.W.; Abdullah, J.; Sadrolhosseini, A.R.; Mahdi, M.A. Design and Optimization of Surface Plasmon Resonance Spectroscopy for Optical Constant Characterization and Potential Sensing Application: Theoretical and Experimental Approaches. *Photonics* **2021**, *8*, 361. [CrossRef]
47. Andam, N.; Refki, S.; Ishitobi, H.; Inouye, Y.; Sekkat, Z. Optical Characterization of Ultra-Thin Films of Azo-Dye-Doped Polymers Using Ellipsometry and Surface Plasmon Resonance Spectroscopy. *Photonics* **2021**, *8*, 41. [CrossRef]
48. Zhang, J.; Feng, H.; Gao, Y. Plasmonic Narrowband Filter Based on an Equilateral Triangular Resonator with a Silver Bar. *Photonics* **2021**, *8*, 244. [CrossRef]
49. Adibzadeh, F.; Olyaee, S. Plasmonic Enhanced InP Nanowire Array Solar Cell through Optoelectronic Modeling. *Photonics* **2021**, *8*, 90. [CrossRef]
50. Gonçalves, M.; Melikyan, A.; Minassian, H.; Makaryan, T.; Petrosyan, P.; Sargsian, T. Interband, Surface Plasmon and Fano Resonances in Titanium Carbide (MXene) Nanoparticles in the Visible to Infrared Range. *Photonics* **2021**, *8*, 36. [CrossRef]
51. Cardoso, M.P.; Silva, A.O.; Romeiro, A.F.; Giraldi, M.T.R.; Costa, J.C.W.A.; Santos, J.L.; Baptista, J.M.; Guerreiro, A. Second-Order Dispersion Sensor Based on Multi-Plasmonic Surface Resonances in D-Shaped Photonic Crystal Fibers. *Photonics* **2021**, *8*, 181. [CrossRef]
52. Ramdzan, N.S.M.; Fen, Y.W.; Liew, J.Y.C.; Omar, N.A.S.; Anas, N.A.A.; Daniyal, W.M.E.M.M.; Fauzi, N.I.M. Exploration on Structural and Optical Properties of Nanocrystalline Cellulose/Poly(3,4-Ethylenedioxythiophene) Thin Film for Potential Plasmonic Sensing Application. *Photonics* **2021**, *8*, 419. [CrossRef]
53. Barchiesi, D.; Gharbi, T.; Cakir, D.; Anglaret, E.; Fréty, N.; Kessentini, S.; Maalej, R. Performance of Surface Plasmon Resonance Sensors Using Copper/Copper Oxide Films: Influence of Thicknesses and Optical Properties. *Photonics* **2022**, *9*, 104. [CrossRef]
54. Humbert, C.; Pluchery, O.; Lacaze, E.; Busson, B.; Tadjeddine, A. Two-Colour Sum-Frequency Generation Spectroscopy Coupled to Plasmonics with the CLIO Free Electron Laser. *Photonics* **2022**, *9*, 55. [CrossRef]
55. Yang, Y.; Creedon, N.; O'Riordan, A.; Lovera, P. Surface Enhanced Raman Spectroscopy : Applications in Agriculture and Food Safety. *Photonics* **2021**, *8*, 568. [CrossRef]
56. Barbillon, G. Applications of Shell-Isolated Nanoparticle-Enhanced Raman Spectroscopy. *Photonics* **2021**, *8*, 46. [CrossRef]
57. Barbillon, G. Nanoplasmonics in High Pressure Environment. *Photonics* **2020**, *7*, 53. [CrossRef]

Article

Design and Optimization of Surface Plasmon Resonance Spectroscopy for Optical Constant Characterization and Potential Sensing Application: Theoretical and Experimental Approaches

Wan Mohd Ebtisyam Mustaqim Mohd Daniyal [1], Yap Wing Fen [1,2,*], Jaafar Abdullah [1], Amir Reza Sadrolhosseini [3] and Mohd Adzir Mahdi [4]

[1] Functional Devices Laboratory, Institute of Advanced Technology, Universiti Putra Malaysia, UPM Serdang 43400, Selangor, Malaysia; gs50207@student.upm.edu.my (W.M.E.M.M.D.); jafar@upm.edu.my (J.A.)

[2] Department of Physics, Faculty of Science, Universiti Putra Malaysia, UPM Serdang 43400, Selangor, Malaysia

[3] Magneto-Plasmonic Lab, Laser and Plasma Research Institute, Shahid Beheshti University, Tehran 1983969411, Iran; a_sadrolhosseini@sbu.ac.ir

[4] Wireless and Photonics Networks, Faculty of Engineering, Universiti Putra Malaysia, UPM Serdang 43400, Selangor, Malaysia; mam@upm.edu.my

* Correspondence: yapwingfen@upm.edu.my

Citation: Daniyal, W.M.E.M.M.; Fen, Y.W.; Abdullah, J.; Sadrolhosseini, A.R.; Mahdi, M.A. Design and Optimization of Surface Plasmon Resonance Spectroscopy for Optical Constant Characterization and Potential Sensing Application: Theoretical and Experimental Approaches. Photonics 2021, 8, 361. https://doi.org/10.3390/photonics8090361

Received: 4 July 2021
Accepted: 2 August 2021
Published: 29 August 2021

Abstract: The best surface plasmon resonance (SPR) signal can be generated based on several factors that include the excitation wavelength, the type of metal used, and the thickness of the metal layer. In this study, the aforementioned factors have been investigated to obtain the best SPR signal. The excitation wavelength of 633 nm and gold metal with thickness of 50 nm were required to generate the SPR signal before the SPR was used for optical constant characterization by fitting of experimental results to the theoretical data. The employed strategy has good agreement with the theoretical value where the real part refractive index, n value, of the gold thin film was 0.1245 while the value for the imaginary part, k, was 3.6812 with 47.7 nm thickness. Besides that, the optical characterization of nanocrystalline cellulose (NCC)-based thin film has also been demonstrated. The n and k values found for this thin film were 1.4240 and 0.2520, respectively, with optimal thickness of 9.5 nm. Interestingly when the NCC-based thin film was exposed to copper ion solution with n value of 1.3333 and k value of 0.0060 to 0.0070 with various concentrations (0.01–10 ppm), a clear change of the refractive index value was observed. This result suggests that the NCC-based thin film has high potential for copper ion sensing using SPR with a sensitivity of 8.0052°/RIU.

Keywords: surface plasmon resonance; nanocrystalline cellulose; optical characterization; copper ion

1. Introduction

Surface plasmon resonance (SPR) is a collective oscillation of electrons that can be generated by light [1,2]. This oscillation occurs at the interface between metal and dielectric material and can be observed using various configurations and the most common configuration is the Kretschmann configuration as shown in Figure 1. This configuration has been used in most practical applications such as clinical diagnostic, gas sensor, biological sensing, and chemical sensing [3–14]. Theoretically in the Kretschmann configuration, the SPR can be generated using the basic principles of physics. The light beam that passes through two different media with a higher to lower refractive index will be totally reflected if the incident angle is higher than the critical angle. The electromagnetic field component of the incident light will penetrate through the prism and reach the interface between the metal and dielectric medium to excite the free electron on the metal surface if the metal layer is thin enough [15–22]. As a result, the electron is excited and plasmon wave is formed that propagates along the metal surface. The plasmon wave is the sensing component that interacts optically with the metal surfaces. At specific angle, when the momentum of the plasmon wave is equivalent to the incident light, resonance occurs and the intensity of the reflected light decreases [23–27]. The plasmon wave is the sensing component of SPR and very sensitive towards the changes of the thin film surrounding properties that includes the excitation wavelength, type of metal used, and the thickness of the metal layer [28–30]. Changes of these properties will change the reflectance curve and will affect the sensitivity, full width half maximum (FWHM), and the accuracy of the SPR, curve which are important parameters to define the performance of the SPR. In this study, the surrounding properties to generate SPR have been investigated using a simulation and automatic fitting program that has been developed using the Matlab program to achieve the best SPR signal.

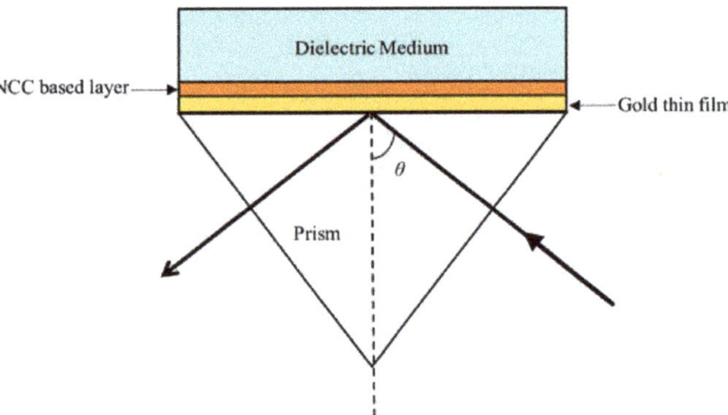

Figure 1. Kretschmann configuration to generate the surface plasmon resonance (SPR) signal.

SPR is an optical technique that has many advantages that include being cost-effective, and having linear properties, low mass, and fast measurement [31–39]. One of the most important findings in SPR is the introduction of the active layer on the surface of the metal thin film. Since the past decades, a wide range of materials have been exploited to be used as an active layer to improve the sensitivity of the SPR and most of the studies only focused on the target analyte's adsorption uptake performance for potential sensing applications [40–55]. Other important information that can be investigated using SPR is the refractive index value of the active layer. The refractive index value of the active layer is very crucial as it can provide information regarding the intermolecular interactions in liquid mixtures [56–60]. To the best of our knowledge, SPR has not yet been used to study the refractive index value of nanocrystalline cellulose (NCC) and graphene oxide (GO) composite. The NCC-based active layer has high potential to improve the SPR sensitivity owing to the excellent properties for metal ion adsorption [61,62]. The optical properties of the NCC-based thin film can provide information on the intermolecular interactions with metal ions for sensing applications. Hence in this present work, SPR has been used to study the optical properties of NCC-based thin film and the changes of the optical properties after interaction with copper ion also was investigated.

2. Theory

To generate an SPR signal, the incident light must first be in a transverse magnetic mode as the electric field is perpendicular to the metal thin film. The electric field of the light source can be described by [63]:

$$\vec{E} = E_o(\widehat{x} + i\widehat{z})e^{i(kx - \omega t)}e^{-k|z|} \tag{1}$$

where E_o is the amplitude, $\hat{x}$ and $\hat{z}$ are the unit vectors, k is the wave vector, and ω are the angular optical frequency of the electrical field. When total internal reflection occurs, the evanescent wave will excite the free electrons that exist on the surface of the thin film forming a surface plasmon that propagates along the surface of the thin film. The wave vector for the surface plasmon, K_{sp} can be described by the following equation [64]:

$$K_{sp} = \frac{\omega}{c} \sqrt{\left(\frac{\varepsilon_1 \varepsilon_2}{\varepsilon_1 + \varepsilon_2} \right)} \tag{2}$$

where ω is the frequency, c is the velocity of light, ε_1 is the complex dielectric constant for the surface active, and ε_2 is the complex dielectric constant for the dielectric media. The dielectric constant can also be described by:

$$\varepsilon = n^2 \tag{3}$$

Using Equation (3), Equation (2) then can be rewritten as:

$$K_{sp} = \frac{\omega}{c} \sqrt{\left(\frac{n_1^2 n_2^2}{n_1^2 + n_2^2} \right)} \tag{4}$$

where n_1 and n_2, is the refractive index of gold layer and sample layer respectively. On the other hand, the component of the incident light vector that is parallel to the prism/metal interface, K_x can be described as [65]:

$$K_x = \left(\frac{\omega}{c} \right) n_p \sin \theta_{SPR} \tag{5}$$

where n_p is the refractive index of the prism. SPR then can be generated when the wave vector for the surface plasmon is equal to the incident light vector and can be described by [66]:

$$K_{sp} = K_x \tag{6}$$

with

$$\sqrt{\left(\frac{n_1^2 n_2^2}{n_1^2 + n_2^2} \right)} = n_p \sin \theta_{SPR} \tag{7}$$

The coupling of these two wave vectors, K_{sp} and K_x result in a sharp dip of the reflectance at a resonance angle, θ_{SPR}. The SPR optical sensor works by detecting the changes of the thin film surface refractive index. Thus, the refractive index of the sample layer is [67]:

$$n_2 = \sqrt{\frac{n_1^2 n_p^2 \sin^2 \theta_{SPR}}{n_1^2 - n_p^2 \sin^2 \theta_{SPR}}} \tag{8}$$

In accordance with the boundary conditions for the electrical and magnetic fields at the interfaces between multilayers, the reflection coefficient, r, can be expressed as [68]:

$$r = \frac{m_{21} + m_{22}\gamma_2 - m_{11}\gamma_0 - m_{12}\gamma_2\gamma_0}{m_{21} + m_{22}\gamma_2 + m_{11}\gamma_0 + m_{12}\gamma_2\gamma_0} \tag{9}$$

m_{ij} are the matrix transfer element with thickness, d given by [69]:

$$m = \begin{pmatrix} \cos\delta & -i\frac{\sin\delta}{\gamma_1} \\ -i\gamma_1 \sin\delta & \cos\delta \end{pmatrix} \tag{10}$$

where,

$$\gamma_1 = \frac{n_1}{\cos\theta_1}\sqrt{\varepsilon_0\mu_0} \text{ and } \delta = \frac{2\pi}{\lambda}dn_1\cos\theta_1 \tag{11}$$

The reflectivity of the multilayer system, R, is defined as the ratio of the energy reflected at the surface to the energy of the incident and can be expressed as [70]:

$$R = rr^* \tag{12}$$

where it is the function of the thickness and refractive index of both metal and sample layer.

3. Materials and Methods

3.1. Surface Plasmon Resonance (SPR) Optimization

Based on theoretical equations in Section 2, a simulation and automatic fitting program has been developed using the Matlab program. The simulation was used to investigate the optimal condition to generate an SPR signal. As the SPR signal is very sensitive towards the surrounding properties, the simulation was used to determine the optimum surrounding condition to generate an SPR signal that includes the type of light source, type of metal, and thickness of the metal used.

3.2. Preparation of Thin Film

After the SPR simulation, thin films were prepared to be studied using SPR. In this study, two thin films were prepared, i.e., gold thin film and NCC based thin film. The gold thin film was prepared using a glass substrate of 0.13–0.16 mm thickness and area of 24 mm × 24. Gold layer was deposited on top of the glass substrate using SC7640 sputter coater (Quorum Technologies, West Sussex, UK).

To prepare the NCC-based thin film, the NCC-based composite solution was initially prepared by mixing 1 mL of NCC that was modified using hexadecyltrimethylammonium bromide, with 1 mL of graphene oxide [71]. All chemicals were analytical grade and deionized water was used during the preparation of the NCC-based composite solution. The NCC-based composite solution was deposited on top of the gold thin film using the spin coating technique. About 1 mL of the solution was dropped at the center of the gold thin film using a micropipette and spun for 30 s at 6000 rpm by a P-6708D spin coater machine (Inc. Medical Devices, Indianapolis, IN, USA) to a homogenous NCC based thin film.

3.3. Surface Plasmon Resonance

After the evaluation of the SPR signal, the SPR setup was designed and prepared in the laboratory as shown in Figure 2. The SPR setup then was used to obtain the SPR curve of the gold thin film and the NCC-based thin film. The thin film was placed between the prism and the dielectric medium. Then, deionized water and copper ion solution with different concentrations was injected before the SPR signal was recorded [72].

3.4. Fitting Experimental to Theoretical

After the SPR experiment, the reflectance curve was analyzed to investigate the optical properties of the gold thin film, copper ion solution, and NCC-based thin film. The reflectance curve for the gold thin film in contact with deionized water was first fitted to the theoretical data. Using Equation (7), the refractive index of gold thin film was determined by taking the refractive index of deionized water as 1.3333 [31]. To investigate the refractive index value of copper ion solution, the reflectance curve for the gold thin film in contact with copper ion solution was fitted. The obtained refractive index value of the gold thin film was used to further determine the optical properties of copper ion using the same equation.

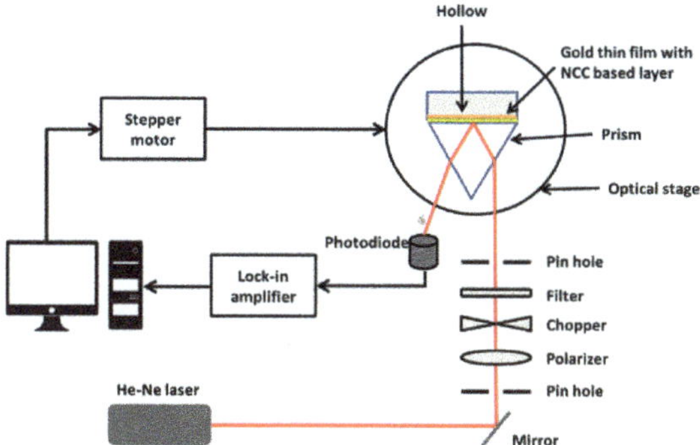

Figure 2. Optical setup of surface plasmon resonance spectroscopy with nanocrystalline cellulose (NCC)-based thin film.

Then, the reflectance curve for the NCC-based thin film in contact with deionized water and copper ion solution was fitted to investigate the optical properties and the changes after in contact with copper ion at different concentrations. The refractive index value for the NCC-based thin film was calculated using Equation (8). The thickness of the thin film was calculated using the reflection coefficient, *r* from Equation (9).

4. Results and Discussion

4.1. Simulation of SPR Signal

Prior to the SPR simulation, the refractive index of copper, gold, and silver were obtained from previous study [73,74]. The refractive index of metal is one of the critical parameters that will affect the SPR signal performance. The real part and imaginary part of the refractive index as a function of wavelength for all three metals are shown in Figure 3a,b respectively. All three metals have different characteristics of refractive index. From the figures, the real part of the refractive index for copper and gold decrease exponentially with the increase of wavelength from 400 nm to 650 nm while silver only decreases gradually. For the imaginary part of the refractive index, all three metals show the same characteristics where the value increases with increasing wavelength.

To obtain the best SPR signal, different light sources and metals were first evaluated through the simulation program that was developed using Matlab based on the theory from Section 2. Using different light sources and the value of refractive index for each metals obtained previously, the SPR signal can be simulated to select the best light sources and best metal to generate SPR signal. The thickness of all metals was kept fixed at 50 nm before the investigation was carried out. Figure 4a,c shows the comparison of SPR signal between copper, gold, and silver respectively using air as the dielectric media and light sources of different wavelength, i.e., blue light (436 nm), violet light (441 nm), green light (546 nm), yellow light (589 nm), and red light (650 nm). It can be observed that SPR signal can be obtained using a green, yellow, and red light source for both copper and gold while for silver, the SPR signal can be obtained using a blue, violet, green, yellow, and red light source. On the other hand, the SPR signal changed when water was used as the dielectric media. Only yellow and red light sources were able to generate an SPR signal for both copper and gold while for silver, the SPR signal can be generated using green, yellow, and red light source as shown in Figure 5a,c respectively. Besides that, blue and violet light sources are not feasible in generating an SPR signal for all three metals. SPR refers to the electromagnetic response that occurs when plasmons are oscillating with the same frequency to the incident light. The existence of surface plasmon on a metal dielectric interface is confined to the wavelengths longer than the critical wavelength, which depends on the plasma frequency and is specific to the metal. The frequency of the incidence light must be equal to the natural frequency of the material or resonance will not occur; thus, SPR signal can be seen was not generated at some wavelengths for each metal [75]. Moreover, the SPR curve also can be observed to shift further to the right when water was used as the dielectric media. SPR is very sensitive towards the changes of the refractive index near the surface of the metal. Thus, changes of the dielectric media from air to water greatly affect the SPR signal owing to the refractive index value, i.e., 1.0003 and 1.3333 for air and water respectively [76]. Furthermore, the use of a higher wavelength is more favorable in generating the SPR signal owing to the appreciable effect of the width of SPR signal [77]. From this simulation, the red-light source has been selected as it is the best light source to generate an SPR signal for all three metal thin films.

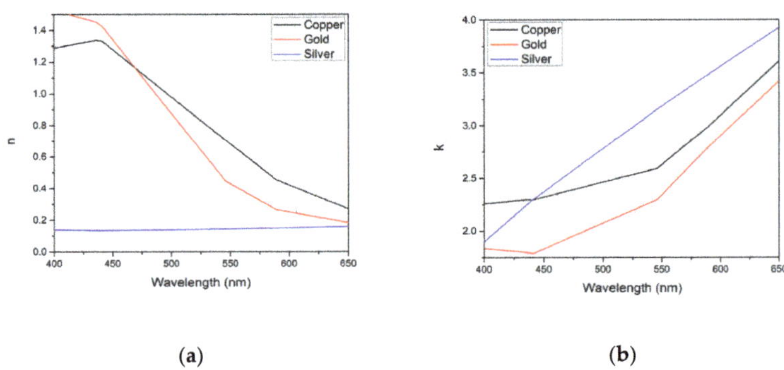

(a) (b)

Figure 3. (a) Real part and (b) imaginary part refractive index for copper, gold, and silver from 400–650 nm wavelength.

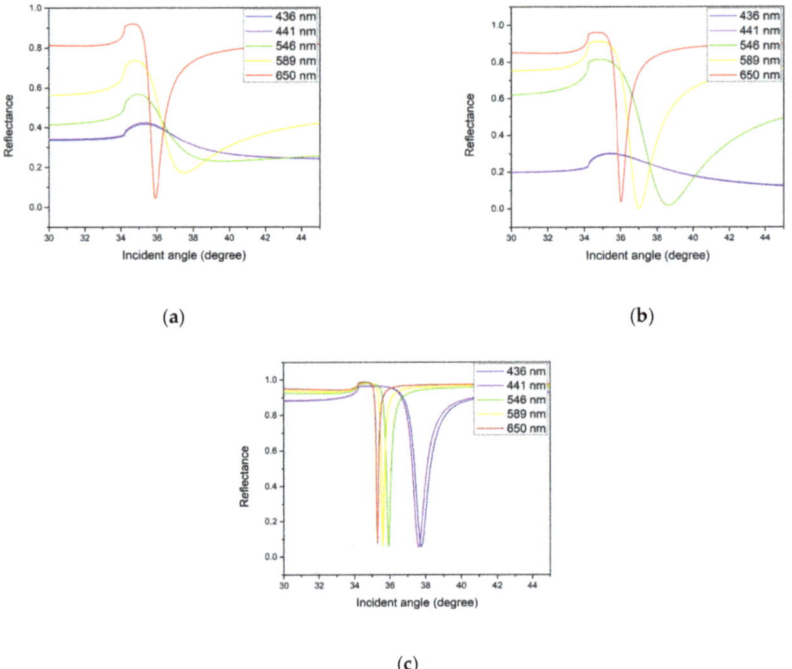

(a) (b)

(c)

Figure 4. SPR signal using different excitation wavelength for (a) copper, (b) gold, and (c) silver (thickness of 50 nm) in contact with air.

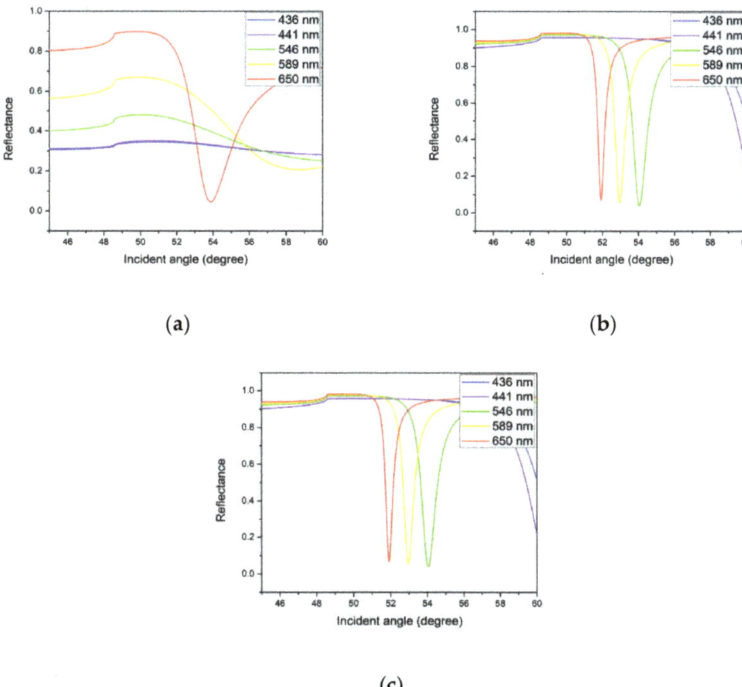

Figure 5. SPR signal using different excitation wavelength for (**a**) copper, (**b**) gold, and (**c**) silver (thickness of 50 nm) in contact with water.

To obtain the best SPR signal, one of the parameters that needs to be optimized is the wavelength of the light source. To determine a more accurate wavelength of the light source to generate the best SPR signal, the wavelength in the red-light range (613–653 nm) was used to simulate the SPR signal for all three metal thin films. The best SPR signal was determined by the lowest reflectance minimum and a smaller value of full width half maximum (FWHM). To obtain the FWHM value, the width of the SPR curve at half of the maximum value was calculated as shown in Figure 6. Then, the accuracy of the SPR signal was calculated using the inverse of the FWHM [78].

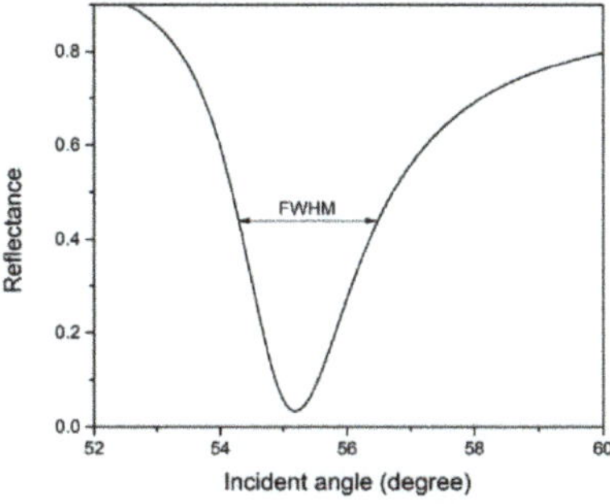

Figure 6. The full width half maximum (FWHM) of the SPR curve.

From the simulation, the optimal SPR signal using copper thin film can be obtained at 653 nm where the FWHM value was the lowest at approximately 2.2° thus, gives the highest accuracy. At 613 nm to 643 nm, the FWHM increased from 3.2° to 2.5° while the lowest reflectance minimum also increased as shown in Figure 7a. Moving on to the gold thin film, the generated SPR signal using a light source with a wavelength of 613 nm and 623 nm has the highest FWHM value at approximately 2.5° and 2.3°, respectively. The FWHM and accuracy of the reflectance curve do not change when a light source with a wavelength of 633, 643, and 653 nm was used. The calculated FWHM value for these wavelengths was 2.2°. Besides that, at 633 nm the SPR signal has the lowest reflectance compared to at 643 and 653 nm as shown in Figure 7b. For the silver thin film, the lowest reflectance minimum for all wavelengths does not show any significant change as shown in Figure 7c. Moreover, only a slight difference is identified in the calculated FWHM value of the reflectance curve for all ranges of wavelength. From 613 nm until 633 nm, the calculated FWHM value was 0.5° while at 643 until 653 nm, 0.4° was obtained. The calculated FWHM value and accuracy for all three metal thin films are recorded in Table 1.

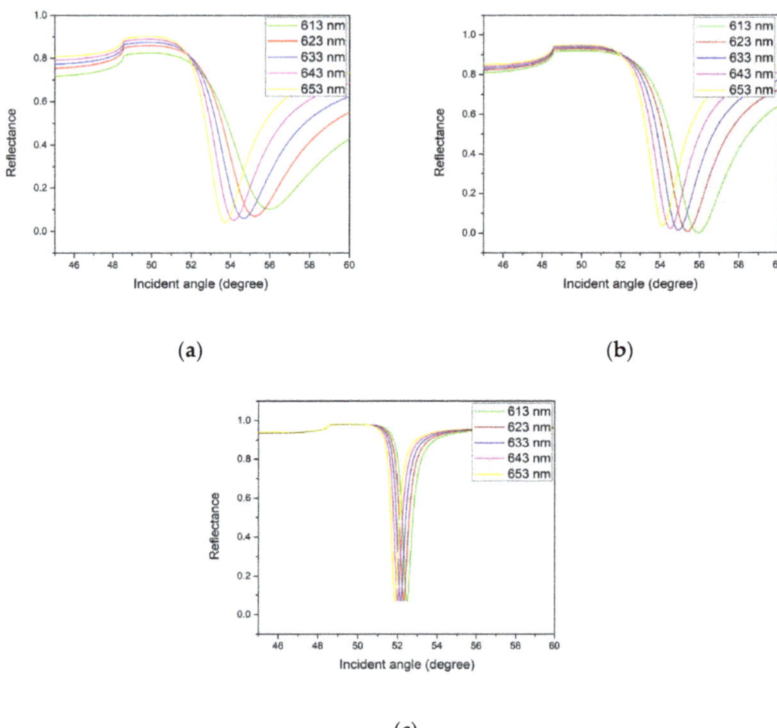

Figure 7. Optimization of the SPR signal based on different wavelengths of light sources for (**a**) copper, (**b**) gold, and (**c**) silver.

Another parameter that needs to be optimized to obtain the best SPR signal for maximum sensitivity is the thickness of the metal thin film. The optimization of the reflectance minimum and FWHM value of the reflectance curve can be achieved by selecting the appropriate thickness of the metals. Different thickness from 40 nm to 60 nm was used to investigate the SPR signal for all three metal thin films at the 633 nm wavelength light source. For the copper thin film, the lowest reflectance minimum of the reflectance curve can be obtained at 44 nm thickness as shown in Figure 8a. It can also be observed that at higher thickness, the reflectance minimum gradually decreases. The FWHM value of the SPR signal for the copper thin film also was the lowest at 44 nm thickness at approximately 3.1°. For the gold thin film, the best SPR signal can be obtained around 50 nm where the reflectance minimum was the lowest as shown in Figure 8b. The FWHM value calculated at 50 nm was also the lowest approximately 2.1° that gives higher accuracy compared to other thicknesses. On the other hand, the best SPR signal using silver thin film can be obtained around 52 nm where the reflectance minimum and the FWHM were the lowest around 0.4° as shown in Figure 8c.

Table 1. The FWHM and accuracy of copper, gold, and silver thin film at different wavelengths.

Metal	Wavelength (nm)	FWHM (Degree)	Accuracy (Degree^{-1})
	613	3.2	0.31
	623	3.0	0.33
Copper	633	2.8	0.36
	643	2.5	0.40
	653	2.2	0.45
	613	2.5	0.40
	623	2.3	0.43
Gold	633	2.2	0.45
	643	2.2	0.45
	653	2.2	0.45
	613	0.5	2.0
	623	0.5	2.0
Silver	633	0.5	2.0
	643	0.4	2.5
	653	0.4	2.5

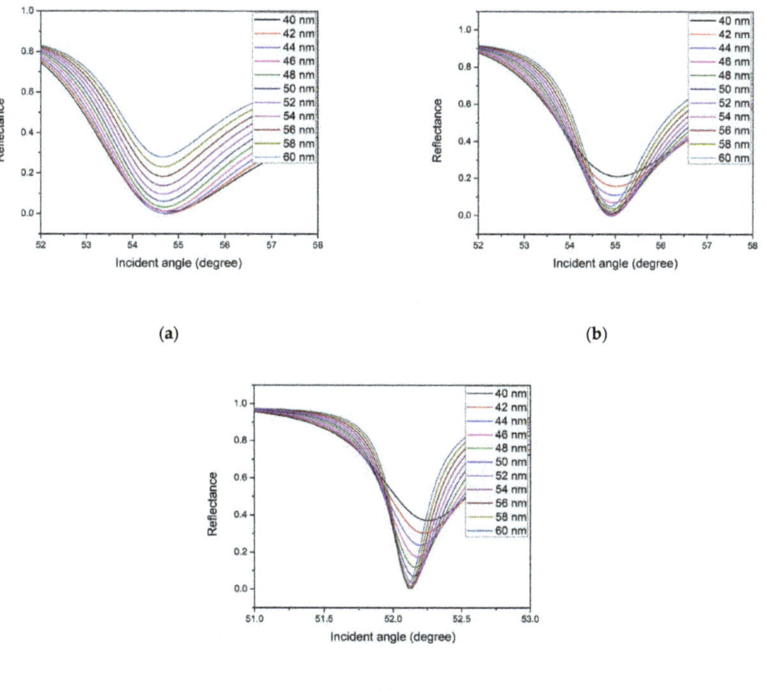

Figure 8. Optimization of the SPR signal based on different thickness of (**a**) copper, (**b**) gold, and (**c**) silver in contact with water and excitation wavelength of 633 nm.

From the simulation results, the red light of 633 nm wavelength has been proven to be the most appropriate light source to generate SPR. For metal selection, copper is not favorable to generate SPR owing to its optical properties that give a higher reflectance minimum and high FWHM value of the reflectance curve, hence decreasing the accuracy of the SPR in sensing application. The SPR signal is the best for both gold and silver thin films. However, silver has poor chemical stability and less inert compared to the gold that limits its application. Therefore, the most favorable metal to generate SPR is gold that gives a low reflectance minimum and smaller width of the reflectance curve [77]. Besides that, gold is also inert and has higher chemical stability that gives advantages for wider application.

4.2. Analysis of Thin Films Properties

Gold thin film has been used to generate SPR using water as the dielectric medium at room temperature. The experimental SPR signal then was fitted to obtain the information on the thin film properties as shown in Figure 9. According to the fitted SPR signal, the optical properties of the prepared gold thin film for the real part, n and imaginary part, k refractive index was 0.1245 and 3.6812, respectively. The thickness of the gold layer was also determined from the fitting where 47.7 nm thickness was revealed.

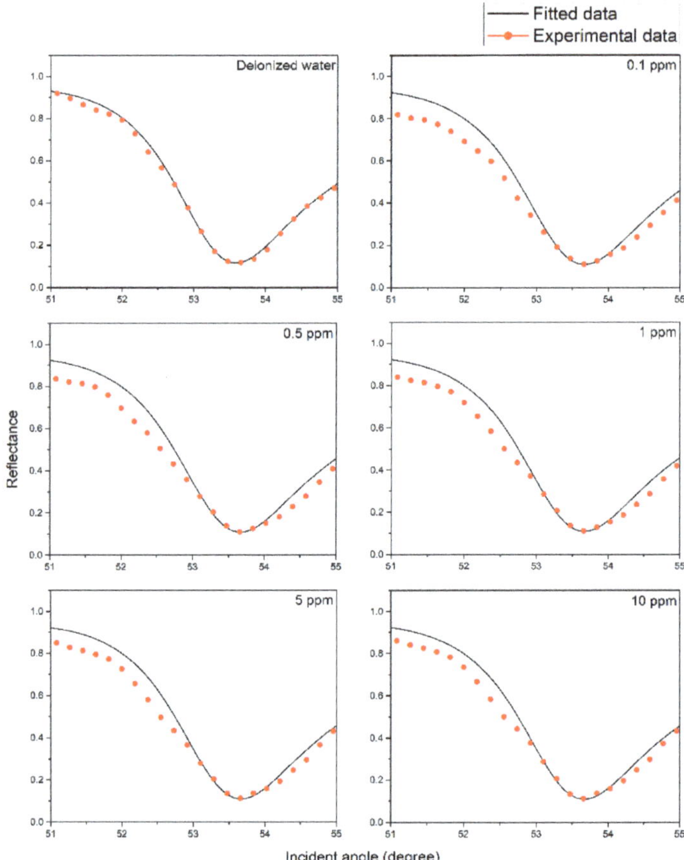

Figure 9. Fitted SPR signal of the gold thin film in contact with deionized water and copper ion.

The SPR signal of the gold thin film in contact with copper ion at low concentration has also been investigated. This result was also fitted to analyze the optical properties of the copper ion solution. From the fitting result, the k value for the copper ion at 0.1 ppm found was 0.0060. At 0.5 ppm, the k value increases to 0.0070 and the value remains the same until 10 ppm. This result has a good agreement with the previous study where the *n* value of copper ion at low concentration is almost equal to the deionized water refractive index at room temperature, i.e., 1.3333 [76].

The SPR experiment then continued using the NCC-based thin film. The SPR experiment using NCC-based thin film with deionized water was first carried out to investigate the optical properties. The obtained reflectance curve then was fitted as shown in Figure 10. The fitting of the SPR curve reveals that the refractive index of the NCC based thin film was 1.4240 and 0.2520 for *n* and *k* respectively. This refractive index value has a good agreement with the refractive index of NCC from the previous study, i.e., 1.499 [79].

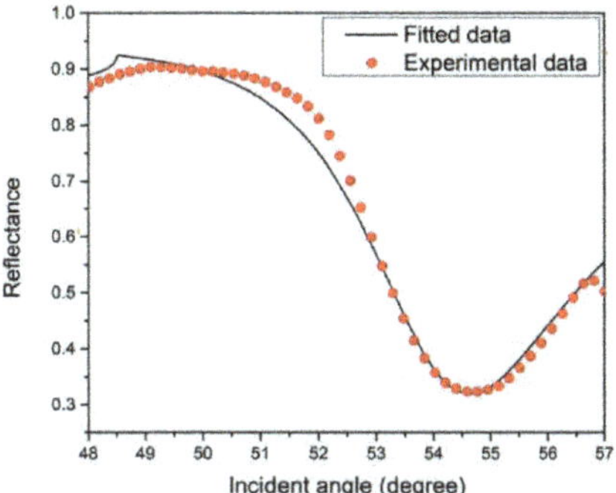

Figure 10. Fitted data and experimental data of SPR signal for NCC-based sensing layer on gold thin film in contact with deionized water.

In further analysis of the SPR results, the thickness of the NCC-based layer was also investigated. The fitting of the SPR curve reveals that the NCC-based layer has a thickness of 9.5 nm. The effect of the NCC-based layer thickness towards the SPR signal can be demonstrated as shown in Figure 11. A thinner NCC-based layer will give a lower FWHM value and sharper SPR curve that has higher accuracy in the determination of resonance angle but the binding interaction with copper ion might be minimal thus reducing the sensitivity of the SPR sensor. For a thicker sensing layer, although more interaction with copper ion can be achieved, a higher FWHM value and broader SPR curve have lower accuracy that can reduce the SPR sensor's effectiveness [80]. Hence, by considering both factors, i.e., the binding interaction of copper ion with the NCC based layer, and to obtain the best SPR result for resonance angle determination and better detection accuracy, the optimal thickness of the sensing layer was fixed around 9 to 10 nm. The FWHM value and the accuracy of the SPR curve for each thickness were recorded in Table 2.

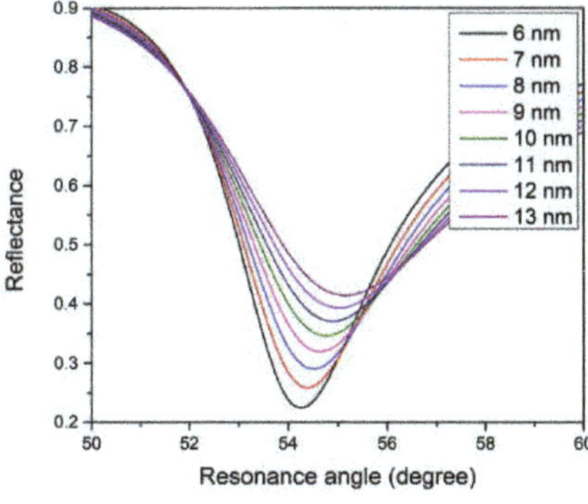

Figure 11. Optimization of the SPR signal based on different thickness of NCC-based layer.

Table 2. The FWHM and accuracy of NCC-based thin film at different thickness.

Thickness (nm)	FWHM (Degree)	Accuracy (Degree^{-1})
6	3.0	0.333
7	3.2	0.313
8	3.4	0.294
9	3.5	0.286
10	3.6	0.277
11	3.7	0.270
12	3.8	0.263
13	3.9	0.256

Then, the SPR experiment for the NCC-based thin film was carried out using copper ion solution. The obtained refractive index value for the gold thin film and copper ion was used to investigate the optical properties of NCC-based thin film through the fitting procedure. The curves presented in Figure 12 show the fitting of the SPR curve for NCC-based thin film in contact with copper ion solution. The refractive index of the NCC-based thin film increased from 1.424 to 1.457 for the n while the value of k decreased from 0.252 to 0.192 after in contact with copper ion at 0.01 ppm. Interestingly, the refractive index of the NCC-based thin film changes gradually for both n and k when a higher concentration of copper ion was used, as shown in Figure 13a,b respectively. Changes in the refractive index might be due to the chemical interaction of the NCC-based thin film with copper ion thus. The refractive index of the NCC-based thin film remains the same when in contact with copper ion above 0.5 ppm due to the saturation of copper ion on the surface of the NCC-based thin film [81]. The refractive index value and the thickness of the NCC-based thin film is recorded in Table 3. The thickness of the NCC-based thin film after interaction with copper ion was also investigated. The thickness of the NCC-based thin film gradually increased after being in contact with the copper ion that was believed due to the interaction that causes changes of the thin film surface. The thickness remains the same at around 11 nm when in contact with copper ion from 0.5 ppm to 10 ppm after the thin film reaches saturation.

Table 3. Refractive index of the NCC-based layer in contact with deionized water and copper ion.

Concentration (ppm)	Refractive Index of Sensing Layer After in Contact with Copper Ions		Thickness of NCC Based Layer (nm)
	Real Part, n	Imaginary Part, k	
0	1.424	0.252	9.5
0.01	1.457	0.192	9.7
0.05	1.467	0.173	9.9
0.08	1.482	0.155	10.3
0.1	1.487	0.136	10.5
0.5	1.504	0.123	11.0
1	1.504	0.123	11.0
5	1.504	0.123	11.0
10	1.504	0.123	11.0

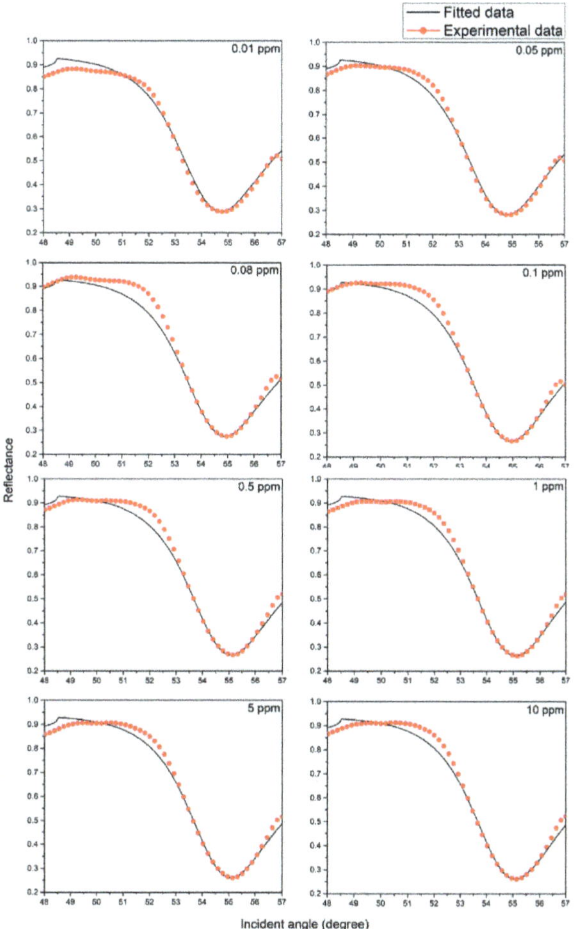

Figure 12. Fitted data and experimental data of SPR signal for NCC-based thin film in contact with copper ion from 0.01–10 ppm.

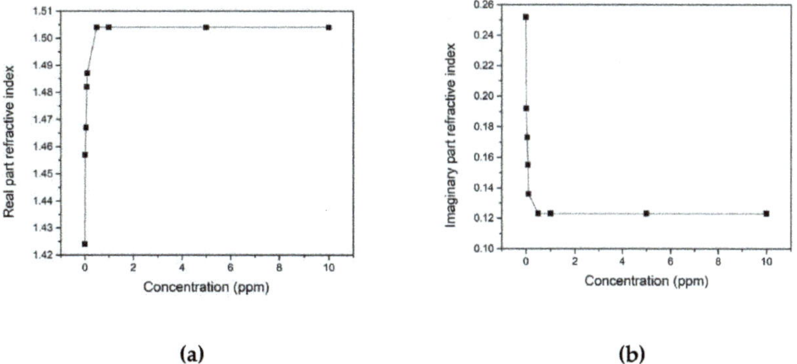

Figure 13. Refractive index of the NCC-based thin film (**a**) real part and (**b**) imaginary part at different concentrations.

After the analysis of the refractive index, determination of the resonance angle for the SPR result was also carried out. The resonance angle was obtained by measuring the angle of the minimum reflectance curve. The changes of the resonance angle, $\Delta\theta$ was calculated using the resonance angle of the NCC-based thin film in contact with deionized water as the reference. The result obtained then can be used to examine the sensitivity of the SPR. The sensitivity, S is a key performance parameter for a sensor and can be defined as [82–85]:

$$S = \frac{\Delta\theta}{\Delta n} \tag{13}$$

Figure 14 shows the $\Delta\theta$ against Δn for the NCC based thin film in contact with copper ion from 0.01 until 0.5 ppm. The calculated slope for the plotted graph was 8.0052°/RIU which represents the S with a correlation coefficient, R^2 of 0.99. From this result, the NCC-based thin film has good sensitivity for potential copper ion detection.

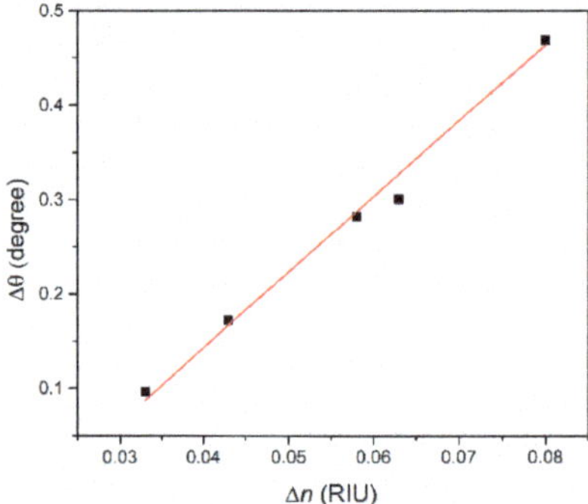

Figure 14. Sensitivity for the NCC-based thin film in contact with copper ion using SPR.

5. Conclusions

In this research, the excitation wavelength of 633 nm and gold metal with thickness of 50 nm has been demonstrated to be the optimal condition to generate the best SPR signal. The SPR has also successfully been applied for optical characterization and has good agreement with the theoretical value where the real part refractive index, n and imaginary part, k for the gold thin film were 0.1245 and 3.6812, respectively, with 47.7 nm thickness. Moreover, the optical constant for NCC-based thin film has also been investigated. The n and k values found for the NCC-based thin film were 1.4240 and 0.2520, respectively, with optimal thickness of 9.5 nm. When the NCC-based thin film was exposed to copper ion solution with n value of 1.3333 and k value of 0.0060 to 0.0070 from 0.01–10 ppm, a clear change of the refractive index value was observed. The studies of the NCC-based thin film shows that it has a good sensitivity of 8.0052°/RIU for copper ion sensing using SPR.

Author Contributions: Conceptualization, Y.W.F.; methodology, Y.W.F. and W.M.E.M.M.D.; validation, Y.W.F., J.A. and M.A.M.; formal analysis, W.M.E.M.M.D.; investigation, W.M.E.M.M.D.; data curation, W.M.E.M.M.D. and A.R.S.; writing—original draft preparation, W.M.E.M.M.D.; writing—review and editing, Y.W.F.; supervision, Y.W.F., J.A. and M.A.M. All authors have read and agreed to the published version of the manuscript.

Funding: This research work was funded by the Ministry of Education, Malaysia, through the Fundamental Research Grant Scheme (FRGS/1/2019/STG02/UPM/02/1).

Institutional Review Board Statement: Not applicable.

Informed Consent Statement: Not applicable.

Data Availability Statement: Not applicable.

Conflicts of Interest: The authors declare no conflict of interest.

References

1.   Kravets, V.G.; Kabashin, A.V.; Barnes, W.L.; Grigorenko, A.N. Plasmonic surface lattice resonances: A review of properties and applications. *Chem. Rev.* **2018**, *118*, 5912–5951. [CrossRef] [PubMed]
2.   Mukhtar, W.M.; Susthitha Menon, P.; Shaari, S.; Malek, M.Z.A.; Abdullah, A.M. Angle shifting in surface plasmon resonance: Experimental and theoretical verification. *J. Phys. Conf. Ser.* **2013**, *431*, 012028. [CrossRef]
3.   Usman, F.; Ojur, J.; Yousif, A.; Cheng, K. Structural characterization and optical constants of p-toluene sulfonic acid doped polyaniline and its composites of chitosan and reduced graphene-oxide. *Integr. Med. Res.* **2019**, *9*, 1468–1476. [CrossRef]
4.   Omar, N.A.S.; Fen, Y.W.; Saleviter, S.; Daniyal, W.M.E.M.M.; Anas, N.A.A.; Ramdzan, N.S.M.; Roshidi, M.D.A. Development of a graphene-based surface plasmon resonance optical sensor chip for potential biomedical application. *Materials* **2019**, *12*, 1928. [CrossRef] [PubMed]
5.   Gerislioglu, B.; Dong, L.; Ahmadivand, A.; Hu, H.; Nordlander, P.; Halas, N.J. Monolithic metal dimer-on-film structure: New plasmonic properties introduced by the underlying metal. *Nano Lett.* **2020**, *20*, 2087–2093. [CrossRef]
6.   Mansouri, M.; Fathi, F.; Jalili, R.; Shoeibie, S.; Dastmalchi, S.; Khataee, A.; Rashidi, M.R. SPR enhanced DNA biosensor for sensitive detection of donkey meat adulteration. *Food Chem.* **2020**, *331*, 127163–127172. [CrossRef]
7.   Firdous, S.; Anwar, S.; Rafya, R. Development of surface plasmon resonance (SPR) biosensors for use in the diagnostics of malignant and infectious diseases. *Laser Phys. Lett.* **2018**, *15*, 18–23. [CrossRef]
8.   Rosddi, N.N.M.; Fen, Y.W.; Anas, N.A.A.; Omar, N.A.S.; Ramdzan, N.S.M.; Daniyal, W.M.E.M.M. Cationically modified nanocrystalline cellulose/carboxyl-functionalized graphene quantum dots nanocomposite thin film: Characterization and potential sensing application. *Crystals* **2020**, *10*, 875. [CrossRef]
9.   Kamaruddin, N.H.; Bakar, A.A.A.; Yaacob, M.H.; Mahdi, M.A.; Zan, M.S.D.; Shaari, S. Enhancement of chitosan-graphene oxide SPR sensor with a multi-metallic layers of Au-Ag-Au nanostructure for lead(II) ion detection. *Appl. Surf. Sci.* **2016**, *361*, 177–184. [CrossRef]
10. Moon, J.; Kang, T.; Oh, S.; Hong, S.; Yi, J. In situ sensing of metal ion adsorption to a thiolated surface using surface plasmon resonance spectroscopy. *J. Colloid Interface Sci.* **2006**, *298*, 543–549. [CrossRef] [PubMed]
11. Eddin, F.B.K.; Fen, Y.W.; Omar, N.A.S.; Liew, J.Y.C.; Daniyal, W.M.E.M.M. Femtomolar detection of dopamine using surface plasmon resonance sensor based on chitosan/graphene quantum dots thin film. *Spectrochim. Acta A Mol. Biomol. Spectrosc.* **2021**, *263*, 120202. [CrossRef]
12. Ock, K.; Jang, G.; Roh, Y.; Kim, S.; Kim, J.; Koh, K. Optical detection of Cu^{2+} ion using a SQ-dye containing polymeric thin-film on Au surface. *Microchem. J.* **2001**, *70*, 301–305. [CrossRef]
13. Abdallah, T.; Abdalla, S.; Negm, S.; Talaat, H. Surface plasmons resonance technique for the detection of nicotine in cigarette smoke. *Sens. Actuators A Phys.* **2003**, *102*, 234–239. [CrossRef]
14. Rezabakhsh, A.; Rahbarghazi, R.; Fathi, F. Surface plasmon resonance biosensors for detection of Alzheimer's biomarkers; an effective step in early and accurate diagnosis. *Biosens. Bioelectron.* **2020**, *167*, 112511–112523. [CrossRef] [PubMed]
15. Anas, N.A.A.; Fen, Y.W.; Omar, N.A.S.; Daniyal, W.M.E.M.M.; Ramdzan, N.S.M.; Saleviter, S. Development of graphene quantum dots-based optical sensor for toxic metal ion detection. *Sensors* **2019**, *19*, 3850. [CrossRef] [PubMed]
16. Ciminelli, C.; Campanella, C.M.; Olio, F.D.; Campanella, C.E.; Armenise, M.N. Label-free optical resonant sensors for biochemical applications. *Prog. Quantum Electron.* **2013**, *37*, 51–107. [CrossRef]
17. Sun, R.J.; Huang, H.J.; Hsiao, C.N.; Lin, Y.W.; Liao, B.H.; Chou Chau, Y.F.; Chiang, H.P. Reusable TiN substrate for surface plasmon resonance heterodyne phase interrogation sensor. *Nanomaterials* **2020**, *10*, 1325. [CrossRef]
18. Wang, L.; Li, T.; Du, Y.; Chen, C.; Li, B.; Zhou, M.; Dong, S. Au NPs-enhanced surface plasmon resonance for sensitive detection of mercury(II) ions. *Biosens. Bioelectron.* **2010**, *25*, 2622–2626. [CrossRef]
19. Lee, K.S.; Son, J.M.; Jeong, D.Y.; Lee, T.S.; Kim, W.M. Resolution enhancement in surface plasmon resonance sensor based on waveguide coupled mode by combining a bimetallic approach. *Sensors* **2010**, *10*, 11390–11399. [CrossRef]
20. Chang, C.C.; Lin, S.; Wei, S.C.; Chen, C.Y.; Lin, C.W. An amplified surface plasmon resonance "turn-on" sensor for mercury ion using gold nanoparticles. *Biosens. Bioelectron.* **2011**, *30*, 235–240. [CrossRef]
21. Fauzi, N.I.M.; Fen, Y.W.; Omar, N.A.S.; Saleviter, S.; Daniyal, W.M.E.M.M.; Hashim, H.S.; Nasrullah, M. Nanostructured chitosan/maghemite composites thin film for potential optical detection of mercury ion by surface plasmon resonance investigation. *Polymers* **2020**, *12*, 1497. [CrossRef] [PubMed]
22. Wu, F.; Thomas, P.A.; Kravets, V.G.; Arola, H.O.; Soikkeli, M.; Iljin, K.; Kim, G.; Kim, M.; Shin, H.S.; Andreeva, D.V.; et al. Layered material platform for surface plasmon resonance biosensing. *Sci. Rep.* **2019**, *9*, 20286. [CrossRef]
23. Zhang, P.; Chen, Y.P.; Wang, W.; Shen, Y.; Guo, J.S. Surface plasmon resonance for water pollutant detection and water process analysis. *Trends Anal. Chem.* **2016**, *85*, 153–165. [CrossRef]
24. Zhang, Y.; Xu, M.; Wang, Y.; Toledo, F.; Zhou, F. Studies of metal ion binding by apo-metallothioneins attached onto preformed self-assembled monolayers using a highly sensitive surface plasmon resonance spectrometer. *Sens. Actuators B Chem.* **2007**, *123*, 784–792. [CrossRef]
25. Shaban, M.; Hady, A.G.A.; Serry, M. A new sensor for heavy metals detection in aqueous media. *IEEE Sens. J.* **2014**, *14*, 436–441. [CrossRef]
26. Zijlstra, P.; Paulo, P.M.R.; Yu, K.; Xu, Q.H.; Orrit, M. Chemical interface damping in single gold nanorods and its near elimination by tip-specific functionalization. *Angew. Chemie Int. Ed.* **2012**, *51*, 8352–8355. [CrossRef] [PubMed]

27. Kim, S.-H.; Han, S.-K.; Jang, G.-S.; Koh, K.-N.; Kang, S.-W.; Keum, S.-R.; Yoon, C.-M. Surface plasmon resonance study on the interaction of a dithiosquarylium dye with metal ions. *Dyes Pigm.* **2000**, *44*, 169–173. [CrossRef]

28. Panta, Y.M.; Liu, J.; Cheney, M.A.; Joo, S.W.; Qian, S. Ultrasensitive detection of mercury (II) ions using electrochemical surface plasmon resonance with magnetohydrodynamic convection. *J. Colloid Interface Sci.* **2009**, *333*, 485–490. [CrossRef] [PubMed]

29. Daniyal, W.M.E.M.M.; Fen, Y.W.; Abdullah, J.; Saleviter, S.; Omar, N.A.S. Preparation and characterization of hexadecyltrimethy-lammonium bromide modified nanocrystalline cellulose/graphene oxide composite thin film and its potential in sensing copper ion using surface plasmon resonance technique. *Optik* **2018**, *173*, 71–77. [CrossRef]

30. Wang, C.; Zhang, C.; Zhao, Y.; Yan, X.; Cao, P. Poisoning Effect of SO_2 on honeycomb cordierite-based $Mn–Ce/Al_2O_3$ catalysts for NO reduction with NH_3 at low temperature. *Appl. Sci.* **2018**, *8*, 95. [CrossRef]

31. Fen, Y.W.; Yunus, W.M.M.; Talib, Z.A. Analysis of Pb(II) ion sensing by crosslinked chitosan thin film using surface plasmon resonance spectroscopy. *Optik* **2013**, *124*, 126–133. [CrossRef]

32. Hong, S.; Kang, T.; Moon, J.; Oh, S.; Yi, J. Surface plasmon resonance analysis of aqueous copper ions with amino-terminated self-assembled monolayers. *Colloid Surf. A Physicochem. Eng. Asp.* **2007**, *292*, 264–270. [CrossRef]

33. Raj, D.R.; Prasanth, S.; Vineeshkumar, T.V.; Sudarsanakumar, C. Surface plasmon resonance based fiber optic sensor for mercury detection using gold nanoparticles PVA hybrid. *Opt. Commun.* **2016**, *367*, 102–107.

34. Eddin, F.B.K.; Fen, Y.W. The principle of nanomaterials based surface plasmon resonance biosensors and its potential for dopamine detection. *Molecules* **2020**, *25*, 2769. [CrossRef] [PubMed]

35. Chah, S.; Yi, J.; Zare, R.N. Surface plasmon resonance analysis of aqueous mercuric ions. *Sensors Actuators B Chem.* **2004**, *99*, 216–222. [CrossRef]

36. Yu, J.C.C.; Lai, E.P.C.; Sadeghi, S. Surface plasmon resonance sensor for Hg(II) detection by binding interactions with polypyrrole and 2-mercaptobenzothiazole. *Sens. Actuators B Chem.* **2004**, *101*, 236–241. [CrossRef]

37. Lee, S.M.; Kang, S.W.; Kim, D.U.; Cui, J.Z.; Kim, S.H. Effect of metal ions on the absorption spectra and surface plasmon resonance of an azacrown indoaniline dye. *Dyes Pigm.* **2001**, *49*, 109–115. [CrossRef]

38. Hur, Y.; Ock, K.; Kim, K.; Jin, S.; Gal, Y.; Kim, J.; Kim, S.; Koh, K. Surface plasmon resonance study on enhanced refractive index change of an Ag^+ ion-sensing membrane containing dithiosquarylium dye. *Anal. Chim. Acta* **2002**, *460*, 133–139. [CrossRef]

39. Forzani, E.S.; Zhang, H.; Chen, W.; Tao, N. Detection of heavy metal ions in drinking water using a high-resolution differential surface plasmon resonance sensor. *Environ. Sci. Technol.* **2005**, *39*, 1257–1262. [CrossRef]

40. Chau, Y.C.; Wang, C.; Shen, L.; Lim, C.M.; Chao, C.C.; Huang, H.J.; Lin, C.; Kumara, N.T.R.N. Simultaneous realization of high sensing sensitivity and tunability in plasmonic nanostructures arrays. *Sci. Rep.* **2017**, *7*, 16817. [CrossRef]

41. Pelossof, G.; Tel-Vered, R.; Willner, I. Amplified surface plasmon resonance and electrochemical detection of Pb^{2+} ions using the Pb^{2+}-dependent DNAzyme and hemin/G-quadruplex as a label. *Anal. Chem.* **2012**, *84*, 3703–3709. [CrossRef] [PubMed]

42. Kim, E.J.; Chung, B.H.; Lee, H.J. Parts per trillion detection of Ni(II) ions by nanoparticle-enhanced surface plasmon resonance. *Anal. Chem.* **2012**, *84*, 10091–10096. [CrossRef]

43. May, L.M.; Russell, D.A. Novel determination of cadmium ions using an enzyme self-assembled monolayer with surface plasmon resonance. *Anal. Chim. Acta* **2003**, *500*, 119–125. [CrossRef]

44. Lin, T.J.; Chung, M.F. Detection of cadmium by a fiber-optic biosensor based on localized surface plasmon resonance. *Biosens. Bioelectron.* **2009**, *24*, 1213–1218. [CrossRef]

45. Wang, S.; Forzani, E.S.; Tao, N. Detection of heavy metal ions in water by high-resolution surface plasmon resonance spectroscopy combined with anodic stripping voltammetry. *Anal. Chem.* **2007**, *79*, 4427–4432. [CrossRef] [PubMed]

46. Verma, R.; Gupta, B.D. Detection of heavy metal ions in contaminated water by surface plasmon resonance based optical fibre sensor using conducting polymer and chitosan. *Food Chem.* **2015**, *166*, 568–575. [CrossRef]

47. Eum, N.-S.; Lee, S.-H.; Lee, D.-R.; Kwon, D.-K.; Shin, J.-K.; Kim, J.-H.; Kang, S.-W. K^+-ion sensing using surface plasmon resonance by NIR light source. *Sens. Actuators B Chem.* **2003**, *96*, 446–450. [CrossRef]

48. Zainuddin, N.H.; Fen, Y.W.; Alwahib, A.A.; Yaacob, M.H.; Bidin, N.; Omar, N.A.S.; Mahdi, M.A. Detection of adulterated honey by surface plasmon resonance optical sensor. *Optik* **2018**, *168*, 134–139. [CrossRef]

49. Fen, Y.W.; Yunus, W.M.M.; Moksin, M.M.; Talib, Z.A.; Yusof, N.A. Surface plasmon resonance optical sensor for mercury ion detection by crosslinked chitosan thin film. *J. Optoelectron. Adv. Mater.* **2011**, *13*, 279–285.

50. Kumar, P.; Kim, K.-H.; Bansal, V.; Lazarides, T.; Kumar, N. Progress in the sensing techniques for heavy metal ions using nanomaterials. *J. Ind. Eng. Chem.* **2017**, *54*, 30–43. [CrossRef]

51. Dhayal, M.; Ratner, D.M. XPS and SPR analysis of glycoarray surface density. *Langmuir* **2009**, *25*, 2181–2187. [CrossRef]

52. Martinis, E.M.; Wuilloud, R.G. Enhanced spectrophotometric detection of Hg in water samples by surface plasmon resonance of Au nanoparticles after preconcentration with vortex-assisted liquid-liquid microextraction. *Spectrochim. Acta A Mol. Biomol. Spectrosc.* **2016**, *167*, 111–115. [CrossRef]

53. Wu, C.M.; Lin, L.Y. Utilization of albumin-based sensor chips for the detection of metal content and characterization of metal-protein interaction by surface plasmon resonance. *Sens. Actuators B Chem.* **2005**, *110*, 231–238. [CrossRef]

54. Yao, F.; He, J.; Li, X.; Zou, H.; Yuan, Z. Studies of interaction of copper and zinc ions with Alzheimer's Aβ(1-16) using surface plasmon resonance spectrometer. *Sens. Actuators B Chem.* **2012**, *161*, 886–891. [CrossRef]

55. Fahnestock, K.J.; Manesse, M.; McIlwee, H.A.; Schauer, C.L.; Boukherroub, R.; Szunerits, S. Selective detection of hexachromium ions by localized surface plasmon resonance measurements using gold nanoparticles/chitosan composite interfaces. *Analyst* **2009**, *134*, 881–886. [CrossRef]

56. Roshidi, M.D.A.; Fen, Y.W.; Daniyal, W.M.E.M.M.; Omar, N.A.S.; Zulholinda, M. Structural and optical properties of chitosan-poly(amidoamine) dendrimer composite thin film for potential sensing Pb^{2+} using an optical spectroscopy. *Optik* **2019**, *185*, 351–358. [CrossRef]

57. Cennamo, N.; Massarotti, D.; Galatus, R.; Conte, L.; Zeni, L. Performance Comparison of Two Sensors Based on Surface Plasmon Resonance in a Plastic Optical Fiber. *Sensors* **2013**, *13*, 721–735. [CrossRef]

58. Qiu, G.Y.; Law, A.H.L.; Ng, S.P.; Wu, C.M.L. Label-free detection of lead(II) ion using differential phase modulated localized surface plasmon resonance sensors. *Procedia Eng.* **2016**, *168*, 533–536. [CrossRef]

59. Castillo, J.; Chirinos, J.; Gutiérrez, H.; La Cruz, M. Surface plasmon resonance sensor based on golden nanoparticles and cold vapour generation technique for the detection of mercury in aqueous samples. *Opt. Laser Technol.* **2017**, *94*, 34–39. [CrossRef]

60. Daniyal, W.M.E.M.M.; Fen, Y.W.; Abdullah, J.; Hashim, H.S.; Fauzi, N.I.M.; Chanlek, N.; Mahdi, M.A. X-ray photoelectron study on gold/nanocrystalline cellulose-graphene oxide thin film as surface plasmon resonance active layer for metal ion detection. *Thin Solid Films* **2020**, *713*, 138340–138350. [CrossRef]

61. Amjadi, M.; Shokri, R.; Hallaj, T. A new turn-off fluorescence probe based on graphene quantum dots for detection of Au(III) ion. *Spectrochim. Acta A Mol. Biomol. Spectrosc.* **2016**, *153*, 619–624. [CrossRef]

62. Balandin, A.A. Thermal properties of graphene and nanostructured carbon materials. *Nat. Mater.* **2011**, *10*, 569–581. [CrossRef]

63. Omar, N.A.S.; Fen, Y.W.; Saleviter, S.; Kamil, Y.M.; Daniyal, W.M.E.M.M.; Abdullah, J.; Mahdi, M.A. Experimental evaluation on surface plasmon resonance sensor performance based on sensitive hyperbranched polymer nanocomposite thin films. *Sensors Actuators A Phys.* **2020**, *303*, 111830–111840. [CrossRef]

64. Daghestani, H.N.; Day, B.W. Theory and applications of surface plasmon resonance, resonant mirror, resonant waveguide grating, and dual polarization interferometry biosensors. *Sensors* **2010**, *10*, 9630–9646. [CrossRef]

65. Jana, J.; Ganguly, M.; Pal, T. Enlightening surface plasmon resonance effect of metal nanoparticles for practical spectroscopic application. *RSC Adv.* **2016**, *6*, 86174–86211. [CrossRef]

66. Liu, C.; Liu, Q.; Hu, X. SPR phase detection for measuring the thickness of thin metal films. *Opt. Express* **2014**, *22*, 7574–7580. [CrossRef]

67. Fen, Y.W.; Yunus, W.M.M. Utilization of chitosan-based sensor thin films for the detection of lead ion by surface plasmon resonance optical sensor. *IEEE Sens. J.* **2013**, *13*, 1413–1418. [CrossRef]

68. Shushama, K.N.; Rana, M.M.; Inum, R.; Hossain, M.B. Sensitivity enhancement of graphene coated surface plasmon resonance biosensor. *Opt. Quantum Electron.* **2017**, *49*, 381. [CrossRef]

69. Yue, C.; Qin, Z.; Lang, Y.; Liu, Q. Determination of thin metal film's thickness and optical constants based on SPR phase detection by simulated annealing particle swarm optimization. *Opt. Commun.* **2019**, *430*, 238–245. [CrossRef]

70. Hasib, M.H.H.; Nur, J.N.; Rizal, C.; Shushama, K.N. Improved transition metal dichalcogenides-based surface plasmon resonance biosensors. *Condens. Matter* **2019**, *4*, 49. [CrossRef]

71. Abitbol, T.; Marway, H.; Cranston, E.D. Surface modification of cellulose nanocrystals with cetyltrimethylammonium bromide. *Nord. Pulp Pap. Res. J.* **2014**, *29*, 46–57. [CrossRef]

72. Fen, Y.W.; Yunus, W.M.M.; Talib, Z.A.; Yusof, N.A. Development of surface plasmon resonance sensor for determining zinc ion using novel active nanolayers as probe. *Spectrochim. Acta A Mol. Biomol. Spectrosc.* **2015**, *134*, 48–52. [CrossRef]

73. Babar, S.; Weaver, J.H. Optical constants of Cu, Ag, and Au revisited. *Appl. Opt.* **2015**, *54*, 477–481. [CrossRef]

74. Johnson, P.B.; Christy, R.W. Optical constant of the nobel metals. *Phys. Rev. B* **1972**, *6*, 4370–4379. [CrossRef]

75. Homola, J. *Surface Plasmon Resonance Based Sensors*; Springer Series on Chemical Sensors and Biosensors; Springer: Berlin/Heidelberg, Germany, 2006; Volume 4, pp. 3–44.

76. Kedenburg, S.; Vieweg, M.; Gissibl, T.; Giessen, H. Linear refractive index and absorption measurements of nonlinear optical liquids in the visible and near-infrared spectral region. *Opt. Mat. Express* **2012**, *2*, 1588–1611. [CrossRef]

77. Usman, F.; Dennis, J.O.; Seong, K.C.; Ahmed, A.Y.; Ferrell, T.L.; Fen, Y.W.; Sadrolhosseini, A.R.; Ayodele, O.B.; Meriaudeau, F.; Saidu, A. Enhanced sensitivity of surface plasmon resonance biosensor functionalized with doped polyaniline composites for the detection of low-concentration acetone vapour. *J. Sensors* **2019**, *2019*, 5786105. [CrossRef]

78. Zynio, S.A.; Samoylov, A.V.; Surovtseva, E.R.; Mirsky, V.M.; Shirshov, Y.M. Bimetallic layers increase sensitivity of affinity sensors based on surface plasmon resonance. *Sensors* **2002**, *2*, 62–70. [CrossRef]

79. Landry, V.; Alemdar, A.; Blanchet, P. Nanocrystalline cellulose: Morphological, physical, and mechanical properties. *For. Prod. J.* **2011**, *61*, 104–112. [CrossRef]

80. Anas, N.A.A.; Fen, Y.W.; Yusof, N.A.; Omar, N.A.S.; Daniyal, W.M.E.M.M.; Ramdzan, N.S.M. Highly sensitive surface plasmon resonance optical detection of ferric ion using CTAB/hydroxylated graphene quantum dots thin film. *J. Appl. Phys.* **2020**, *128*, 083105. [CrossRef]

81. Saleviter, S.; Fen, Y.W.; Daniyal, W.M.E.M.M.; Abdullah, J.; Sadrolhosseini, A.R.; Omar, N.A.S. Design and analysis of surface plasmon resonance optical sensor for determining cobalt ion based on chitosan-graphene oxide decorated quantum dots-modified gold active layer. *Opt. Express* **2019**, *27*, 32294–32307. [CrossRef]

82. Fouad, S.; Sabri, N.; Jamal, Z.A.Z.; Poopalan, P. Surface plasmon resonance sensor sensitivity enhancement using gold-dielectric material. *Int. J. Nanoelectron. Mater.* **2017**, *10*, 147–156.
83. Mudgal, N.; Saharia, A.; Agarwal, A.; Singh, G. ZnO and Bi-metallic (Ag–Au) layers based surface plasmon resonance (SPR) biosensor with BaTiO$_3$ and graphene for biosensing applications. *IETE J. Res.* **2020**, *1*, 1–8. [CrossRef]
84. Agarwal, S.; Giri, P.; Prajapati, Y.K.; Chakrabarti, P. Effect of surface roughness on the performance of optical spr sensor for sucrose detection: Fabrication, characterization, and simulation study. *IEEE Sens. J.* **2016**, *16*, 8865–8873. [CrossRef]
85. Hoseinian, M.S.; Bolorizadeh, M.A. Design and simulation of a highly sensitive SPR optical fiber sensor. *Photonic Sens.* **2019**, *9*, 33–42. [CrossRef]

Article

Optical Characterization of Ultra-Thin Films of Azo-Dye-Doped Polymers Using Ellipsometry and Surface Plasmon Resonance Spectroscopy

Najat Andam [1,2], Siham Refki [2], Hidekazu Ishitobi [3,4], Yasushi Inouye [3,4] and Zouheir Sekkat [1,2,4,*]

[1] Department of Chemistry, Faculty of Sciences, Mohammed V University, Rabat BP 1014, Morocco; najat.andam@um5r.ac.ma
[2] Optics and Photonics Center, Moroccan Foundation for Advanced Science, Innovation and Research, Rabat BP 10100, Morocco; s.refki@mascir.ma
[3] Frontiers Biosciences, Osaka University, Osaka 565-0871, Japan; ishitobi@ap.eng.osaka-u.ac.jp (H.I.); ya-inoue@ap.eng.osaka-u.ac.jp (Y.I.)
[4] Department of Applied Physics, Osaka University, Osaka 565-0871, Japan
[*] Correspondence: z.sekkat@mascir.ma

Abstract: The determination of optical constants (i.e., real and imaginary parts of the complex refractive index (n_c) and thickness (d)) of ultrathin films is often required in photonics. It may be done by using, for example, surface plasmon resonance (SPR) spectroscopy combined with either profilometry or atomic force microscopy (AFM). SPR yields the optical thickness (i.e., the product of n_c and d) of the film, while profilometry and AFM yield its thickness, thereby allowing for the separate determination of n_c and d. In this paper, we use SPR and profilometry to determine the complex refractive index of very thin (i.e., 58 nm) films of dye-doped polymers at different dye/polymer concentrations (a feature which constitutes the originality of this work), and we compare the SPR results with those obtained by using spectroscopic ellipsometry measurements performed on the same samples. To determine the optical properties of our film samples by ellipsometry, we used, for the theoretical fits to experimental data, Bruggeman's effective medium model for the dye/polymer, assumed as a composite material, and the Lorentz model for dye absorption. We found an excellent agreement between the results obtained by SPR and ellipsometry, confirming that SPR is appropriate for measuring the optical properties of very thin coatings at a single light frequency, given that it is simpler in operation and data analysis than spectroscopic ellipsometry.

Keywords: optical properties of ultra-thin dielectric films; surface plasmon spectroscopy; spectroscopic ellipsometry

Citation: Andam, N.; Refki, S.; Ishitobi, H.; Inouye, Y.; Sekkat, Z. Optical Characterization of Ultra-Thin Films of Azo-Dye-Doped Polymers Using Ellipsometry and Surface Plasmon Resonance Spectroscopy. *Photonics* **2021**, *8*, 41. https://doi.org/10.3390/photonics8020041

Received: 28 December 2020
Accepted: 27 January 2021
Published: 5 February 2021

Publisher's Note: MDPI stays neutral with regard to jurisdictional claims in published maps and institutional affiliations.

1. Introduction

Surface plasmons (SPs) are electromagnetic waves that are bound to metal/dielectric interfaces and are capable, among other things, of yielding the optical properties of very thin films (i.e., down to few angstroms) deposited on metal layers [1–3]. Optical characterization (i.e., determination of the complex refractive index (n_c) and thickness (d)) of thin films is required inasmuch as these films find applications in optoelectronics and photonics. Azo-dye-containing polymers are no exception in this regard. Besides SPs, other techniques can be applied to thin films, such as spectroscopic ellipsometry (SE) and photothermal deflection (PTD) spectroscopy. The latter can be applied to dye-doped polymers to measure optical absorption, requiring optical modeling such as Kramer–Kronig to get the refractive response. Unlike PTD, surface plasmon resonance (SPR) spectroscopy yields a direct measurement of the complex refractive index (i.e., n and κ) without the need for data transformation. PTD may be useful, however, where there is a necessity to differentiate between scattering and absorption losses [4,5]. In this paper, we report on the optical characterization of very thin films (thickness: $\sim$58 nm) of guest–host azo-dye-doped polymers at different dye/polymer concentrations. We use SP resonance (SPR) spectroscopy and spectroscopic ellipsometry (SE) to independently measure n_c and d of our samples by both techniques. The experimental SPR spectra are theoretically compared to Fresnel's reflectivity calculations of the layers system used in our study, and SE results are rationalized by the Lorentz equation for the resonance characteristics (i.e., the absorption of the azo dye, the host polymer being transparent in the frequency range studied, and Bruggeman's effective medium approximation for the dye polymer system, which is considered as a composite material).

SPs and localized surface plasmons (LSPs) were reported long ago [6,7], and their use is of the utmost importance in, for example, the field of bio-sensing [8–10], enhanced spectroscopies [11], dye-sensitized solar cells [12], and photocatalysis [13]. Important reviews summarize the developments in the field [9,14–19]. The research discussed in this paper is part of a series of works published by our group in the field of plasmonics, combining metals and dielectric organic materials. This includes work on surface-enhanced visible absorption of dye molecules by LSPs at gold nanoparticles [11], optical characterization of nano-thin layers of dielectric and metals [1,2,20–22], sensing of photoreactions in molecularly thin layers, plasmons coupling at metal-insulator-metal (MIM) structures, glucose sensing using such structures [23–25], plasmons and waveguides (WGs), WG–WG coupling generating Fano resonances (FR), external control of FR by light action, and so on (for examples, see [26,27]).

Azo-dye-containing materials have been studied extensively in past decades for applications in holography and optical data storage [28,29], nonlinear optics (NLO), electro-optic modulation (EO), second-harmonic generation (SHG) [30–33], photomechanical actuation and matter motion, and so on [34–40]. Azo dyes also have potential applications in dye chemistry and bio-photonics [41,42], driving a natural interest in their optical characterization, namely the determination of the real and imaginary parts of n_c as a function of the wavelength of light. For example, Prêtre et al. developed a method for calculating the real and imaginary parts of n_c and the first order molecular hyperpolarizability of NLO azo-polymers for SHG based on UV–vis absorption spectroscopy and SE [43], and Bodarenko et al. determined n_c for an azo-dye-doped polymer by SE, driven by the need for application of EO polymers in silicon-hybrid-organic photonics [44]. Several other researchers determined the optical properties of azo-dye-containing materials by SE [43,45,46]. The investigation of the optical properties of the DR1/PMMA nanocomposite material, which is considered as a mixture of two constituents (i.e., DR1 and PMMA), requires the use of a mixing law describing this material as an effective medium. Maxwell–Garnett (MG) and Bruggeman (BR) theories are usually employed to model the observations for composite materials [47,48]. It is not the purpose of this paper to discuss the detailed differences between these two theories, and we are only stating their usefulness in describing dye-doped polymer systems. For the DR1/PMMA material studied in this paper, we found that BR's theory better fits our observations compared to MG. Other authors compared the usefulness of the two theories to model the same type of materials and found that BR is more appropriate than MG for modeling azo-dye-doped PMMA [45], and that BR theory is appropriate for modeling NLO guest–host systems [46,49].

In this paper, we determine the real and imaginary parts of n_c for Disperse Red One (DR1), an azo dye that is well known for its photoisomerization and NLO properties [50], incorporated into a poly-methyl-methacrylate (PMMA) matrix at different dye/polymer concentrations by using SPR and SE, and we show that there is an excellent agreement between SPR and SE in the determination of the real and imaginary parts of n_c for quite thin (i.e., 58 nm) DR1/PMMA films. For this range of thicknesses of the DR1/PMMA dielectric layer, only SP is observed (vide infra), and waveguides occur for thicker films [51].

2. Materials and Methods

For UV–Visible absorption, SPR, and SE measurements, we prepared DR1/ PMMA thin films in the following manner. Commercially available DR1 and PMMA (product reference 182230), with an average molecular weight of M_w ~120,000 and a glass transition temperature Tg of 110 °C, were purchased from Sigma-Aldrich, and were used without purification. The statistical segment length of the PMMA (density: ~1.188 g·cm^{-3}) used in our study is not available to us, however the statistical segment length of PMMA of a density of 1.13 g·cm^{-3} is 6.5 Å [52]. A set of mixture solutions of DR1 and PMMA with different concentrations was prepared. First, we dissolved PMMA in chloroform with 0.83 weight percent (% w/w) concentration, and we stirred the solution for 12 h with a magnetic stirrer. Then we added the DR1 powder into the PMMA solution with 8%, 10%, and 12% w/w relative to PMMA. The DR1/PMMA solutions were also stirred for 12 h. The prepared solutions were then spin-coated onto cleaned glass substrates with appropriate speed to achieve the desired film thickness. Spin coating was performed with 5500 rpm to obtain thin films with thicknesses of ~58 nm. Finally, the films were dried at 80 °C for 30 min, followed by 115 °C for 1 min to remove the solvent remaining after spin-coating. We measured the film's thickness, d, by a profilometer (Bruker), at four different positions for each sample, and we found that our samples were homogeneous with $d = 58$ nm $\pm$ 4 nm.

3. Results and Discussion

3.1. UV-Visible Spectroscopy

The absorption spectra of the film samples were measured by using a UV–visible spectrophotometer (Perkin Elmer-Lambda 1050). Figure 1a shows the absorption spectra of our film samples (e.g., of DR1/PMMA thin films) with different weight concentrations of DR1 versus PMMA C (% w/w), and the same thickness (i.e., ~58 nm). For all the films prepared, the maximum absorption is observed at 488 nm and is assigned to the $\pi \rightarrow \pi^*$ transition of trans-DR1, as is well known from the literature [50].

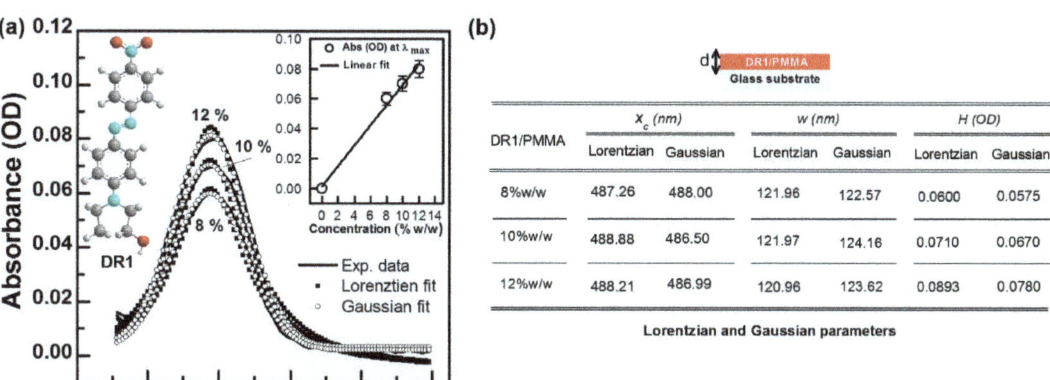

DR1/PMMA	x_c (nm)		w (nm)		H (OD)	
	Lorentzian	Gaussian	Lorentzian	Gaussian	Lorentzian	Gaussian
8%w/w	487.26	488.00	121.96	122.57	0.0600	0.0575
10%w/w	488.88	486.50	121.97	124.16	0.0710	0.0670
12%w/w	488.21	486.99	120.96	123.62	0.0893	0.0780

Figure 1. (a) Absorption spectra of DR1/PMMA thin films with different weight concentrations of DR1 versus PMMA C (% w/w). The inset shows a linear evolution of the maximum of absorption, at $\lambda_{max}^{DR1} = 488$ nm of the film, with its concentration, for a fixed film thickness d ~58 nm; and the absorbance extrapolate to the origin. The solid curves are the experimental data, and the close circles' curve is a Lorentzian theoretical fit $y = y_0 + (2A/\pi)(w/\left(4(x - x_c)^2 + w^2\right)$, while the open circles' curve is Gaussian theoretical fit $y = y_0 + Ae^{-\ln(2)\,(x-x_c)^2/w^2}/w\sqrt{\pi/4ln(2)}$, where y_0 is the baseline. (b) The values of the characteristic parameters of the Lorentzian and Gaussian are given in the table. The Lorentzian function does not fit the data as well as the Gaussian function at the infrared tail ~700 nm.

The inset to Figure 1a shows that the maximum absorption increases linearly with the increased concentration C (% w/w) of the dye according to $Abs\,(OD) = \varepsilon_t \cdot d \cdot C$ (i.e., the Lambert-Beer law), where ε_t is the molecular extinction coefficient of trans-DR1, d the thickness of the DR1/PMMA film, and C the concentration of DR1. Aggregation may occur when dye concentration exceeds 12% in weight [53,54]; however, in our case, we did not observe aggregation of DR1 at 12% w/w, since the shape of the absorption spectrum of the 12% sample is the same as those at lower concentrations, and the linear increase of the absorbance with the concentrations extrapolates to the origin (see inset to Figure 1a). This indicates that the chromophores can be considered as isolated for all concentrations studied.

Lorentz and Gaussian functions curve fitting are applied to the UV–visible spectra of the DR1/PMMA films by using the equations shown in the caption to Figure 1a for each function. Where x_c, is the center of the peak, w is the width of the peak at half height, and H is the amplitude (in OD units) defined by $H = 2A/w\pi$ for the Lorentzian and by $H = A/w\sqrt{\pi/4\ln(2)}$ for the Gaussian (where A is the area under the absorption band). These parameters (i.e., x_c, w, and H) are the most representative of each spectrum in Lorentz and Gaussian fitting. Each of the DR1/PMMA spectrums has unique Lorentz parameters as shown in Figure 1b. For all concentrations, we found that x_c is located at $\sim$488 nm (corresponding to central energy of $\sim$2.54 eV). H has different values corresponding to the optical densities of the different concentrations of DR1. Figure 1a shows that the Gaussian function fits the absorption spectra nicely as well and yields nearly the same theoretical spectra as those obtained by the Lorentzian fit, except at the tail $\sim$700 nm, where the Gaussian better fits the experimental data than the Lorentzian. The characteristic parameters of both functions, summarized in the table in Figure 1b, are close enough. We extracted the extinction coefficient of our sample using both sets of parameters, and we found nearly the same values of the extinction coefficients of the samples at the three concentrations studied. Indeed, using $\kappa = (\lambda\,\alpha/4\pi)$ for the extinction coefficient of the sample, with $\alpha = 2.303H/d$ being the optical density of the sample in units of [43], we found the values of 0.0925 (0.088), 0.1094 (0.1033), and 0.1279 (0.1202) at 488 nm, and 0.0065 (0.0069), 0.0108 (0.0093), and 0.0115 (0.0105) at 632.8 nm, for κ for the 8%, 10%, and 12% samples, respectively. The values between parentheses are those corresponding to the Gaussian function. The values of κ found by UV–vis spectroscopy are nearly the same as those we found by SPR spectroscopy and SE (vide infra).

3.2. SPR Spectroscopy

We will briefly recall the principle of SPR spectroscopy, and how it is used to calculate the optical properties of very thin coatings (up to a few tens of nanometers). Then, we will discuss our experimental results of SPR observations in films of the same thickness and different dye concentrations. SPs are electromagnetic waves that are bound to and propagate along the interface between metal and dielectric. They penetrate to different extents into the metal and the dielectric, with penetration depths given by their field amplitudes decaying exponentially perpendicular to the interface [1,55–57]. The dispersion relation of SPs is given by:

$$k_{SP} = \frac{\omega}{c}\sqrt{\frac{\varepsilon_m(\omega).\varepsilon_d(\omega)}{\varepsilon_m(\omega) + \varepsilon_d(\omega)}} \tag{1}$$

where k_{SP}^0 is the longitudinal component of the SP wave vector and is proportional to the wave frequency ω, and c is the speed of light in vacuum. $\varepsilon_d(\omega)$ and $\varepsilon_m(\omega)$ are the complex dielectric constants of the dielectric and the metal, respectively, and they are light-frequency-dependent. A free wave of wave-vector k and frequency ω propagating in the dielectric, at an incidence angle θ, is characterized by:

$$k_{photon}^0 = \frac{\omega}{c}\sqrt{\varepsilon_d(\omega)}\sin\theta \tag{2}$$

Equations (1) and (2) indicate that a free wave, characterized by a line (i.e., the light line on Figure 2a), cannot couple to an SP wave unless its k vector is augmented, by, for example, adding a prism, or a grating that brings the additional longitudinal k vector required for coupling [1]. When coupling takes place, a resonance phenomenon is observed in the attenuated total reflection (ATR) spectrum, as can be seen from, for example, an angular scan (Figure 2b). In such a scan, θ_c is a critical angle at which light incident on the prism/metal is totally reflected from the base of the prism, a phenomenon referred to as total internal reflection (TIR). Below θ_c, the metal film acts as a mirror, and reflectivity is still high with little transmission, while above θ_c, TIR takes place, and ATR occurs via coupling to a resonance (i.e., SPR). A dip occurs in the reflectivity curve at θ_0. The coupling angle (i.e., θ_0) is given by the energy and momentum matching condition between surface plasmons and photons. When a dielectric coating is applied to the metal layer, the resonance condition changes and θ_0 shifts to larger angles θ_1 (Figure 3a). From this shift and Fresnel's calculations, one can calculate the optical thickness of the coating. This is what is discussed next for our DR1/PMMA thin films.

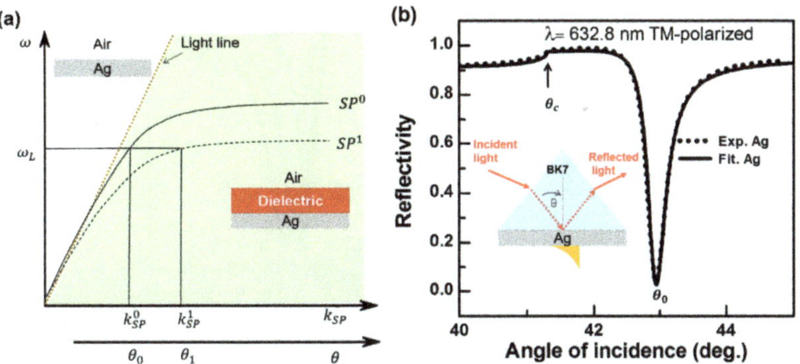

Figure 2. (a) Dispersion relation of *SPs* (i.e., ω vs. k_{SP} (e.g., Equation (1)), referring to the *SP* at an Ag-air interface (SP^0), and when a thin dielectric coating (DC) is deposited on the Ag layer (SP^1). The light line is also indicated in this figure. A light frequency ω_L crosses the dispersion curves at k_{SP}^0 and k_{SP}^1, and it determines the coupling angle θ_0 and θ_1, for k_{SP}^0 and k_{SP}^1, respectively. The thin DC shifts the SP dispersion curve to higher coupling ($k_{SP}^1 = k_{SP}^0 + \Delta k_{SP}$, i.e., resonance, angle, and momentum). (b) Experimentally observed SPR at a prism/Ag (thickness~56 nm) system with a 638 nm laser light. The scatters are experimental data points, and the solid curve is a Fresnel's theoretical fit.

As schematically depicted in Figure 3a, the bilayer stack used for our SPR measurements consisted of an Ag layer (thickness: ~56 nm) coated with a 58-nm thin layer of DR1/PMMA. The stack was deposited onto cleaned BK7 glass slides (15×22 mm with a refractive index of 1.5151 at $\lambda = 632.8$ nm [58]), which were put in contact with a 90° BK7 glass prism by an index-matching oil. The DR1/PMMA dielectric layer was deposited on Ag by the spin-coating process. To measure the angle-scan ATR spectra (i.e., SPR in Kretschmann configuration), we used a custom-made optical setup, which is described in detail in [59]. To do so, the sample/prism system was mounted on a (θ, 2θ) rotating stage, where θ is the internal angle of incidence (see Figure 3a). The sample/prism system was illuminated by P-polarized (e.g., transverse magnetic (TM)-polarized) light from a Helium–Neon (He–Ne) laser operating at the wavelength of 632.8 nm, with few µW of power to avoid absorption by the sample, even though at this wavelength the sample presents negligible absorption. The reflected light was measured by using a Si photodiode connected to a lock-in-amplifier as a function of θ. The precision of the measurement of θ was 0.018°. The error that this precision can cause on reflectivity measurements is $\Delta R = 7 \times 10^{-3}$.

Figure 3a shows the experimental and the theoretical fit ATR angular spectra for an Ag layer (~56 nm) without and with DR1/PMMA (~58 nm) at different weight concentrations C (8%, 10%, and 12% *w/w*). Electromagnetic (EM) calculations of the reflectivity of light at multilayer structures based on Fresnel reflection were performed using a freely available software package (Winspall) to reproduce experimental ATR spectra. We used a value of the refractive index at the wavelength $\lambda = 632.8$ nm, for BK7 prism of $n_g = 1.5151$ from a database [58], and we extracted the refractive index and the thicknesses of Ag, which led to the theoretical spectrum that fits the experimental one well. Theoretical fits to the experimental data yielded the values $n_{Ag} = 0.0679 + i4.087$ of the refractive index and the thicknesses (i.e., 56 nm) of the Ag layer. The refractive index of silver determined in our work (0.0679 + i4.087) is close to that found in the literature (0.0562 + i4.276) [58,60]. The thickness of the Ag layer obtained from the theoretical SPR curves was identical to that measured by profilometry.

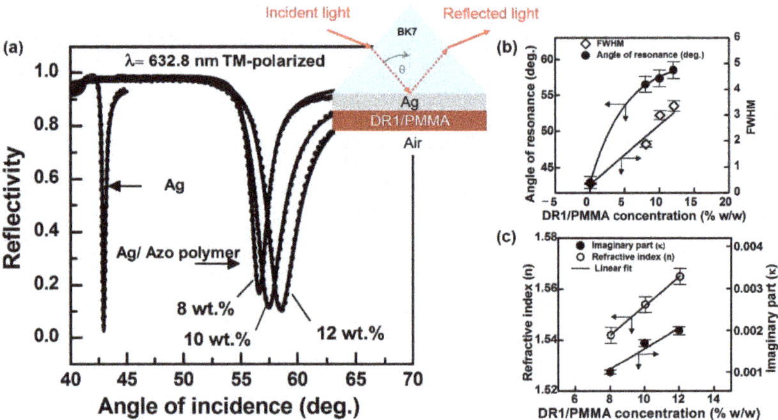

Figure 3. Angle-scan ATR spectra in a Kretschmann configuration for a single Ag layer (56 nm) at a wavelength of 632.8 nm. (**a**) Experimental (dots) and theoretical (solid curve) ATR spectra of an Ag layer sample without and with DR1/PMMA thin films ($d \sim 58$ nm) at 8%, 10%, and 12% w/w as indicated on the figure. (**b**) Dependence of SPR angle and full width at half maximum (FWHM) versus DR1 concentration in PMMA. (**c**) Refractive index (real, n, and imaginary, κ, parts) of DR1/PMMA versus DR1 concentration. Note the linear dependence of n and κ of DR1/PMMA on the DR1 concentration. The scatters are experimental data adapted from the SPR curves at the corresponding concentrations, and the full lines are guides to the eye.

In Figure 3a, we see that the ATR scan at the prism/Ag system exhibits a sharp dip at 42.96°, due to the excitation of SP at the Ag/Air interface. The full width at half maximum (FWHM) of this dip was about 0.34°. After adding the outermost layer of 58 nm of DR1/PMMA, the resonance shifts towards higher incidence angles: 56.59°, 58.13°, and 59.66° for the DR1 concentration of 8%, 10%, and 12% w/w, respectively (Figure 3b). This figure also shows that the FWHM of the mode increases linearly with the dye concentration owing to the increase of losses (i.e., linear increase of κ with the DR1 concentration). Indeed, by fitting the experimental data of SPR observed with DR1/PMMA, we extracted the optical properties (n and κ) of the outermost layer (i.e., DR1/PMMA) for different concentrations of DR1 (Figure 3c). We found a linear relationship between the dye concentration and the mode shift, as well as the refractive index (n and κ) of the material. This result confirms the fact that both the refractive index and the extinction coefficient of material are proportional to the molecules concentration in agreement with the UV–vis data of the previous section (see also [61]). Next, we discuss the determination of $n_c = n + i\kappa$, i.e., n and κ, by SE.

3.3. Spectroscopic Ellipsometry

Ellipsometry data were acquired using a commercially available variable angle SE system (VASETM, J.A. Woollam). The principle of operation of SE is schematically depicted in Figure 4. A light beam is first polarized by passing through a polarizer, and is then reflected from the sample surface at an angle of incidence φ. After reflection, light passes through a second polarizer, which is called an analyzer, and it falls onto the detector. Our sample consists of a thin film of DR1/PMMA (vide infra) deposited on the top of an ITO glass (~147 nm). All measurements were performed in air at room temperature for three angles of incidence (60°, 70°, and 80°) and in the wavelength range 400–1000 nm. The data analysis was made using Complete EASE software (version 5.04). The SE method is based on Fresnel reflection coefficients measurement [62], which is given by:

$$\rho = \frac{r_p}{r_s} = \tan(\psi)e^{i\Delta} \tag{3}$$

where r_p and r_s are the complex Fresnel reflection coefficients for parallel (P-) and perpendicular (S-) polarized (i.e., transverse electric (TE)) light to the plane of incidence, respectively, and delta (Δ) and psi (Ψ) are the ellipsometry angles [62,63]. We determined the optical properties, n and κ, of DR1/PMMA at different dye concentrations, as functions of the wavelength, by fitting the Δ and Ψ using the SE software (Figure 4). The most appropriate way of simulating the dielectric functions of the DR1/PMMA thin films is to treat the films as a mixture of two constituents. Generally, Bruggeman effective medium approximation (B-EMA) is frequently used to calculate the dielectric function of the composite layer based on the volume ratios (f) and the dielectric functions of each constituent [62–64]. The complex dielectric constant of DR1/ PMMA was fitted using B-EMA, given by:

$$f_v^{DR1}\frac{\widetilde{\varepsilon}_{DR1} - \widetilde{\varepsilon}}{\widetilde{\varepsilon}_{DR1} + 2\widetilde{\varepsilon}} + f_v^{PMMA}\frac{\widetilde{\varepsilon}_{PMMA} - \widetilde{\varepsilon}}{\widetilde{\varepsilon}_{PMMA} + 2\widetilde{\varepsilon}} = 0 \tag{4}$$

where $\widetilde{\varepsilon} = \varepsilon' + i\varepsilon''$ is the effective complex dielectric function of the mixture, with $\varepsilon' = n^2 - \kappa^2$ and $\varepsilon'' = 2n\kappa$, and $\widetilde{\varepsilon}_{DR1}$ and $\widetilde{\varepsilon}_{PMMA}$ are the complex dielectric functions of DR1 and PMMA, respectively. f_v^{DR1} and f_v^{PMMA} are the volume fractions of DR1 and PMMA with $(f_v^{DR1} + f_v^{PMMA} = 1)$. The values of volume fractions can be calculated by using Equation (5).

$$f_v^{DR1} = \frac{f_w^{DR1}}{f_w^{DR1} + (1 - f_w^{DR1}).\frac{\rho_{DR1}}{\rho_{PMMA}}} \tag{5}$$

where f_w^{DR1} is the weight fraction of DR1, and $\rho_{DR1} = 1.1523$ g·cm^{-3} and $\rho_{PMMA} = 1.1880$ g·cm^{-3} are the densities of DR1 and PMMA, respectively. We determined the optical properties of the DR1/PMMA mixture as follows: for the PMMA constituent, we used the dielectric functions parameterized in the database of the software, and for the DR1 constituent (i.e., an absorbing compound), we used the Lorentz model describing absorption. The Lorentz parameters are obtained by fitting the experimental data of Δ and Ψ of the DR1/PMMA mixture for each concentration, allowing us to extract n and κ of DR1 and the DR1/PMMA composite.

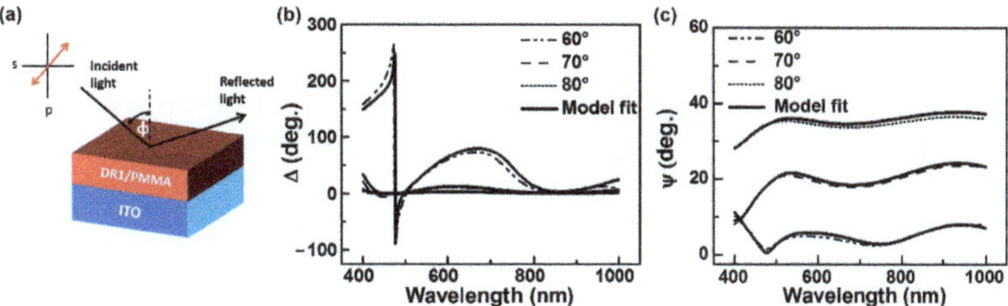

Figure 4. (a) Schematic of light reflection on a layered stack (i.e., polymer film on top of an ITO layer), which we used in our SE experiments. The experimental data, measured at three different angles (i.e., 60°, 70°, and 80° incidence angles) are (b) delta (Δ) angle and (c) psi (ψ). Experiments were performed on DR1/PMMA thin films (thickness ~ 58 nm) at 8%, 10%, and 12% w/w concentrations (only the data of 8% w/w are shown). Dotted curves are experimental data and the theoretical fits are represented by full curves. See text for more details.

It is well known that Lorentz oscillators are primarily useful for describing resonant absorption peaks due to doping of a transparent dielectric material with an absorbing dye [63]. DR1 is one of such dyes, and it absorbs light in the UV–visible range. It has an absorption peak centered at 488 nm. When DR1 is introduced as a guest into a PMMA matrix, the absorption of the DR1 can still be described by the Lorentz formula, which gives the dimensionless complex dielectric function of DR1 as a function of the photon energy E (Equation (6)).

$$\varepsilon(E) = \varepsilon' + i\varepsilon'' = \varepsilon'(\infty) + \sum_{k=1}^{N} \frac{A_k}{E_k^2 - E^2 - iB_k E} \tag{6}$$

where ε' and ε'' are the real and imaginary parts of the complex dielectric function of DR1, $\varepsilon'(\infty)$ is its real dielectric function at infinite energy (offset term), and N is the number of oscillators. Each oscillator is described by three parameters: A_κ is the amplitude of the k_{th} oscillator in units of $(eV)^2$, B_κ is the broadening of the k_{th} oscillator in units of eV, and E_κ is the center energy of the k_{th} oscillator [45]. For DR1, a single oscillator centered at 2.5 eV yields the best fit value, corresponding to 495 nm. This value is close to (i.e., shifted by 7 nm) the experimentally observed wavelength (i.e., 488 nm), corresponding to the maximum absorption of DR1 in the visible region of the spectrum (Figure 1a). SE data allows for the determination of n and κ of all constituents of a composite material, in our case DR1 and PMMA, as well as their composition, and we could extract n and κ of neat DR1. To do so, we used both the model of Brugemann for the composite material and the Lorentz model to account for the absorption of DR1 in the composite (vide infra). Data for PMMA is available from the literature and from the Complete EASE software. We did not measure neat PMMA independently, but the theoretical dispersion curve of PMMA extracted from the database. (i.e., from EASE software) and used in our Bruggeman effective medium match well with that of the literature [44]. Indeed, the refractive index of PMMA extracted from the EASE software at 632.8 nm is $n_{632.8\ nm}^{PMMA} = 1.4888$, and is close enough to the value of $n_{632.8\ nm}^{PMMA} = 1.4889$ that we found in our previous works using plasmonic structures [59]. The fit parameters of DR1 corresponding to Equation (6) (i.e., $\varepsilon'(\infty)$, A_1, E_1, B_1) are shown on the table as an inset to Figure 5a. These fitting parameters allow for the determination of the dispersion curve of n and κ as a function of the wavelength using the data of Δ and Ψ. We found the same values of the parameters of the table in the inset of Figure 5a for all 3 concentrations (i.e., 8%, 10%, and 12% of w/w) of DR1 compared to PMMA. These values yielded the best theoretical fits of the experimental data.

Figure 5 shows the dispersive curves of optical constants (n and κ) of neat DR1 (Figure 5a) and DR1/PMMA composite at different concentrations (Figure 5b–d) as obtained from SE experiments. Figure 5a shows that for neat DR1, the value of κ at the wavelength of maximum extinction (i.e., 495 nm) is 1.066. This is in good agreement with the value of κ (i.e., 1.077) which we extracted from [43] using the data of this reference.

The extinction coefficient κ, which is related to the absorption coefficient α by $\kappa = (\lambda\alpha/4\pi)$, is calculated to be 1.077 using $\lambda = 478$ nm and $H/d = 123 \times 10^3 \mathrm{cm}^{-1}$, with H/d calculated in [43], using 478 nm corresponding to the maximum absorption of DR1 measured in chloroform. Comparing n and κ of DR1/PMMA in Figure 5b–d with those for neat DR1 in Figure 5a, we see that the values of n and κ of DR1/PMMA are lower than those of neat DR1, and that n and κ increase with the increased concentration of DR1 in the DR1/PMMA composite. The molecular polarizability of DR1 is larger than that of PMMA. For both SPR and ellipsometry, the measurements at each concentration were repeated three times and the value of n and κ was calculated as the average of the three measurements for each sample. Table 1 summarizes the results of n and κ of all the samples measured with UV–vis spectroscopy, SE, and SPR, together with data from the literature [43,44]. A good agreement between the data from our paper and the literature is obtained, and interestingly enough, n and κ obtained from SE measurements are reasonably close to those obtained by SPR spectroscopy.

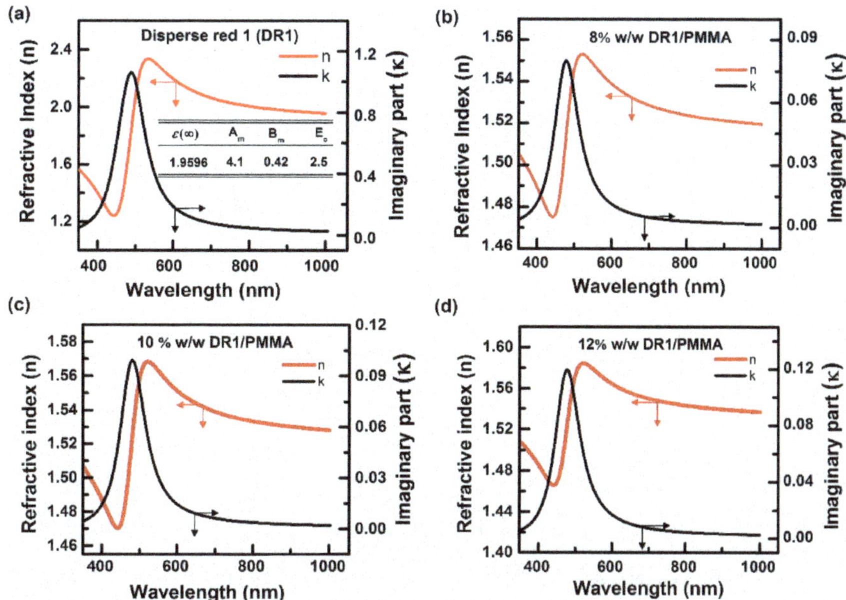

Figure 5. n and κ as a function of the wavelength for (**a**) neat DR1, (**b**) 8% w/w DR1/PMMA, (**c**) 10% w/w DR1/PMMA, and (**d**) 12% w/w DR1/PMMA. The film's thickness was d ∼58 nm.

Table 1. Optical Constants of neat DR1 and thin films of DR1/PMMA at 488 nm and 632.8 nm. The precision of the measured data by spectroscopic ellipsometry and SPR is 2×10^{-3} and 4×10^{-4} for n, and 4×10^{-3} and 1×10^{-4} for κ, respectively.

DR1/PMMA Concentration	Literature		SPR [a]		Ellipsometry [a]				UV-Vis [a]	
	632.8 nm		632.8 nm		488 nm		632.8 nm		488 nm	632.8 nm
	n	κ	n	κ	n	κ	n	κ	κ	κ
Neat DR1	2.0536 [b]	-	-	-	1.835	1.058	2.136	0.126	-	-
8% w/w	-	-	1.5401	0.0075	1.534	0.082	1.537	0.008	0.0925	0.0065
10% w/w	1.5450 [c]	0.008 [c]	1.5540	0.0101	1.545	0.098	1.550	0.011	0.1094	0.0108
12% w/w	-	-	1.5650	0.0120	1.553	0.120	1.561	0.012	0.1279	0.0115

[a] This paper; [b] From [43]; [c] From [44].

4. Conclusions

The optical properties, n and κ, of azo-dye-contacting materials are often researched for photonics applications, especially for NLO including EO modulation and SHG. In this paper, we determined n and κ of DR1-doped PMMA for different dye concentrations, a feature which is useful for researchers working in the fields of, for example, NLO and plasmonics. We also demonstrate the potential of SPR spectroscopy in determining such properties for ultrathin films of the NLO dye-doped polymer studied, especially that SPR mode shifts were observed at different concentrations of DR1. Both the determination of n and κ and the observed SPR modes shifts for different dye concentrations constitute the originality of this work. We also used UV–vis spectroscopy and SE with Lorentz theory and Bruggeman effective medium approximation to determine n and κ of the films studied. The results obtained by SPR measurements are in good agreement with those obtained by UV–vis and ellipsometric spectroscopies.

Author Contributions: Z.S. conceived the idea and wrote the paper; N.A. and S.R. did the experiments; H.I. and Y.I. helped with the spectroscopic ellipsometry experiments and discussed the results; Z.S. gave overall guidance. All authors have read and agreed to the published version of the manuscript.

Funding: This work was funded by the priority research area project "Next Generation Optical Media Recordable Via Nano-photonics" (MEFCRS PPR/2015/69) funded by the Moroccan Ministry of Higher Education and Research.

Data Availability Statement: The data is available from the corresponding author upon reasonable request.

Conflicts of Interest: The authors declare no conflict of interest.

References

1. Sekkat, Z.; Wood, J.; Geerts, Y.; Knoll, W. Surface Plasmon Investigations of Light-Induced Modulation in the Optical Thickness of Molecularly Thin Photochromic Layers. *Langmuir* **1996**, *12*, 2976–2980. [CrossRef]
2. Bartual-Murgui, C.; Salmon, L.; Akou, A.; Thibault, C.; Molnár, G.; Mahfoud, T.; Sekkat, Z.; Real, J.A.; Bousseksou, A. High quality nano-patterned thin films of the coordination compound {Fe(pyrazine)[Pt(CN)4]} deposited layer-by-layer. *New J. Chem.* **2011**, *35*, 2089–2094. [CrossRef]
3. Shah, D.; Reddy, H.; Kinsey, N.; Shalaev, V.M.; Boltasseva, A. Optical Properties of Plasmonic Ultrathin TiN Films. *Adv. Opt. Mater.* **2017**, *5*, 1700065. [CrossRef]
4. Huang, H.T.; Huang, C.Y.; Ger, T.R.; Wei, Z.H. Anti-integrin and integrin detection using the heat dissipation of surface plasmon resonance. *Appl. Phys. Lett.* **2013**, *102*, 111109. [CrossRef]
5. Krasilnikova, S.A. Measurement of optical constants of thin films by non conventional ellipsometry, photothermal deflection spectroscopy and plasmon resonance spectroscopy. In Proceedings of the Advances in Optical Thin Films III, 71010R, Glasgow, Scotland, UK, 2–5 September 2008; Volume 7101. [CrossRef]
6. Wood, R.W. On a Remarkable Case of Uneven Distribution of Light in a Diffraction Grating Spectrum. *Proc. Phys. Soc. Lond.* **1902**, *18*, 269–275. [CrossRef]
7. Mie, G. Beiträge zur Optik trüber Medien, speziell kolloidaler Metallösungen. *Ann. Phys.* **1908**, *330*, 377–445. [CrossRef]
8. Nguyen, H.; Park, J.; Kang, S.; Kim, M. Surface Plasmon Resonance: A Versatile Technique for Biosensor Applications. *Sensors* **2015**, *15*, 10481–10510. [CrossRef] [PubMed]
9. Brolo, A.G. Plasmonics for future biosensors. *Nat. Photon.* **2012**, *6*, 709–713. [CrossRef]
10. Kashyap, R.; Chakraborty, S.; Zeng, S.; Swarnakar, S.; Kaur, S.; Doley, R.; Mondal, B. Enhanced Biosensing Activity of Bimetallic Surface Plasmon Resonance Sensor. *Photonics* **2019**, *6*, 108. [CrossRef]
11. Elhani, S.; Ishitobi, H.; Inouye, Y.; Ono, A.; Hayashi, S.; Sekkat, Z. Surface Enhanced Visible Absorption of Dye Molecules in the Near-Field of Gold Nanoparticles. *Sci. Rep.* **2020**, *10*, 3913. [CrossRef] [PubMed]
12. Jeong, N.C.; Prasittichai, C.; Hupp, J.T. Photocurrent Enhancement by Surface Plasmon Resonance of Silver Nanoparticles in Highly Porous Dye-Sensitized Solar Cells. *Langmuir* **2011**, *27*, 14609–14614. [CrossRef]
13. Zhou, X.; Liu, G.; Yu, J.; Fan, W. Surface plasmon resonance-mediated photocatalysis by noble metal-based composites under visible light. *J. Mater. Chem.* **2012**, *22*, 21337. [CrossRef]
14. Knoll, W. Interfaces and thin films as seen by bound electromagnetic waves. *Annu. Rev. Phys. Chem.* **1998**, *49*, 569–638. [CrossRef] [PubMed]
15. Maier, S.A. *Plasmonics: Fundamentals and Applications*; Springer: New York, NY, USA, 2007.
16. Brongersma, M.L.; Shalaev, V.M. The Case for Plasmonics. *Science* **2010**, *328*, 440–441. [CrossRef] [PubMed]
17. Kawata, S. *Near-Field Optics and Surface Plasmon Polaritons*; Springer: Berlin, Germany; New York, NY, USA, 2001.
18. Kawata, S.; Inouye, Y.; Verma, P. Plasmonics for near-field nano-imaging and superlensing. *Nat. Photon.* **2009**, *3*, 388–394. [CrossRef]
19. Stockman, M.I. Nanoplasmonics: Past, present, and glimpse into future. *Opt. Express* **2011**, *19*, 22029. [CrossRef]
20. Rehman, S.; Rahmouni, A.; Mahfoud, T.; Nesterenko, D.V.; Sekkat, Z. Determination of the Optical Thickness of sub 10-nm Thin Metal Films by SPR Experiments. *Plasmonics* **2014**, *9*, 381–387. [CrossRef]
21. Liu, C.; Liu, Q.; Hu, X. SPR phase detection for measuring the thickness of thin metal films. *Opt. Express* **2014**, *22*, 7574–7580. [CrossRef]
22. Gonzalez-vila, A.; Debliquy, M.; Lahem, D.; Zhang, C.; Mégret, P.; Caucheteur, C. Molecularly imprinted electropolymerization on a metal-coated optical fiber for gas sensing applications. *Sens. Actuators B Chem.* **2017**, *244*, 1145–1151. [CrossRef]
23. Refki, S.; Hayashi, S.; Rahmouni, A.; Nesterenko, D.V.; Sekkat, Z. Anticrossing Behavior of Surface Plasmon Polariton Dispersions in Metal-Insulator-Metal Structures. *Plasmonics* **2016**, *11*, 433–440. [CrossRef]
24. Refki, S.; Hayashi, S.; Nesterenko, D.V.; Sekkat, Z.; Inoue, Y.; Kawata, S. Metal-insulator-metal structures for high-resolution sensing. In *JSAP-OSA Joint Symposia 2014 Abstracts*; Hokkaido: Sapporo, Japan, 2014; p. 19p_C3_5. [CrossRef]

25. Sekkat, Z.; Hayashi, S.; Nesterenko, D.V.; Rahmouni, A.; Refki, S.; Ishitobi, H.; Inouye, Y.; Kawata, S. Plasmonic coupled modes in metal-dielectric multilayer structures: Fano resonance and giant field enhancement. *Opt. Express* **2016**, *24*, 20080. [CrossRef] [PubMed]

26. Hayashi, S.; Nesterenko, D.V.; Rahmouni, A.; Sekkat, Z. Observation of Fano line shapes arising from coupling between surface plasmon polariton and waveguide modes. *Appl. Phys. Lett.* **2016**, *108*, 051101. [CrossRef]

27. Hayashi, S.; Nesterenko, D.V.; Rahmouni, A.; Ishitobi, H.; Inouye, Y.; Kawata, S.; Sekkat, Z. Light-tunable Fano resonance in metal-dielectric multilayer structures. *Sci. Rep.* **2016**, *6*, 33144. [CrossRef] [PubMed]

28. Parthenopoulos, D.A.; Rentzepis, P.M. Three-Dimensional Optical Storage Memory. *Science* **1989**, *245*, 843–845. [CrossRef] [PubMed]

29. Maeda, M.; Ishitobi, H.; Sekkat, Z.; Kawata, S. Polarization storage by nonlinear orientational hole burning in azo dye-containing polymer films. *Appl. Phys. Lett.* **2004**, *85*, 351–353. [CrossRef]

30. Sekkat, Z.; Dumont, M. Photoassisted poling of azo dye doped polymeric films at room temperature. *Appl. Phys. B* **1992**, *54*, 486–489. [CrossRef]

31. Sekkat, Z.; Dumont, M. Poling of polymer films by photoisomerisation of azodye chromophores. *Mol. Chryst. Liq. Chryst. Sci. Technol. B Nonlinear Opt.* **1992**, *2*, 359–362.

32. Xu, H.J.; Liu, F.; Elder, D.L.; Johnson, L.E.; De Coene, Y.; Clays, K.; Robinson, B.H.; Dalton, L.R. Ultrahigh electro-optic coefficients, High index of refraction, and long term stability from Diels-Alder Cross-Linkable Molecular Glasses. *Chem. Mater.* **2020**, *32*, 1408–1421. [CrossRef]

33. Kalinin, A.A.; Islamova, L.N.; Shmelev, A.G.; Fazleeva, G.M.; Fominykh, O.D.; Dudkina, Y.B.; Vakhonina, T.A.; Levitskaya, A.I.; Sharipova, A.V.; Mukhtarov, A.S.; et al. D-pi-A chromophores with a quinoxaline core in the pi-bridge and bulky aryl groups in the acceptor: Synthesis, properties, and femtosecond nonlinear optical activity of the chromophore/PMMA guest-host materials. *Dye. Pigment.* **2021**, *184*, 108801. [CrossRef]

34. Kuzyk, M.G.; Dawson, N.J. Photomechanical materials and applications: A tutorial. *Adv. Opt. Photon.* **2020**, *12*, 847. [CrossRef]

35. Zhou, B.; Bernhardt, E.; Bhuyan, A.; Ghorbanishiadeh, Z.; Rasmussen, N.; Lanska, J.; Kuzyk, M.G. Theoretical and experimental studies of photomechanical materials [Invited]. *J. Opt. Soc. Am. B* **2019**, *36*, 1492. [CrossRef]

36. Sekkat, Z. Optical tweezing by photomigration. *Appl. Opt.* **2016**, *55*, 259. [CrossRef] [PubMed]

37. Sekkat, Z. Vectorial motion of matter induced by light fueled molecular machines. *OSA Contin.* **2018**, *1*, 668. [CrossRef]

38. Moujdi, S.; Rahmouni, A.; Mahfoud, T.; Nesterenko, D.V.; Halim, M.; Sekkat, Z. Surface relief gratings in azo-polymers revisited. *J. Appl. Phys.* **2018**, *124*, 213103. [CrossRef]

39. Sekkat, Z. Model for athermal enhancement of molecular mobility in solid polymers by light. *Phys. Rev. E* **2020**, *102*, 032501. [CrossRef] [PubMed]

40. Ishitobi, H.; Akiyama, T.; Sekkat, Z.; Inouye, Y. Optical Trapping of Photosoftened Solid Polymers. *J. Phys. Chem. C* **2020**, *124*, 26037–26042. [CrossRef]

41. Atkins, S.; Chueh, A.; Barwell, T.; Nunzi, J.-M.; Seroude, L. Capture and light-induced release of antibiotics by an azo dye polymer. *Sci. Rep.* **2020**, *10*, 3267. [CrossRef] [PubMed]

42. Carmen, R.; Ursu, C.; Dascalu, M.; Asandulesa, M.; Tiron, V.; Bele, A.; Tugui, C.; Teodoroff-Onesim, S. Multi-stimuli responsive free-standing films of DR1- grafted silicones. *Chem. Eng. J.* **2020**, *401*, 126087. [CrossRef]

43. Prêtre, P.; Wu, L.-M.; Knoesen, A.; Swalen, J.D. Optical properties of nonlinear optical polymers: A method for calculation. *J. Opt. Soc. Am. B* **1998**, *15*, 359. [CrossRef]

44. Bondarenko, S.; Villringer, C.; Steglich, P. Comparative Study of Nano-Slot Silicon Waveguides Covered by Dye Doped and Undoped Polymer Cladding. *Appl. Sci.* **2018**, *9*, 89. [CrossRef]

45. Taqatqa, O.; Al Attar, H. Spectroscopic ellipsometry investigation of azo dye and azo dye doped polymer. *Eur. Phys. J. Appl. Phys.* **2007**, *37*, 61–64. [CrossRef]

46. Steglich, P.; Villringer, C.; Dietzel, B.; Mai, C.; Schrader, S.; Casalboni, M.; Mai, A. On-Chip Dispersion Measurement of the Quadratic Electro-Optic Effect in Nonlinear Optical Polymers Using a Photonic Integrated Circuit Technology. *IEEE Photonics J.* **2019**, *11*, 1–10. [CrossRef]

47. Niklasson, G.A.; Granqvist, C.G.; Hunderi, O. Effective medium models for the optical properties of inhomogeneous materials. *Appl. Opt.* **1981**, *20*, 26–30. [CrossRef]

48. Boyd, R.W.; Gehr, R.J.; Fischer, G.L.; Sipe, J.E. Nonlinear optical properties of nanocomposite materials. *Pure Appl. Opt.* **1996**, *5*, 505–512. [CrossRef]

49. Lee, K.J.; Kang, T.D.; Lee, H.S.; Lee, H.K. Ellipsometric study of polymer thin films: Nonlinear optical guest-host system. *J. Appl. Phys.* **2005**, *97*, 083543. [CrossRef]

50. Sekkat, Z.; Knoll, W. (Eds.) *Photoreactive Organic Thin Films*; Academic Press: Amsterdam, The Netherlands; Boston, MA, USA, 2002.

51. Sekkat, Z.; Prêtre, P.; Knoesen, A.; Volksen, W.; Lee, V.Y.; Miller, R.D.; Wood, J.; Knoll, W. Correlation between polymer architecture and sub-glass-transition-temperature light-induced molecular movement in azo-polyimide polymers: Influence on linear and second- and third-order nonlinear optical processes. *J. Opt. Soc. Am. B* **1998**, *15*, 401. [CrossRef]

52. Zeroni, I.; Lodge, T.P. Chain Dimensions in Poly(ethylene oxide)/ Poly(methyl methacrylate) Blends. *Macromolecules* **2008**, *41*, 1050–1052. [CrossRef]

53. Swalen, J.D.; Bjorklund, G.C.; Fleming, W.W.; Hung, R.Y.; Jurich, M.C.; Lee, V.Y.; Miller, R.D.; Moerner, W.E.; Morichere, D.Y.; Skumanich, A.; et al. NLO Polymeric Waveguide Electro-Optic Phase Modulator. *Proc. SPIE Int. Soc. Opt. Eng.* **1993**, *1775*. [CrossRef]

54. Kuzyk, M.G.; Sohn, J.E.; Dirk, C.W. Mechanisms of quadratic electro-optic modulation of dye-doped polymer systems. *J. Opt. Soc. Am. B* **1990**, *7*, 842–858. [CrossRef]

55. Kretschmann, E. Decay of non radiative surface plasmons into light on rough silver films. Comparison of experimental and theoretical results. *Opt. Commun.* **1972**, *6*, 185–187. [CrossRef]

56. Aust, E.F.; Sawodny, M.; Ito, S.; Knoll, W. Surface plasmon and guided optical wave microscopies. *Scanning* **1994**, *16*, 353–362. [CrossRef]

57. Borstel, G.; Falge, H.J. Surface polaritons in semi-infinite crystals. *Appl. Phys.* **1978**, *16*, 211–223. [CrossRef]

58. Polyanskiy, M. Refractive Index. 2014. Available online: https://refractiveindex.info/?shelf=glass&book=BK7&page=SCHOTT (accessed on 19 January 2021).

59. Refki, S.; Hayashi, S.; Ishitobi, H.; Nesterenko, D.V.; Rahmouni, A.; Inouye, Y.; Sekkat, Z. Resolution Enhancement of Plasmonic Sensors by Metal-Insulator-Metal Structures. *Ann. Phys.* **2018**, *530*, 1700411. [CrossRef]

60. Johnson, P.B.; Christy, R.W. Optical Constants of the Noble Metals. *Phys. Rev. B* **1972**, *6*, 4370. [CrossRef]

61. Elhani, S.; Maouli, I.; Refki, S.; Halim, M.; Hayashi, S.; Sekkat, Z. Quantitative analyses of optically induced birefringence in azo dye containing polymers. *J. Opt.* **2019**, *21*, 115401. [CrossRef]

62. Jellison, G.E. Data analysis for spectroscopic ellipsometry. *Thin Solid Film.* **1993**, *234*, 416–422. [CrossRef]

63. Woollam, J.A. *VASE Manual*; Woollam, J.A. Inc.: Lincoln, NE, USA, 1997.

64. Bruggeman, D.A.G. Berechnung verschiedener physikalischer Konstanten von heterogenen Substanzen. I. Dielektrizitätskonstanten und Leitfähigkeiten der Mischkörper aus isotropen Substanzen. *Ann. Phys.* **1935**, *416*, 636–664. [CrossRef]

 *photonics*

Communication

Plasmonic Narrowband Filter Based on an Equilateral Triangular Resonator with a Silver Bar

Jingyu Zhang, Hengli Feng and Yang Gao *

School of Electronic Engineering, Heilongjiang University, Harbin 150080, China; 2201611@s.hlju.edu.cn (J.Z.); 2201622@s.hlju.edu.cn (H.F.)
* Correspondence: 2013012@hlju.edu.cn

Abstract: A kind of plasmonic structure consisted of an equilateral triangle-shaped cavity (ETSC) and a metal-insulator-metal (MIM) waveguide is proposed to realize triple Fano resonances. Numerically simulated by the finite difference time domain (FDTD) method, Fano resonances inside the structure are also explained by the coupled mode theory (CMT) and standing wave theory. For further research, inverting ETSC could dramatically increase quality factor to enhance resonance wavelength selectivity. After that, a bar is introduced into the ETSC and the inverted ETSC to increase resonance wavelengths through changing the structural parameters of the bar. In addition, working as a highly efficient narrowband filter, this structure owes a good sensitivity (S = 923 nm/RIU) and a pretty high-quality factor (Q = 322) along with a figure of merit (FOM = 710). Additionally, a narrowband peak with 1.25 nm Full-Width-Half-Maximum (FWHM) can be obtained. This structure will be used in highly integrated optical circuits in future.

Keywords: waveguide; SPPs; FDTD; bandstop filter; CMT

Citation: Zhang, J.; Feng, H.; Gao, Y. Plasmonic Narrowband Filter Based on an Equilateral Triangular Resonator with a Silver Bar. *Photonics* **2021**, *8*, 244. https://doi.org/10.3390/photonics8070244

Received: 25 May 2021
Accepted: 24 June 2021
Published: 29 June 2021

Publisher's Note: MDPI stays neutral with regard to jurisdictional claims in published maps and institutional affiliations.

1. Introduction

Surface plasmon polaritons (SPPs) perform as special electromagnetic (EM) waves that propagate along with the metal–dielectric interface and decay exponentially in the perpendicular direction [1]. SPPs are of great potential value in setting up optical circuits [2] because of their outstanding properties, such as overcoming the conventional optical diffraction limit [3] and achieving the transmission and manipulation of the optical signal within the subwavelength scale [4].

Under these characteristics, SPPs are widely used among optical communication, physical chemistry, biological and biochemical sensing [5,6], etc. For example, plasmonic biosensing is used for the fast, real-time, and label-free probing of biologically relevant analytes, where the main challenges are to detect small molecules at ultralow concentrations and produce compact devices for point-of-care analysis [7–9]. Guatha et al. detected biomonolayers and streptavidin-conjugated semiconducting quantum dots by employing the arrays of gold nanoantennas covered by an ultrathin silicon layer. Additionally, they found surface lattice resonances (SLRs) with high sensitivities to small changes of refractive index [10]. Sadeghi et al. found that narrow SLRs could be shifted by changing the environmental refractive index based on the arrays of large nanodisks for chemical and biological sensing [11]. Additionally, plasmonic biosensing offers a new way to detect coronavirus with low-cost and rapid detection. Liping Huang et al. employed a spike protein specific nanoplasmonic resonance sensor to measure SARS-CoV-2 virus particle with only one step [12].

For Fano resonance, the physical essence is that the interference of a discrete state with a continuum gives rise to characteristically asymmetric peaks in excitation spectra. In this way, the scattering of the structure is limited well in a narrow wavelength scale and has strong localized electromagnetic characteristics in the near field. As the curve of Fano resonance is asymmetric and steep, its subtle wavelength shift is easy to distinguish.

Fano resonance reflected on the metal nanostructures is the interference between the bright mode and the dark mode of the surface plasmons. The bright mode can be directly excited by the incident light, while the dark mode needs to be excited by the coupling effects of the bright mode. The bright mode has a relatively broad band and the dark mode has a relatively narrow band. The resonance of the surface plasmons is the result of the interference between the bright mode and the dark mode, and this kind of resonance can be recognized as Fano resonance [13,14].

As for the metal-insulator-metal (MIM) waveguide, it has the merits of a strong local field enhancement characteristic for SPPs [15] and ease of integration. Unique advantages enable the MIM waveguide to be made into different optical devices, such as filters [16], sensors [17] and slow light devices [18]. Furthermore, Fano effects in the MIM waveguide coupled with resonators owe high figure of merit (*FOM*) and Sensitivity (*S*) and structural parameters' change accompanied by a sharp asymmetric spectral profile. Consequently, more and more researchers are paying attention to its applications in the modulator, optical switches, wavelength division multiplexing (WDM), Mach–Zehnder, and splitters [19–21]. It is worth noting that recent studies have aimed to design different structures for higher *FOM* and *S* [22–25], and through changing the external size of resonators to make the whole structure tunable [26,27]. Additionally, a lot of structures are designed to obtain higher quality factor (*Q*). Quynh et al. demonstrated that local surface plasmon resonances (LSPRs) can be coupled to the $(0, +/- 1)$ diffraction orders to achieve high Q-factor (SLRs) with low dispersion. These ultrahigh Q-factors can be increased to above 1500 without an adhesive layer [28]. Saad Bin-Alam et al. proposed a modified anapole resonator based on Hg with a high Q-factor and large thermal sensitivity of 17.14 MHz/degrees to realize high-precision temperature sensing [29]. Peter A et al. illustrated a metasurface that consisted of hollow dielectric cuboids with an experimental Q-factor of 728 at 1505 nm by using silicon as the high index dielectric [30]. Liang et al. demonstrated a plasmonic metasurface with a Q-factor of 2340 in the telecommunication C band by exploiting SLRs, which is the highest one in the record [31].

In this paper, based on equilateral triangle-shaped cavity (ETSC), we designed a highly selective and tunable plasmonic structure. The spectral characteristics and magnetic distributions in the z-direction are numerically analyzed by finite difference time domain (FDTD). Additionally, the structural transmission properties are calculated by the temporal coupled mode theory (CMT) [32]. Subsequently, simply inverting the ETSC obtains a higher *Q*. In addition, Fano resonances can be adjusted by introducing a bar without sacrificing performance. Furthermore, with a good sensitivity of 923 nm/RIU and a high *Q* of 322, the results manifest that our designed structure could perform as an exceptional narrowband filter. In brief, the devised structure, with highly selective and tunable multiple Fano resonances, has promising applications in the future.

2. Materials and Methods

As can be seen from Figure 1a, the proposed structure includes an ETSC and one MIM waveguide. Specifically, the insulator and the metal material (white and blue areas) are, respectively, set as air and silver. For the structural parameters of the filter, they consist of the side length and height of the ETSC (B = 484 nm, H = 420 nm), coupling distance between the ETSC and the waveguide (G = 200 nm), and the waveguide width (W = 50 nm) enabling only the fundamental mode (TM_0) to propagate through whole structure. Inside this simulation, characteristics of silver are described by the Johnson and Christy model. The computation volume is 800 nm × 1100 nm and all its four boundaries are ended with perfectly matched layers in order to absorb incident light with minimal reflections. Additionally, mesh step is set as dx = dy = 1 nm. A two-dimensional model (xoy plane) is chosen to dramatically decrease computation time.

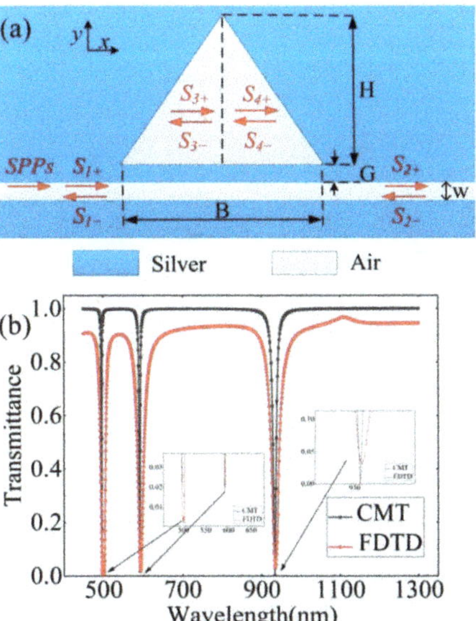

Figure 1. (**a**) The schematic of the equilateral triangle-shaped cavity (ETSC); (**b**) the transmission spectra of the ETSC by using coupled mode theory (CMT) and Johnson and Christy model.

The resonance wavelength of the ETSC can be deduced by standing wave theory as:

$$\lambda_r = \frac{2n_{eff}H}{r - (\phi_1 + \phi_2)/2\pi}, r = 1, 2, 3 \ldots \tag{1}$$

Here, r is equal to the number of the resonance wavelength. ϕ_1, ϕ_2 denotes the phase delay when the SPPs couple into the ETSC and the SPPs return to the waveguide after multiple reflections inside the ETSC, respectively. The effective reflective index n_{eff} could be conducted by combining the following expressions:

$$\varepsilon_d k_d tanh(\frac{Wk_d}{2}) = 0 \tag{2}$$

$$k_{d,m} = \sqrt{\beta_{spp} - \varepsilon_{d,m}k_0^2} \tag{3}$$

$$\triangle \phi = \frac{4\pi n_{eff}H}{\lambda} + \phi_1 + \phi_2 \tag{4}$$

$$n_{eff} = \frac{\beta_{spp}}{k_0} \tag{5}$$

Here, ε_d is the permittivity of dielectric and ε_m is the permittivity of metal. k_d is the propagation constant of dielectric and k_m is propagation constant of metal. β_{spp} is the complex propagation constant of SPPs. $k_0 = \frac{2\pi}{\lambda}$ is the wave vector inside the waveguide. $\triangle \phi$ is the total phase delay which SPPs propagate in the ETSC within one cycle. Especially when the resonance condition is satisfied with $\triangle \phi = 2m\pi$, Equation (1) is attainable. The transmission spectra of the ETSC using both CMT and FDTD simulation are shown in Figure 1b with three resonances at 489 nm, 581 nm and 927 nm.

CMT is utilized to study the optical properties of the proposed structure. As shown in Figure 1a, the input and output energy are expressed by $S_{j\pm}$ (j = 1, 2, 3 and 4), a_1 and a_2 display the energy amplitudes which meet the following conditions:

$$\frac{da_1}{dt} = (iw_1 - \frac{1}{\tau_{i1}} - \frac{1}{\tau_{c1}})a_1 + S_{1+}\sqrt{\frac{1}{\tau_{c1}}} + S_{2-}\sqrt{\frac{1}{\tau_{c1}}} \tag{6}$$

$$\frac{da_2}{dt} = (iw_2 - \frac{1}{\tau_{i2}} - \frac{1}{\tau_{c2}})a_2 + S_{3+}\sqrt{\frac{1}{\tau_{c2}}} + S_{4-}\sqrt{\frac{1}{\tau_{c2}}} \tag{7}$$

w_1 and w_2 are resonant angular frequency of the incident light, and match the order of nth (n = 1 and 2) resonance separately, $\frac{1}{\tau_{in}} = \frac{w_n}{(2Q_{in})}$ (n = 1 and 2) stands for the decay rate of internal loss of cavity, and $\frac{1}{\tau_{cn}} = \frac{w_n}{(2Q_{cn})}$ (n = 1 and 2) denotes the cavity loss which coupled to the ETSC. Q_{in} and Q_{cn} are internal and cavity quality factors. Additionally, we can obtain energy conversion equations:

$$S_{2+} = S_{1+} - \sqrt{\frac{1}{\tau_{c1}}}a_1, \; S_{4+} = S_{3+} - \sqrt{\frac{1}{\tau_{c2}}}a_2, \; S_{3-} = S_{4-} - \sqrt{\frac{1}{\tau_{c2}}}a_2 \tag{8}$$

$$S_{3+} = S_{2+}Ce^{i(\phi_1+\phi_2)}, \; S_{2-} = S_{3-}Ce^{i(\phi_1+\phi_2)}, \; S_{4-} = 0 \tag{9}$$

C is the related damping coefficient. Combining the Equations (6)–(9), we calculate the model's transmission function, and T is representative of the transmission coefficient.

$$t = \frac{S_{4-}}{S_{1+}} = (1 + \frac{\frac{\mu_1}{\tau_{c2}} + \frac{\mu_2}{\tau_{c1}} + \sqrt{\frac{\beta^2}{\tau_{i1}\tau_{c2}}}}{\mu_1\mu_2 - \beta^2})Ce^{i(\phi_1+\phi_2)} + \frac{\sqrt{\frac{\beta^2}{\tau_{i1}\tau_{c2}}}}{\mu_1\mu_2 - \beta^2} \tag{10}$$

$$\mu_1 = i(w - w_1) - \frac{1}{\tau_{i1}} - \frac{1}{\tau_{c1}}, \; \mu_2 = i(w - w_2) - \frac{1}{\tau_{i2}} - \frac{1}{\tau_{c2}}, \; \beta = \sqrt{\frac{C^2 e^{2i(\phi_1+\phi_2)}}{\tau_{c1}\tau_{c2}}} \tag{11}$$

$$T = |t|^2 \tag{12}$$

The FDTD simulation result has the same tendency as the CMT method, which is depicted by the black dot line in Figure 1b. It owes the parameters C = 1, $\phi_1 = \phi_2 = 0$ and when the resonant angular frequency w_1 = 3.80 fs^{-1} and w_2 = 3.9 fs^{-1}, τ_{i1} = 48.2 ps, τ_{i2} = 72.3 ps, τ_{c1} = 5.1ps, τ_{c2} = 7.9 ps. and τ_{i1} = 722.5 ps, τ_{i2} = 100.3 ps, τ_{c1} = 7.9 ps, τ_{c2} = 9.6 ps when w_1 = 3.2 fs^{-1} and w_2 = 2.0 fs^{-1}.

Figure 2a–c reveal the magnetic field distribution ($|H_z|$) of different surface plasmon wavelengths inside the ETSC. Three resonances mainly happen in the three vertices of the ETSC. Figure 2a indicates that intense resonances exist at the bottom left and the top of the ETSC, and Figure 2b,c show that resonances happen at the bottom of the ETSC. Additionally, Figure 2b denotes there is also a little energy at the top of the cavity. To further illustrate the excitation method of the resonance modes, Figure 2d is depicted. In Figure 2d, the vectorial charge mainly exists between the bottom of the ETSC and upper side of the MIM waveguide, and along with every side of the ETSC.

To further study the multiple Fano resonances based on this structure and achieve a high Q narrowband filter, an inverted ETSC schematic is proposed.

Figure 3a shows the inverted ETSC schematic. In detail, the distance between the vertex of the inverted ETSC and the waveguide is D = 30 nm. Additionally, the transmittance spectrum of the inverted one is shown in Figure 3b. It displays quintuple Fano resonances at 406 nm, 419 nm, 491 nm, 582 nm and 930 nm.

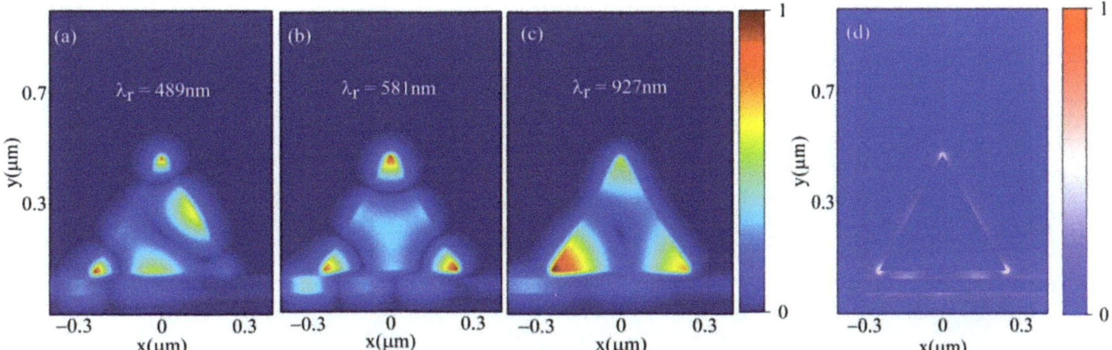

Figure 2. The magnetic distribution profile of the ETSC at: (**a**) $\lambda_r = 489$ nm; (**b**) $\lambda_r = 581$ nm; (**c**) $\lambda_r = 927$ nm; (**d**) the vectorial charge distribution of the ETSC at $\lambda_r = 489$ nm.

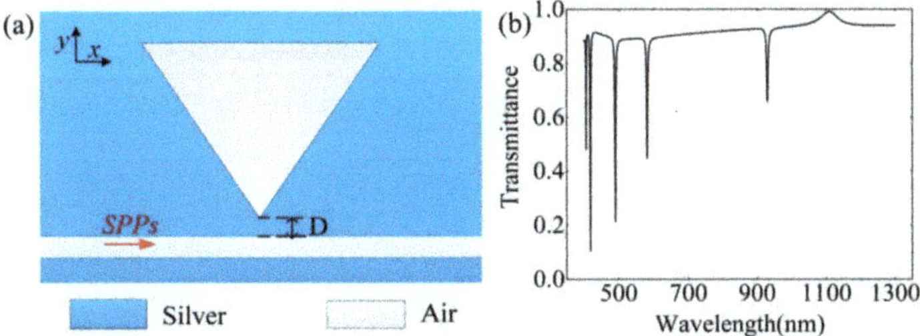

Figure 3. (**a**) The schematic of the inverted ETSC; (**b**) the transmittance spectrum of the inverted ETSC.

Figure 4a–e manifest the magnetic field profile of the inverted ETSC at five resonance wavelengths, separately. A little energy at the top of the inverted ETSC is exhibited in Figure 4a, while it does not exist for both Figure 4b,c. Referring to Figure 4d, intense resonances occur at the bottom and the top compared with the three resonance modes above. Additionally, the strong resonance occurs at the bottom of the inverted ETSC in Figure 4e. It is apparent when comparing Figure 4a–e with Figure 4f,g that the physics mechanisms of FR1, FR2, FR3, FR4 and FR5 are the same as the Fano resonances. It can be observed that the inverted ETSC and the waveguide cause destructive interference so that most of the input energy cannot pass through the bus waveguide and are restricted in the inverted ETSC from Figure 4a–e. On the contrary, in Figure 4f,g, most of the energy passes through the waveguide and little energy is coupled into the inverted ETSC, causing constructive interference between two excitations.

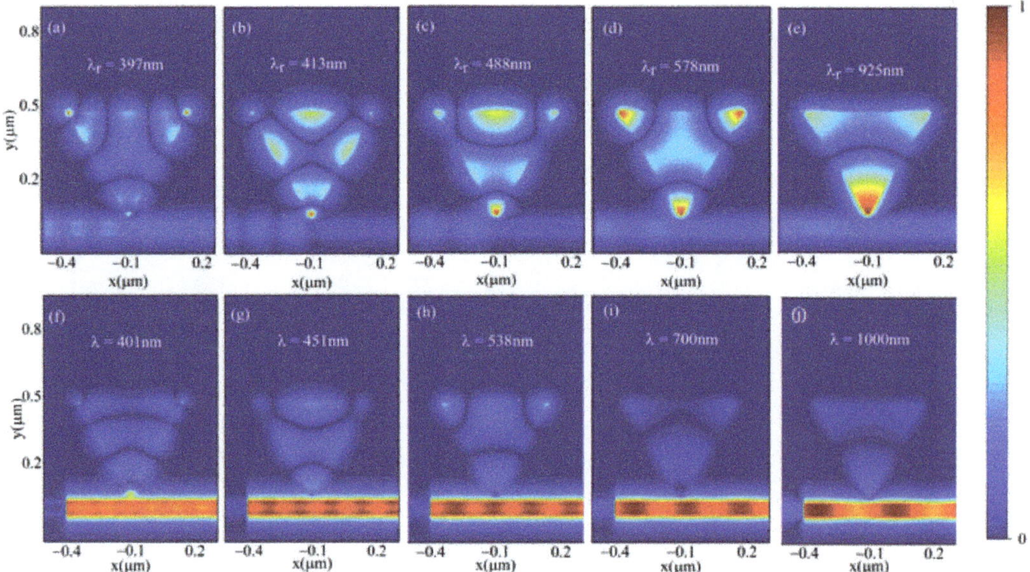

Figure 4. The magnetic distribution profile of the inverted ETSC at: (**a**) λ_r = 397 nm; (**b**) λ_r = 413 nm; (**c**) λ_r = 488 nm; (**d**) λ_r = 578 nm; (**e**) λ_r = 925 nm; (**f**) λ = 401 nm; (**g**) λ = 451 nm; (**h**) λ = 538 nm; (**i**) λ = 700 nm; (**j**) λ = 1000 nm.

3. Results

On the one hand, the former designed structure can realize tunability by adding a bar. In this way, visible and near-infrared resonance wavelengths can be controlled via changing the geometrical parameters of bar. On the other hand, inverting the ETSC results in quality factors of all resonance wavelengths increasing dramatically with peak at 322, and FWHM decreasing significantly with minimum at 1.25 nm when λ_r = 413 nm.

3.1. Bar Embedded ETSC Configuration

Figure 5a shows the ETSC with a bar. The height h_1 and width w_1 of bar are 100 nm and 20 nm by default. Additionally, Figure 5b indicates the transmittance spectrum of the ETSC with a bar. From Figure 5b, adding a bar into the ETSC is almost the same as the ETSC without a bar.

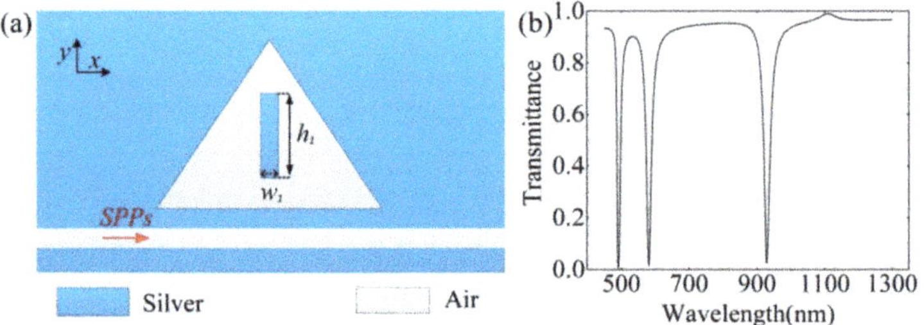

Figure 5. (**a**) The schematic of the ETSC with a bar; (**b**) the transmittance spectrum of the ETSC with a bar.

The transmission spectra of the structure with various heights are shown in Figure 6a. From Figure 6a, by increasing height of the bar, FR1 shifts obviously to a higher wavelength without a significant reduction in T_{min}, while a new resonance mode called FR2 emerges when h_1 range from 50 nm to 100 nm, and after that, FR2 does not change with the growth of h_1.

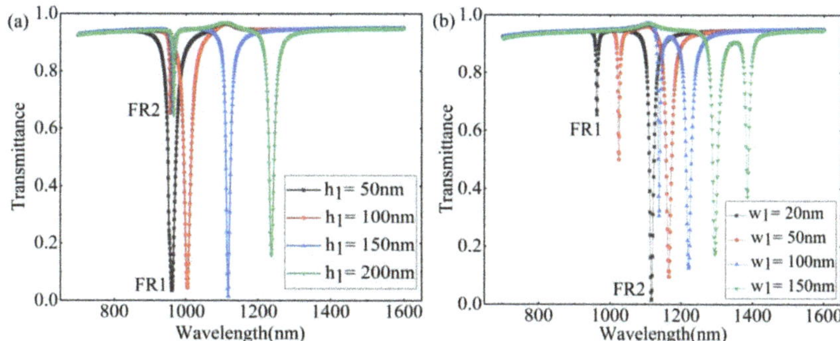

Figure 6. (a) The transmission spectra of different h_1; (b) the transmission spectra of different w_1.

From Figure 6b, based on the ETSC, the effects of w_1 which increases from 20 nm to 150 nm on transmittance spectra are studied. There is an apparent tendency in Figure 6b that both FR1 and FR2 have evident redshifts, and two resonance wavelengths became closer since the increase in width, when w_1 = 100 nm the distance is at the minimum.

3.2. Bar Embedded the Inverted ETSC Configuration

Figure 7a denotes that the derived structure consists of the inverted ETSC and a bar, which is introduced into the inverted ETSC to realize the increase in λ_r. The bar is described by two structural parameters h_2 and w_2. h_2 and w_2 of the bar are defined as 150 nm and 20 nm by default. the structural parameters are altered independently, and the influences on the transmittance spectra are investigated. As for Figure 7b, it exhibits the transmittance spectrum of the inverted ETSC with a bar, which has little difference with the inverted ETSC only.

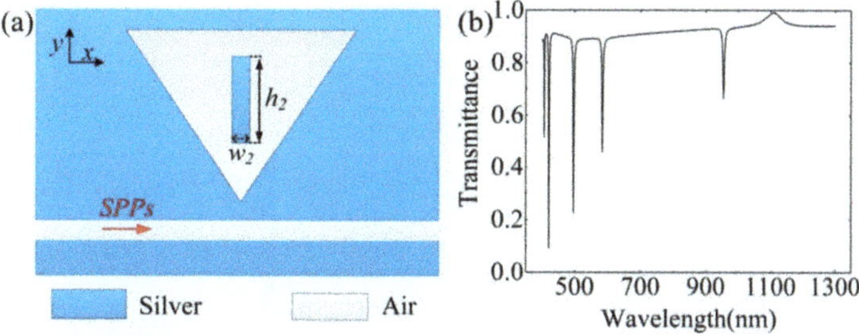

Figure 7. (a) The schematic of the inverted ETSC added with a bar; (b) the transmittance spectrum of the inverted ETSC with a bar.

On the foundation of the inverted ETSC, the effects of h_2, which climbs from 0 nm to 200 nm on the transmittance spectra, are studied, and the outcomes are depicted in Figure 8a. It is distinct that, without a significant reduction in T_{min}, the resonance wavelengths of the FR1, FR2, FR3 and FR4 are almost fixed, while the FR5 has a redshift by altering h_2. To further certify the linear performance, the resonance wavelengths of FR1, FR2, FR3, FR4 and FR5 are researched, and the results do manifest this linear relation, as is shown in Figure 8b.

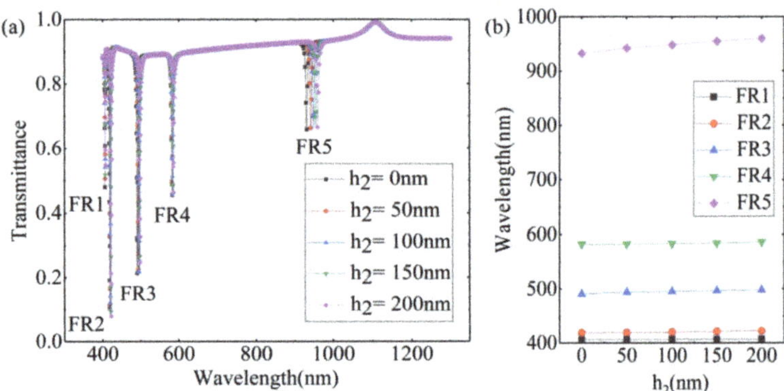

Figure 8. (**a**) The transmittance spectra for different h_2; (**b**) the resonance wavelength of FR1, FR2, FR3, FR4 and FR5 with different h_2.

After that, the transmission spectra of the structure with different widths are shown in Figure 9a. By increasing the width of the bar from 20 nm to 150 nm and keeping the other parameter constant, FR1 and FR4 are nearly fixed at their original location, while λ_r of FR2 and FR3 shift to higher wavelengths. For FR5, the movement to a higher wavelength is significant. In the same way, the resonance wavelengths of FR1, FR2, FR3, FR4 and FR5 are depicted in Figure 9b. Through the range of different parameters, transmittance spectra can be better controlled to meet various demands in practice.

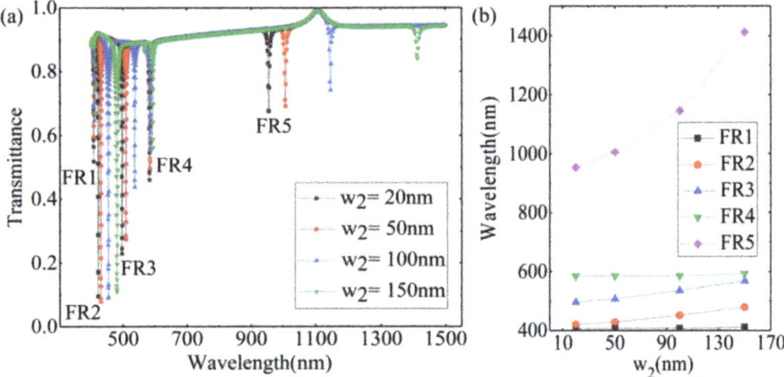

Figure 9. (**a**) The transmittance spectra of different w_2; (**b**) the resonance wavelength of FR1, FR2, FR3, FR4 and FR5 with different w_2.

The sensitivity of the proposed structure in Figure 7a depends on the response degree to the change of the refractive index (RI) inside the white area of the inverted ETSC. With all the structural parameters in Figure 7a unchanged, the n increases from 1.00 to 1.16 at the step of 0.04. The performance of the proposed structure in a nanometer scale is described in Figure 10a, which indicates that these quadruple Fano resonances owe redshifts. Since the sensitivity (S) of narrowband filter can be explained as the wavelength shift per unit refractive index change, expressed as $S = \frac{d\lambda}{dn}$ (nm/RIU), the sensitivities are 923 nm/RIU, 542 nm/RIU, 427 nm/RIU, 301 nm/RIU and 250 nm/RIU at FR5, FR4, FR3, FR2 and FR1 separately, which are concluded in Figure 10b.

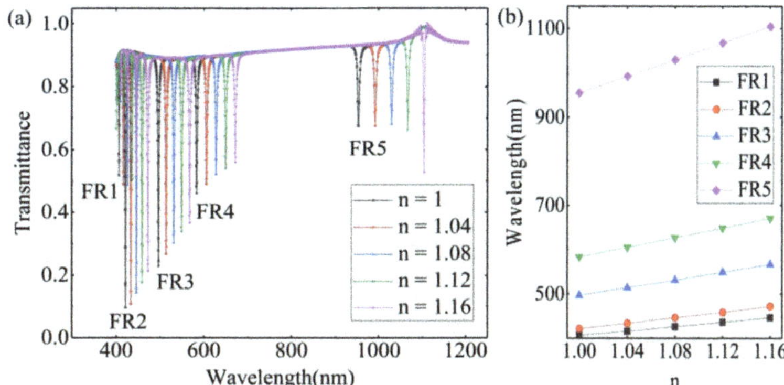

Figure 10. (**a**) The transmittance spectra of the designed structure with n varying from 1 to 1.16; (**b**) the resonance wavelength of FR1, FR2, FR3, FR4 and FR5 with different n.

Moreover, *FOM* can be represented as $FOM = \frac{S}{FWHM}$, where *FWHM* is the full width at half-maximum of the transmittance spectrum. From Figure 7b, different resonance wavelengths' *FOMs* are calculated, in which the maximum can reach 710 at 933 nm.

Another essential factor of a filter measuring the ability to give wavelength selectivity is represented as a quality factor (*Q*). The quality factor can be expressed as $Q = \frac{\lambda_r}{FWHM}$. Additionally, the dephasing time of the inverted ETSC is a critical parameter that can be defined by taking into account the resonance narrowness as follows: $t_d = \frac{2\hbar}{\delta}$ [33]. Here, $\hbar$ is the reduced Planck's constant and $\delta = \left| w_r^{max} - w_r^{min} \right|$. w_r^{max} and w_r^{min} are corresponding to λ_r^{max} and λ_r^{min}. Therefore, the dephasing times are 5.4×10^{-3} fs, 5.7×10^{-3} fs, 1.2×10^{-2} fs, 9.7×10^{-3} fs, and 2.5×10^{-2} fs.

Another important factor is V_{eff} which represents the effective mode volume of the confined electromagnetic field in the cavity. Additionally, the effective mode volume is calculated according to $V_{eff} = \frac{\int_V \varepsilon(r)|E(r)|^2 dV}{max\left[\varepsilon(r)|E(r)|^2\right]}$ [34–46], where $\varepsilon(r)$ is the dielectric constant, $|E(r)|$ is the electric field strength and V is the volume encompassing the resonator with a boundary in the radiation zone of the cavity mode. As for Figure 11a, the minimum of V_{eff} is 0.02×10^5 μm when $h_2 = 50$ nm. By increasing the height of the silver bar, V_{eff} has a slight redshift in the near-infrared band. In Figure 11b, the minimum of V_{eff} is 0.06×10^5 μm when $w_2 = 20$ nm. Increasing the width of the silver bar, V_{eff} has an obvious redshift in the near-infrared band.

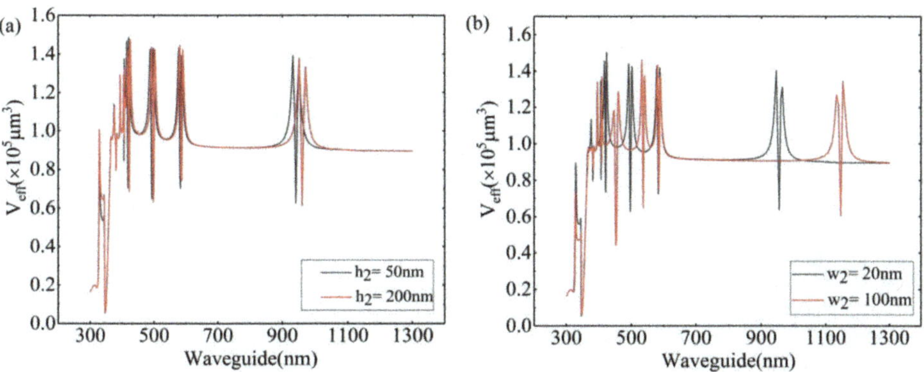

Figure 11. (**a**) Effective mode volume (V_{eff}) for different heights of the inverted ETSC; (**b**) V_{eff} for different widths of the inverted ETSC.

To manifest the advantages of the designed structure performed as a narrowband filter, Table 1 is formed, which compares the performance characteristics of our structure with other likely designs which can produce multiple Fano resonances based on SPPs published in recent years. From Table 1, we could conclude that our structure has relatively good sensitivity and it is superior to other designs in the maximal Q.

Table 1. Comparison of the proposed structure with the latest relative designs.

References	Type	Q	S	FOM	t_d	V_{eff}
[4]	Single	29.67	2080	29.92	N	N
[15]	Dual	Max 31.1	Max 2300	Max 31.5	N	N
[16]	Dual	106/123	900/1700	Max 1.38×10^5	N	N
This work	Quintuple	Max 322	923/542/427/301/250	Max 710	Min 5.4×10^{-3} fs	Min 0.02×10^5 μm

Q: represents quality factor. S: represents sensitivity. FOM: represents figure of merit. t_d: represents dephasing time.

4. Conclusions

Several plasmonic bandstop filters based on the ETSC are designed in this work. Additionally, triple and quintuple resonance modes occur when light passes through the proposed structure, including the ETSC and the inverted ETSC. By introducing a bar into the ETSC and the inverted one, structural resonant wavelengths can be increased by changing the height and width of the bar. Moreover, inverting the ETSC could reduce the FWHM of every resonance wavelength to increase Q efficiently. The main advantage of the proposed structures compared with previous methods is the better quality factor, narrow bandwidth, good sensitivity and tunability. It is believed that the proposed structures will be used in high integration plasmonic devices in future.

Author Contributions: Conceptualization, J.Z. and H.F.; data curation, J.Z.; formal analysis, J.Z.; investigation, J.Z.; methodology, J.Z.; project administration, Y.G.; resources, J.Z.; software, J.Z.; supervision, Y.G.; validation, J.Z., H.F. and Y.G.; writing—original draft, J.Z.; writing—review and editing, Y.G. All authors have read and agreed to the published version of the manuscript.

Funding: This research was funded in part by University Nursing Program for Young Scholars with Creative Talents in Heilongjiang Province (Grant No. UNPYSCT-2015016), Natural Science Foundation of Heilongjiang Province (Grant No. LH2019F047) and Project of the central government supporting the reform and development of local colleges and universities (Grant No. 2020YQ01).

Data Availability Statement: The available data have already been stated in the article.

Acknowledgments: Structure numerical simulation was provided by Lumerical Solutions, Inc.

Conflicts of Interest: The authors declare no conflict of interest.

References

1. Chen, J.; Sun, C.; Gong, Q. Fano resonances in a single defect nanocavity coupled with a plasmonic waveguide. *Opt. Lett.* **2014**, *39*, 52–55. [CrossRef]
2. Chou Chao, C.T.; Chou Chau, Y.F.; Huang, H.J.; Kumara, N.; Kooh, M.R.R.; Lim, C.M.; Chiang, H.P. Highly Sensitive and Tunable Plasmonic Sensor Based on a Nanoring Resonator with Silver Nanorods. *Nanomaterials* **2020**, *10*, 1399. [CrossRef]
3. Farahani, M.; Granpayeh, N.; Rezvani, M. Improved plasmonic splitters and demultiplexers. *Photonics Nanostruct. Fundam. Appl.* **2013**, *11*, 157–165. [CrossRef]
4. Lu, H.; Wang, G.; Liu, X. Manipulation of light in MIM plasmonic waveguide systems. *Chin. Sci. Bull.* **2013**, *58*, 3607–3616. [CrossRef]
5. Li, Z.; Wen, K.; Chen, L.; Lei, L.; Zhou, J.; Zhou, D.; Fang, Y.; Wu, B. Control of Multiple Fano Resonances Based on a Subwavelength MIM Coupled Cavities System. *IEEE Access* **2019**, *7*, 59369–59375. [CrossRef]
6. Wen, K.; Yan, L.; Pan, W.; Luo, B.; Guo, Z.; Guo, Y.; Luo, X. Electromagnetically Induced Transparency-Like Transmission in a Compact Side-Coupled T-Shaped Resonator. *J. Lightwave Technol.* **2014**, *32*, 1701–1707. [CrossRef]
7. Kravets, V.G.; Kabashin, A.V.; Barnes, W.L.; Grigorenko, A.N. Plasmonic Surface Lattice Resonances: A Review of Properties and Applications. *Chem. Rev.* **2018**, *118*, 5912–5951. [CrossRef]
8. Mejia-Salazar, J.R.; Oliveira, O.N., Jr. Plasmonic Biosensing. *Chem. Rev.* **2018**, *118*, 10617–10625. [CrossRef]

9. Ahmadivand, A.; Gerislioglu, B.; Ahuja, R.; Kumar Mishra, Y. Terahertz plasmonics: The rise of toroidal metadevices towards immunobiosensings. *Mater. Today* **2020**, *32*, 108–130. [CrossRef]
10. Gutha, R.R.; Sadeghi, S.M.; Sharp, C.; Wing, W.J. Biological sensing using hybridization phase of plasmonic resonances with photonic lattice modes in arrays of gold nanoantennas. *Nanotechnology* **2017**, *28*, 355504. [CrossRef]
11. Sadeghi, S.M.; Wing, W.J.; Campbell, Q. Tunable plasmonic-lattice mode sensors with ultrahigh sensitivities and figure-of-merits. *J. Appl. Phys.* **2016**, *119*, 244503. [CrossRef]
12. Huang, L.; Ding, L.; Zhou, J.; Chen, S.; Chen, F.; Zhao, C.; Xu, J.; Hu, W.; Ji, J.; Xu, H.; et al. One-step rapid quantification of SARS-CoV-2 virus particles via low-cost nanoplasmonic sensors in generic microplate reader and point-of-care device. *Biosens. Bioelectron.* **2021**, *171*, 112685. [CrossRef]
13. Fano, U. Effects of Configuration Interaction on Intensities and Phase Shifts. *Phys. Rev.* **1961**, *124*, 1866–1878. [CrossRef]
14. Anderson, P.W. Localized Magnetic States in Metals. *Phys. Rev.* **1961**, *124*, 41–53. [CrossRef]
15. Akhavan, A.; Ghafoorifard, H.; Abdolhosseini, S.; Habibiyan, H. Metal–insulator–metal waveguide-coupled asymmetric resonators for sensing and slow light applications. *IET Optoelectron.* **2018**, *12*, 220–227. [CrossRef]
16. Chou Chau, Y.-F. Mid-infrared sensing properties of a plasmonic metal–insulator–metal waveguide with a single stub including defects. *J. Phys. D Appl. Phys.* **2020**, *53*, 115401. [CrossRef]
17. Butt, M.A.; Kazanskiy, N.L.; Khonina, S.N. Nanodots decorated asymmetric metal–insulator–metal waveguide resonator structure based on Fano resonances for refractive index sensing application. *Laser Phys.* **2020**, *30*, 076204. [CrossRef]
18. Fang, Y.; Sun, M. Nanoplasmonic waveguides: Towards applications in integrated nanophotonic circuits. *Light Sci. Appl.* **2015**, *4*, e294. [CrossRef]
19. Wang, S.; Yu, S.; Zhao, T.; Wang, Y.; Shi, X. A nanosensor with ultra-high FOM based on tunable malleable multiple Fano resonances in a waveguide coupled isosceles triangular resonator. *Opt. Commun.* **2020**, *465*, 125614. [CrossRef]
20. Zhang, Z.; Wang, J.; Zhao, Y.; Lu, D.; Xiong, Z. Numerical Investigation of a Branch-Shaped Filter Based on Metal-Insulator-Metal Waveguide. *Plasmonics* **2011**, *6*, 773–778. [CrossRef]
21. Zhang, Z.; Yang, J.; He, X.; Han, Y.; Zhang, J.; Huang, J.; Chen, D. Plasmonic Filter and Demultiplexer Based on Square Ring Resonator. *Appl. Sci.* **2018**, *8*, 462. [CrossRef]
22. Gramotnev, D.K.; Bozhevolnyi, S.I. Plasmonics beyond the diffraction limit. *Nat. Photonics* **2010**, *4*, 83–91. [CrossRef]
23. Butt, M.A.; Kazanskiy, N.L.; Khonina, S.N. Highly Sensitive Refractive Index Sensor Based on Plasmonic Bow Tie Configuration. *Photonic Sens.* **2020**, *10*, 223–232. [CrossRef]
24. Yu, S.; Wang, S.; Zhao, T.; Yu, J. Tunable Plasmonic System Based on a Slotted Side-Coupled Disk Resonator and Its Multiple Applications on Chip-Scale Devices. *Optik* **2020**, *212*, 164748. [CrossRef]
25. Yan, S.; Shi, H.; Yang, X.; Guo, J.; Wu, W.; Hua, E. Study on the Nanosensor Based on a MIM Waveguide with a Stub Coupled with a Horizontal B-Type Cavity. *Photonics* **2021**, *8*, 125. [CrossRef]
26. Yang, X.; Hu, X.; Yang, H.; Gong, Q. Ultracompact all-optical logic gates based on nonlinear plasmonic nanocavities. *Nanophotonics* **2017**, *6*, 365–376. [CrossRef]
27. Wang, Y.; Xue, B.; Mao, J.; Lu, M. Plasmonic-induced transparency in a metallic stub with two cuts and transmission line model. *J. Mod. Opt.* **2018**, *65*, 2301–2307. [CrossRef]
28. Le-Van, Q.; Zoethout, E.; Geluk, E.J.; Ramezani, M.; Berghuis, M.; Rivas, J.G. Enhanced Quality Factors of Surface Lattice Resonances in Plasmonic Arrays of Nanoparticles. *Adv. Opt. Mater.* **2019**, *7*, 8. [CrossRef]
29. Bin-Alam, M.S.; Reshef, O.; Mamchur, Y.; Alam, M.Z.; Carlow, G.; Upham, J.; Sullivan, B.T.; Menard, J.M.; Huttunen, M.J.; Boyd, R.W.; et al. Ultra-high-Q resonances in plasmonic metasurfaces. *Nat. Commun.* **2021**, *12*, 974. [CrossRef]
30. Jeong, P.A.; Goldflam, M.D.; Campione, S.; Briscoe, J.L.; Vabishchevich, P.P.; Nogan, J.; Sinclair, M.B.; Luk, T.S.; Brener, I. High Quality Factor Toroidal Resonances in Dielectric Metasurfaces. *ACS Photonics* **2020**, *7*, 1699–1707. [CrossRef]
31. Ma, L.; Zheng, W.X.; Li, J.; Chen, D.X.; Wang, W.J.; Liu, Y.F.; Zhou, Y.D.; Yang, Y.J.; Huang, Y.J.; Wen, G.J. High-Q Hg-anapole resonator with microstrip line coupling for high-precision temperature sensing applications. *Results Phys.* **2021**, *24*, 5. [CrossRef]
32. Dong, L.; Xu, X.; Sun, K.; Ding, Y.; Ouyang, P.; Wang, P. Sensing analysis based on fano resonance in arch bridge structure. *J. Phys. Commun.* **2018**, *2*, 105010. [CrossRef]
33. Gupta, M.; Singh, R. Terahertz Sensing with Optimized Q /Veff Metasurface Cavities. *Adv. Opt. Mater.* **2020**, *8*, 1902025. [CrossRef]
34. Ahmadivand, A.; Gerislioglu, B.; Ramezani, Z. Gated graphene island-enabled tunable charge transfer plasmon terahertz metamodulator. *Nanoscale* **2019**, *11*, 8091–8095. [CrossRef]
35. Min, B.; Ostby, E.; Sorger, V.; Ulin-Avila, E.; Yang, L.; Zhang, X.; Vahala, K. High-Q surface-plasmon-polariton whispering-gallery microcavity. *Nature* **2009**, *457*, 455–458. [CrossRef] [PubMed]
36. Gierak, J.; Madouri, A.; Biance, A.L.; Bourhis, E.; Patriarche, G.; Ulysse, C.; Lucot, D.; Lafosse, X.; Auvray, L.; Bruchhaus, L.; et al. Sub-5nm FIB direct patterning of nanodevices. *Microelectron. Eng.* **2007**, *84*, 779–783. [CrossRef]
37. Li, G.; Winick, K.A.; Griffin, H.C.; Joseph, S. Hayden Systematic modeling study of channel waveguide fabrication by thermal silver ion exchange. *Appl. Opt.* **2006**, *45*, 743–755. [CrossRef]
38. Stewart, G.; Millar, C.A.; Laybourn, P.J.R.; Wilkinson, C.D.W.; Delarue, R.M. Planar Optical Waveguides Formed by Silver-Ion Migration in Glass. *IEEE J. Quantum Electron.* **1977**, *13*, 192–200. [CrossRef]

39. Asgari, S.; Pooretemad, S.; Granpayeh, N. Plasmonic refractive index sensor based on a double concentric square ring resonator and stubs. *Photonics Nanostruct. Fundam. Appl.* **2020**, *42*, 100857. [CrossRef]
40. Moradiani, F.; Farmani, A.; Mozaffari, M.H.; Seifouri, M.; Abedi, K. Systematic engineering of a nanostructure plasmonic sensing platform for ultrasensitive biomaterial detection. *Opt. Commun.* **2020**, *474*, 126178. [CrossRef]
41. Rakhshani, M.R. Optical refractive index sensor with two plasmonic double-square resonators for simultaneous sensing of human blood groups. *Photonics Nanostruct. Fundam. Appl.* **2020**, *39*, 100768. [CrossRef]
42. Rakhshani, M.R. Wide-angle perfect absorber using a 3D nanorod metasurface as a plasmonic sensor for detecting cancerous cells and its tuning with a graphene layer. *Photonics Nanostruct. Fundam. Appl.* **2021**, *43*, 100883. [CrossRef]
43. Rakhshani, M.R.; Mansouri-Birjandi, M.A. A high-sensitivity sensor based on three-dimensional metal–insulator–metal racetrack resonator and application for hemoglobin detection. *Photonics Nanostruct. Fundam. Appl.* **2018**, *32*, 28–34. [CrossRef]
44. To, N.; Juodkazis, S.; Nishijima, Y. Detailed Experiment-Theory Comparison of Mid-Infrared Metasurface Perfect Absorbers. *Micromachines* **2020**, *11*, 409. [CrossRef] [PubMed]
45. Nishijima, Y.; Balčytis, A.; Naganuma, S.; Seniutinas, G.; Juodkazis, S. Kirchhoff's metasurfaces towards efficient photo-thermal energy conversion. *Sci. Rep.* **2019**, *9*, 8284. [CrossRef]
46. Nishijima, Y.; Balčytis, A.; Naganuma, S.; Seniutinas, G.; Juodkazis, S. Tailoring metal and insulator contributions in plasmonic perfect absorber metasurfaces. *ACS Appl. Nano Mater.* **2018**, *1*, 3557–3564. [CrossRef]

 *photonics*

Article

Plasmonic Enhanced InP Nanowire Array Solar Cell through Optoelectronic Modeling

Farzaneh Adibzadeh [1] and Saeed Olyaee [2],*

[1] Faculty of Electrical Engineering, Shahid Rajaee Teacher Training University, Tehran 16788-15811, Iran; f.adibzadeh@sru.ac.ir

[2] Nano-Photonics and Optoelectronics Research Laboratory (NORLab), Shahid Rajaee Teacher Training University, Tehran 16788-15811, Iran

* Correspondence: s_olyaee@sru.ac.ir; Tel.: +98-21-2297-0030

Abstract: Vertical nanowire (NW) arrays are a promising candidate for the next generation of the optoelectronics industry because of their significant features. Here, we investigated the InP NW array solar cells and obtained the optoelectronic properties of the structure. To improve the performance of the NW array solar cells, we placed a metal layer of Au at the bottom of the NWs and considered their top part to be a conical-shaped parabola. Using optical and electrical simulations, it has been shown that the proposed structure improves the absorption of light in normal incidence, especially at wavelengths near the bandgap of InP, where photons are usually not absorbed. Under inclined radiation, light absorption is also improved in the middle part of the solar spectrum. Increased light absorption in the cell led to the generation of more electron–hole pairs, resulting in an increase in short circuit current density from 24.1 mA/cm^2 to 27.64 mA/cm^2, which is equivalent to 14.69% improvement.

Keywords: nanowires; back reflector; solar cells; plasmonic; III-V semiconductor

---

Citation: Adibzadeh, F.; Olyaee, S. Plasmonic Enhanced InP Nanowire Array Solar Cell through Optoelectronic Modeling. *Photonics* **2021**, *8*, 90. https://doi.org/10.3390/photonics8040090

Received: 1 March 2021
Accepted: 23 March 2021
Published: 25 March 2021

Publisher's Note: MDPI stays neutral with regard to jurisdictional claims in published maps and institutional affiliations.

1. Introduction

Photovoltaics, first introduced in 1958, is a process in which light is converted directly into electricity. Various materials and techniques are used to make solar cells based on the cost and efficiency of conversion. The first material used was silicon, which had disadvantages such as low efficiency. The second generation of solar cells was thin-film solar cells for producing electrical power, which received a lot of attention due to the use of thin layers in their structure [1]. Then, following further research in the field of photovoltaics technology to achieve new structures, the vertical arrays of III-V direct band gap semiconductor nanowire (NWs) were considered due to the reduction in the volume of material consumed compared to planar structures.

The low cost and other features of semiconductor NW structures have led to great interest in their use in a variety of applications, including photodetectors [1–3], light-emitting diodes [4–6], and lasers [7,8]. In NW array solar cells, proper design and improved absorption of excited optical modes are effective factors in optimizing photovoltaic efficiency. One of the common strategies to achieve these points is the proper design of the structure geometry. For example, increasing the length of the NW, which reduces the light transmission at the NW/substrate interface, increases the volume of consumables, and decreases the efficiency due to the increase in the dark current of the solar cell. Furthermore, the reduction of the pitch leads to a stronger absorption but increases the top insertion reflection losses. Another strategy to improve the absorption characteristic of NWs is to break the symmetry of the incident light/NW system. Oblique radiation stimulates new polarization-dependent optical modes. However, the improvement is limited to a narrow band of the solar spectrum. One way to increase light absorption in NWs is to reduce light transmission at the NW-substrate interface and reduce light reflection at the

NW/superstrate interface. Another way to improve the absorption in solar cell structures is to use the plasmonic effect. In this effect with light irradiation, charge carriers at the metal and the dielectric boundary with positive or negative permittivity begin collective oscillation and produce surface plasmon resonance (SPR).

Surface plasmon resonances are divided into two categories, localized surface plasmon resonance (LSPR) and surface plasmon polariton (SPP). LSPRs are observed when surface plasmon is confined in the subwavelength nanostructure, and SPPs are caused by the propagation of charge carriers along the planar surface. Upon excitation, both forms of SPRs can confine incident light on a low-subwavelength scale. The confinement of some carriers improves the local field and allows the manipulation of light below the diffraction limit. These remarkable SPR features allow plasmonic materials to be used in a variety of fields, including photonics [9], energy [10–12], sensors [13], and more. For example, metal nanoparticles inside thin-film solar cells act as antennas that capture the incident light and store its energy in LSPR modes [14–16]. This energy can be transferred to the semiconductor layer by near-field coupling or scattered to the active layer by metal nanoparticles. Nanoparticles can also be used as front scatterers or back reflectors [17–21].

The metal nanostructures at the front trap light in the solar cell beneath by forward scattering the incident light at their LSPR wavelengths with a scattering cross-section greater than their geometric cross-section. Plasmonic nanoparticles at the back of the solar cell increase the optical thickness and improve performance by scattering the far-field. In this theoretical study, we present a structure in which a layer of Au metal was placed at the bottom of the NW to prevent the transmission of light into the substrate and reflect the light that is not absorbed in the single pass into the NW. Increasing the optical path length of excited modes led to improved absorption and photovoltaic efficiency without increasing the volume of material used. More recently, with the development of nanotechnology, the use of metamaterials in many applications such as biosensors [22], photodetectors [23], and absorbers [24,25] was considered. This method can be beneficial in solar harvesting as well.

Furthermore, by considering the conical-shaped parabola on the top of the NW, we reduced leakage of the light into the space outside the NW from the top. The proposed structure is a new method to improve the performance of the NW array solar cell and achieve absorption of over 90% at wavelengths close to the band gap wavelength of InP.

2. Materials and Methods

In this work, we investigated a square array of vertical InP NWs with Au layer at the bottom and conical-shaped parabola at the top. Cone-shaped nanowires grown in vapor-liquid-solid (VLS) mode are a good candidate for the fabrication of low-cost high-performance solar cells. Thus, this shape of nanowires can be beneficial in solar cells. Methods of fabrication of this kind of nanowire are presented in [26]. According to [27], we can place an Au layer at the bottom of the nanowire. Figure 1 shows a schematic of the proposed structure (with Au layer at the bottom and conical-shaped parabola at the top of NWs).

Here, we calculate the absorption of incident photons and the separation of electron–hole pairs photogenerated over an axial n-i-p junction for the InP NW array. The optical response of the NW array is calculated by solving the Maxwell's equations using tabulated data for the refractive indices of InP [28]. Electron–hole transfer in NWs is solved using the drift–diffusion formulas. We use the finite element method (FEM) for this numerical analysis.

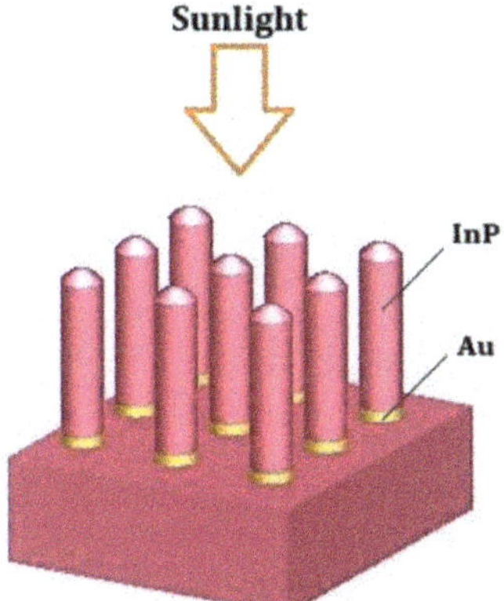

Figure 1. 3D configuration of the proposed InP nanowire (NW) array solar cell with Au layer at the bottom and conical-shaped parabola on the top.

The solutions of Maxwell's equations show how light is distributed inside and outside of the NWs. By obtaining the electric field of the system, $E(r)$, the number of electron–hole pairs photogenerated per unit volume per unit time, $G(r)$, can be obtained for a given incident intensity spectrum. The absorption at each wavelength on the volume V of the NW can be calculated by:

$$A(\lambda) = \frac{1}{2}\omega\varepsilon_0 \int |E(\lambda)|^2 \mathrm{Im}\left(n^2(\lambda)\right) dV \tag{1}$$

where ε_0 is the vacuum permittivity, ω is the angular frequency, E is the electric field, and n is the complex refractive index.

After calculating the optical generation rate, $G(r)$, and recombination mechanisms, R, the electron–hole transfer is analyzed by solving the drift–diffusion equations presented below:

$$\begin{aligned}
\nabla.(-\varepsilon\nabla\varphi) &= q(p - n + N_D - N_A), \\
\nabla.J_n = \nabla.(-q\mu_n n\nabla\varphi_n) &= q(R - G), \\
\nabla.J_p = \nabla.\left(-q\mu_p p\nabla\varphi_p\right) &= -q(R - G),
\end{aligned} \tag{2}$$

R is the net recombination rate in the bulk of the NW obtained from:

$$R = R_{SRH} + R_{Rad} + R_{Aug} = \frac{A}{n + p + 2n_i} + B + C(n + p)\left(np - n_i^2\right), \tag{3}$$

where A, B, and C are the recombination coefficient of the SRH/Radiative/Auger recombination, and n_i is the intrinsic carrier concentration. We neglect radiative recombination and we use $B = 0$ in Equation (3). We assume that the Shockley–Read–Hall (SRH) lifetime and the Auger recombination coefficient for both electron and holes are equal. The surface recombination at the surface of the NW is also considered through the term:

$$R_{surface} = \frac{v_{sr}}{n + p + 2n_i}\left(np + n_i^2\right), \tag{4}$$

where v_{sr} is the surface recombination velocity [29].

Assuming that all the photogenerated carriers contribute to the current, we can obtain the short circuit current density from:

$$J_{sc} = e \int \frac{\lambda}{hc} \frac{P_{abs}(\lambda)}{P_{in}(\lambda)} I_{AM1.5}(\lambda) d\lambda, \tag{5}$$

The geometric structure used for the drift–diffusion model has a length of 100 nm in top n-segment and 300 nm in bottom p-segment with 10^{18} cm^{-3} ionized doping concentration for both of them. The i-segment with 10^{15} cm^{-3} p-doping concentration is located between the n- and p-segments.

The calculations are performed using the finite-difference time-domain (FDTD) method at a wavelength range of 320–1000 nm for unpolarized incident light parallel to the axis of the NWs, taking into account the periodic boundary conditions in the x-, y- directions and perfectly matched layer (PML) in the z-direction. The specific values of the parameters used in the drift–diffusion model are presented in Table 1. We consider here, surface recombination velocity for the bare cell without Au layer at the bottom and conical-shaped parabola at the top equals to 20,000 cms^{-1} [30]. Placing the Au layer at the bottom of the NWs causes a change in the motion of the electrons, which is considered in modeling by changing the surface recombination velocity. The value of this surface recombination velocity is selected according to reference [31]. Reflection is monitored with a power monitor placed behind the radiation source; transmission is monitored with a power monitor placed behind the structure. Electric and magnetic fields are detected within the frequency profile monitors. For a solar cell, in addition to the efficiency of converting sunlight into electrical energy, η, three other parameters are usually considered: (1) the short circuit current density J_{sc}, (2) the open circuit voltage V_{oc}, and (3) the fill factor FF defined by $\eta = \frac{J_{sc} V_{oc}\ FF}{P_{in}}$.

Table 1. The parameters used in the drift–diffusion model [32,33].

Parameters	InP
Dielectric constant (ε_r)	12.5
SRH recombination coefficient (A)	10^7 s^{-1}
Auger recombination coefficient (C)	9×10^{-31} cm^{6}s^{-1}
Electron mobility (μ_n)	5400 cm^{2}V^{-1}s^{-1}
Hole mobility (μ_p)	250 cm^{2}V^{-1}s^{-1}
Band gap	1.34 eV

3. Results and Discussion

As mentioned in previous works [34–37], light absorption in NWs depends on the geometrical parameters, and with proper design, light absorption can be improved in them. One way to improve the absorption of light to achieve almost complete absorption in NWs is to increase their length. Increasing this parameter reduces the transmission of light to the substrate, but in solar cells it increases the dark current, which results in decreased structural efficiency. Furthermore, with increasing length, the volume of materials used increases. Increasing the radius of the NWs increases the absorption of light in them, but due to the increase in light transmission between the NWs and the substrate, the current of the structure decreases. In addition, by reducing the pitch of the array, the absorption of coupled light into the array increases. However, the problem is the increase in top-insertion reflection losses.

In this study, to reduce the transfer of guided modes into the substrate without changing the geometrical parameter of the NWs, we placed a layer of Au metal at the bottom of the NWs which are placed in a square lattice with geometrical values of radius of 100 nm, pitch of 500 nm, and length of 1400 nm. The geometrical values of the radius and the pitch are chosen so that the NWs form a sparse array and the results are independent of the lattice arrangement [38]. Furthermore, the length of 1400 nm for the total length (with or without Au layer) is chosen so that the absorption saturation that is seen in long NWs does not occur [30,39,40].

Because most of the light in the NWs is absorbed in the top parts and the absorption decreases as it moves towards the bottom, we placed Au layer at the bottom to act as a reflector at optical frequencies and reflect light into the NWs. This reflection increases the optical path length of the excited modes and improves the absorption of the guided modes. In addition, when the light reaches to the metal, plasmon is excited in metal and it generates dipole. Due to dipole–dipole coupling between the plasmon excited metal and the adjacent semiconductor in near-field, plasmonic energy in the metal is transferred to semiconductor and it generates more electron–hole pairs in the semiconductor. This can promote performance of the solar cell. Therefore, placing the Au layer at the bottom of the NW has a better result than other parts. By placing the Au layer on the top part of the NW, the light cannot reach the lower part of the NW and the light absorption is reduced.

Decreased light absorption also reduces the number of photogenerated carriers compared to the bare structure. To determine the appropriate thickness of the Au layer, one must sweep its thickness from 0 to 250 nm, and obtain the J_{sc} each time. According to the results presented in Figure 2, sweeping the thickness of this layer has little effect on the J_{sc}. Among the various values, 80 nm thickness gives the maximum improvement in J_{sc}. In this case, the J_{sc} increases from 24.1 mA/cm^2 for the bare cell to 26.75 mA/cm^2, which is equivalent to a 10% improvement.

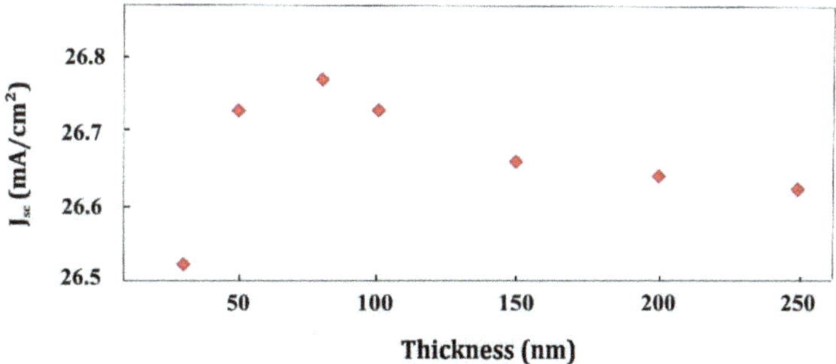

Figure 2. The short circuit current density (in mA/cm^2) of the InP NW array solar cell for varying thickness of Au layer for the fixed radius of 100 nm and pitch of 500 nm.

In NWs with the same geometry, we can improve the absorption if we can reduce the reflection of light from the NW/superstrate. To achieve this goal, we consider the top part of the NW, which is the interface between air and NW, in the form of a conical-shaped parabola. With this assumption, as can be seen in Figure 3, the light reflection from the top part decreases and the absorption in the NW increases. This increase in absorption increases J_{sc} from 24.1 mA/cm^2 for the bare cell to 24.931 mA/cm^2 for the NW with conical-shaped parabola on the top.

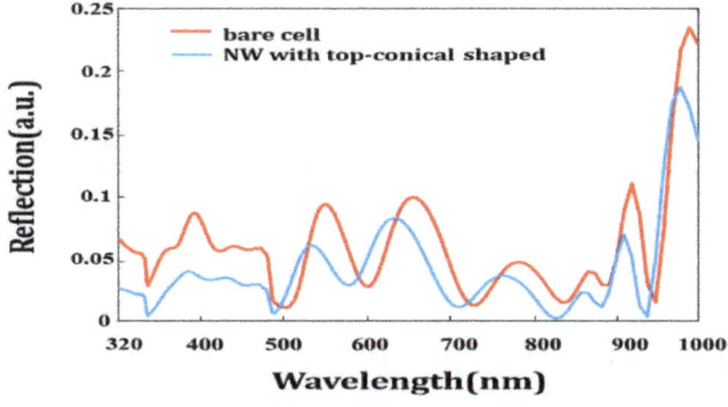

Figure 3. Reflection spectra for the bare cell and for the NWs with conical-shaped top.

Considering that the use of an Au layer at the bottom of the NW and the conical-shaped parabola on the top part both cause more light to be trapped and more electrons and holes are generated by the light, in the following we will examine the effect of applying both of them on the performance of the array. According to the results presented in Figure 4, the absorption increases at all wavelengths, and close to the band gap wavelength of InP is approximately 1 and is limited by the reflection losses at the top side.

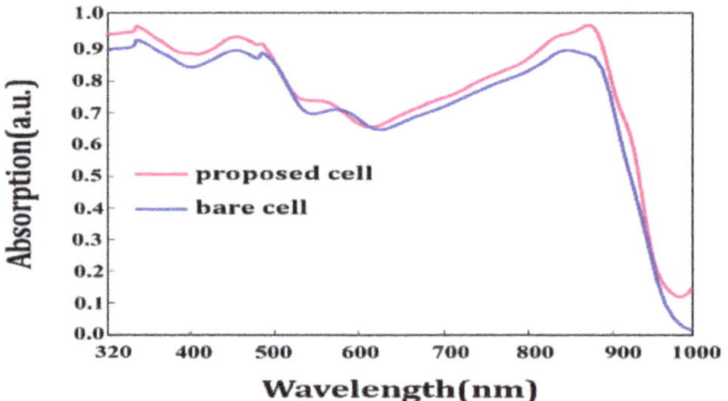

Figure 4. Absorption spectra for bare cell and proposed cell.

Observation of the electric field distribution profile within the NWs in Figure 5 for the bare cell and proposed cell shows that the Au layer prevents the transmission of light into the substrate and reflects most of the light into the NW. Thus, the modes are broadening and the absorption at the bottom of the NW is improved. By improving the absorption within the NW, more carriers contribute to the production of current and the J_{sc} increases from 24.1 mA/cm^2 to 27.64 mA/cm^2.

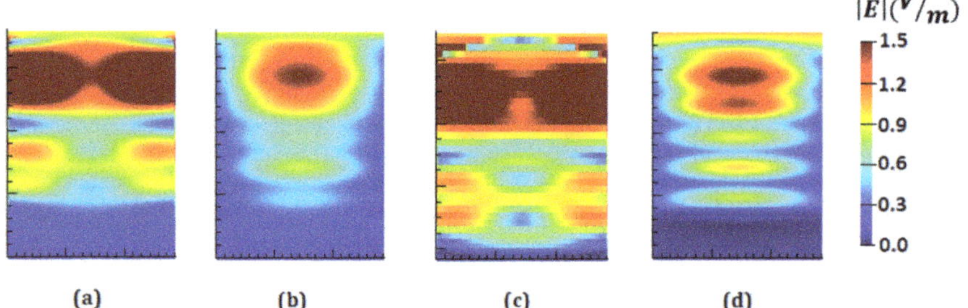

Figure 5. The electric field profile for transverse magnetic (TM) and transverse electric (TE) modes respectively for (**a**,**b**) the bare InP NW array solar cell, and (**c**,**d**) the proposed InP NW array solar cell.

Since the performance of the NWs depends on the geometric parameters of the structure, we sweep the radius and pitch parameters and examine their effects. As shown in Figure 6, sweeping the radius up to about 50 nm does not cause a significant change in the J_{sc} of the proposed cell relative to the bare cell. As can be seen in Figure 7, as the radius increases, the absorption spectrum of the proposed cell expands to longer wavelengths, and the location of the wavelength of HE$_{11}$ mode, which is equivalent to the absorption peak at the longest wavelength and close to the InP band gap wavelength, red-shifts.

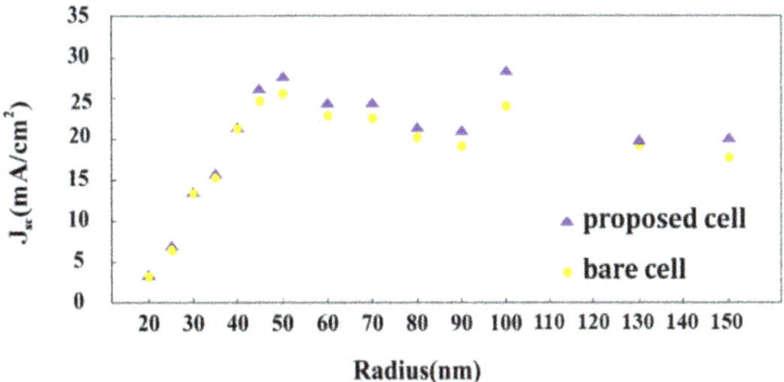

Figure 6. The short circuit current density (in mA/cm^2) of the bare and proposed cell of InP NW array solar cell for varying radius.

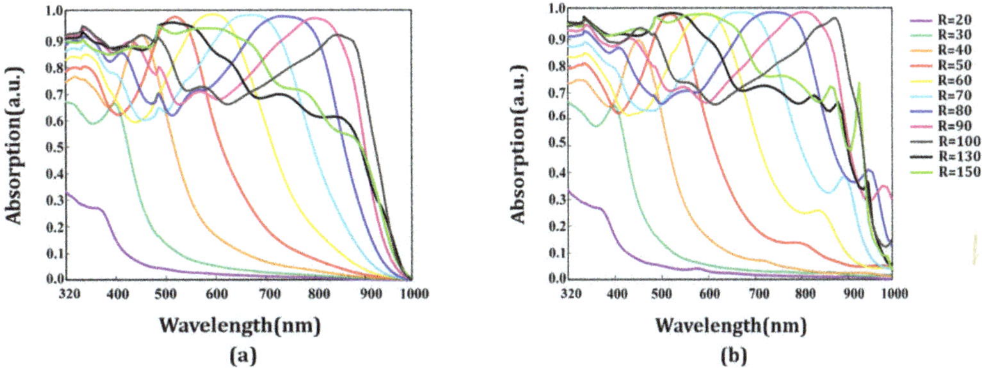

Figure 7. Absorption spectra for varying radius for (**a**) bare cell and (**b**) proposed cell.

In Figure 8, the J_{sc} is plotted as a function of pitch of the array. Increasing this parameter reduces the semiconductor material in the NW, which leads to a decrease in the absorption in the NW, as can be seen in Figure 9.

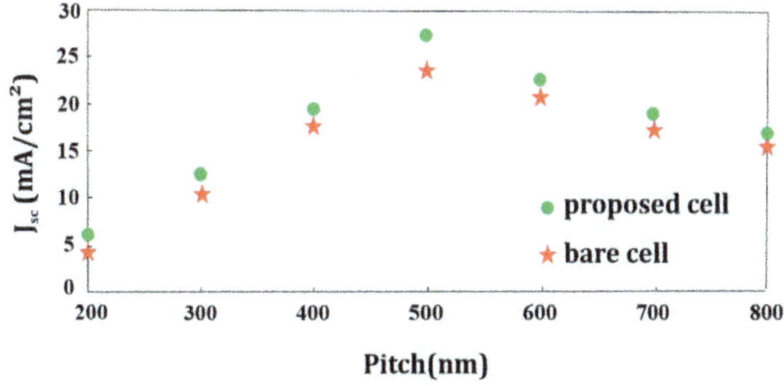

Figure 8. The short circuit current density (in mA/cm^2) of the bare cell and proposed cell of InP NW array solar cell for varying pitch.

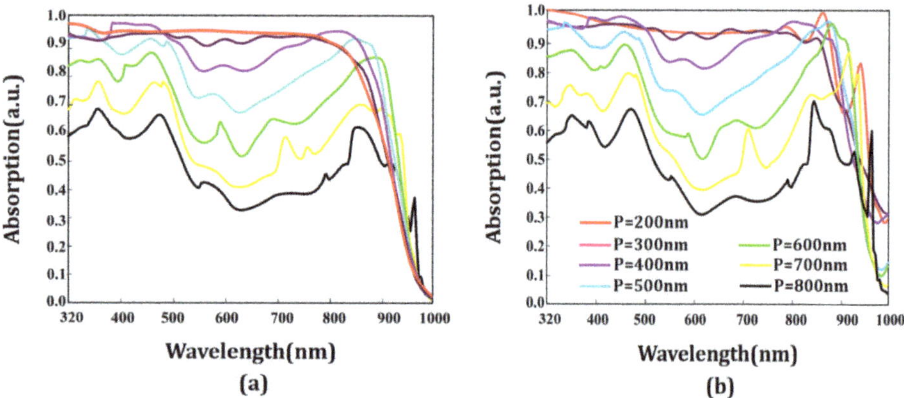

Figure 9. Absorption spectra for varying pitch for (**a**) bare cell and (**b**) proposed cell.

According to Figure 10, by increasing pitch of the array, the insertion reflection losses at the top air/NW interface, which are an upper limit on the J_{sc} decrease and the absorption in the NWs can be improved. Therefore, it is necessary to choose the pitch of the array in such a way that the volume of material consumed is sufficient for absorption and the insertion reflection losses are reduced.

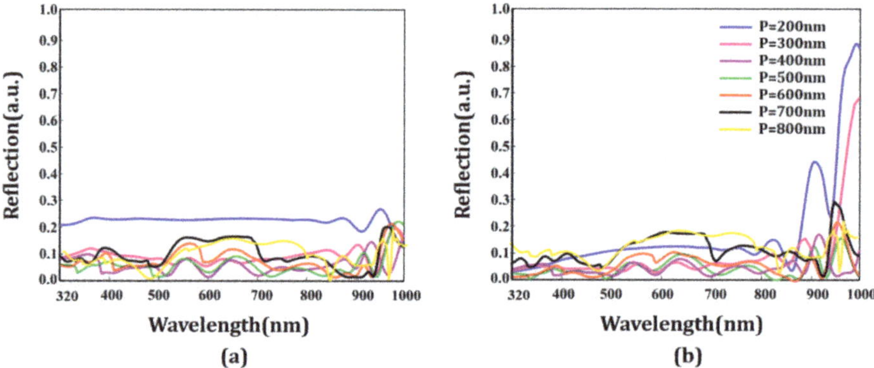

Figure 10. Reflection spectra for varying pitch for (**a**) bare cell and (**b**) proposed cell.

Another way to improve the absorption of NWs, especially in the middle part of the solar spectrum, is to break the symmetry of the NW array/light incident system, which can be achieved by obliquely incident light on the vertical NWs. In addition to the HE_{1m}-guided modes, which depend on symmetry and are excited in the NWs, oblique radiation also stimulates additional Mie resonances. In this way we can achieve broadband absorption.

As can be seen in Figure 11, the absorption improvement occurs in a broad spectral range for inclined radiation compared to normal radiation for both bare and proposed structures. Improved absorption due to the excitation of additional modes indicates strong light trapping in the NWs and prevents its transmission to the substrate. Figure 12 shows the effect of inclined radiation on the transmission spectrum of both bare and proposed cells.

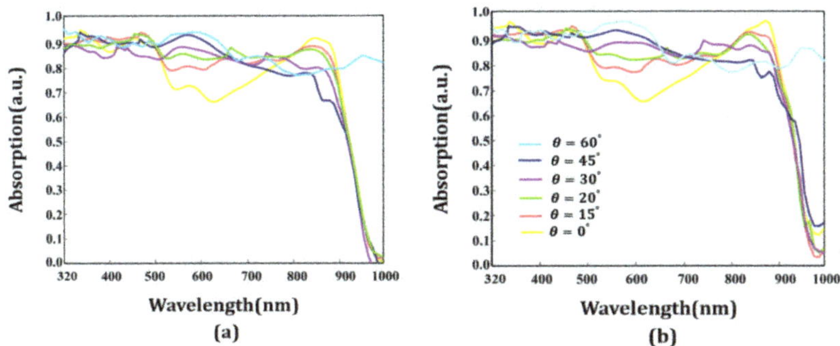

Figure 11. Absorption spectra for inclined radiation for (**a**) bare cell and (**b**) proposed cell.

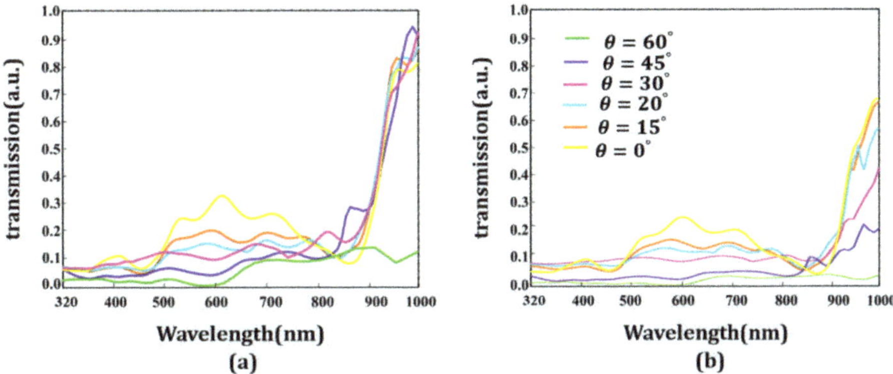

Figure 12. Transmission spectra for inclined radiation for (**a**) bare cell (**b**) proposed cell.

Figure 13 shows the changes in the Jsc by changing the radiation angle. Increasing the absorption of light in oblique radiation compared to normal radiation has increased the generation of the optical carriers, which leads to an increase in Jsc. Among the various radiation angles studied, the 60 ° angle achieves the greatest improvement in Jsc.

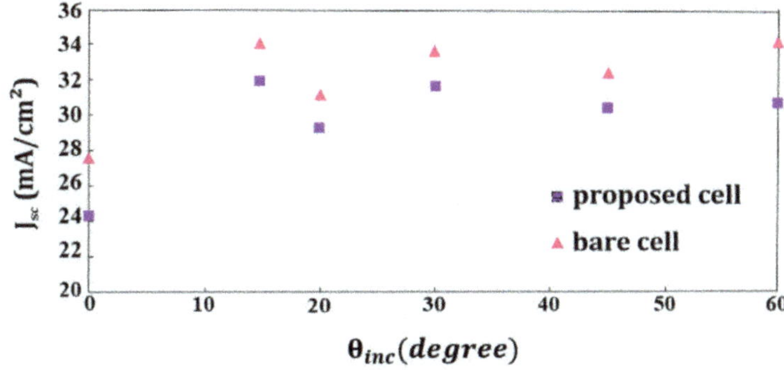

Figure 13. The short circuit current density (in mA/cm^2) of the bare and proposed cell of InP NW array solar cell for varying radiation angle.

Quantitative performance of the cell can be estimated according to the current density–voltage curve obtained by solving the Poisson and drift–diffusion equations with bulk conditions and surface recombination and parameters listed in Table 1. The proposed structure which consists of a layer of Au at the bottom of the NW and a conical-shaped parabola at the top improves the J_{sc} by 14.69%. The power conversion efficiency varies from 14.7 for the bare cell to 17 for the proposed cell.

If we replace the ohmic contacts with carrier selective contact, the J_{sc} will increase from 24.1 mA/cm^2 to 24.71 mA/cm^2 for the bare cell and 27.64 mA/cm^2 to 28.2 mA/cm^2 for the proposed cell due to neglecting the recombination of minority carriers in the contacts [41]. The results of the electrical simulation for the ohmic and carrier selective contacts shown in Figure 14 are summarized in Table 2.

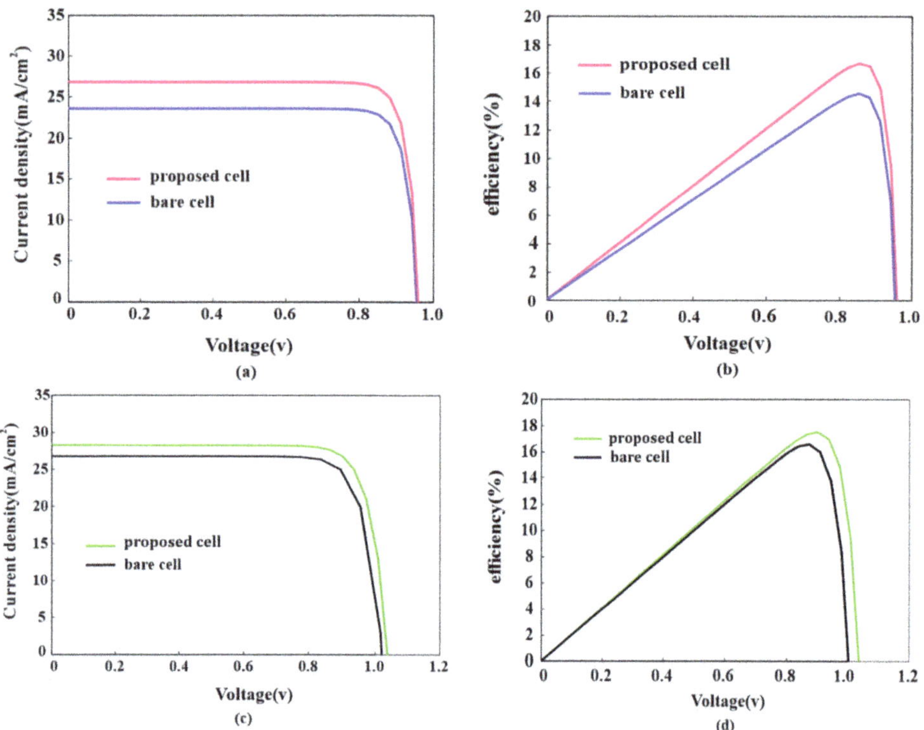

Figure 14. Applied voltage vs. current density for the bare InP NW array solar cell and for proposed cell, applied voltage vs. photovoltaic efficiency for the bare InP NW array solar cell and for proposed cell for (**a–b**) ohmic contacts (**c–d**) perfect carrier selective contacts.

Table 2. Electrical result for the bare and proposed InP NW array solar cells.

Structure	V_{oc} (Volt)	J_{sc} (mA/cm^2)	FF (%)	Efficiency (%)	J_{sc} Enhancement (%)
Ohmic Contacts					
Bare cell	0.958	24.1	63.67	14.7	————
Proposed cell	0.963	27.64	63.87	17	14.69
Perfect Carrier Selective Contacts					
Bare cell	1.015	24.71	66.03	16.56	————
Proposed cell	1.02	28.2	67.97	19.55	14.12

4. Conclusions

We theoretically investigated an InP NW array solar cell with an Au layer at the bottom and a conical-shaped parabola at the top with an axial n-i-p junction on each NW. Among the various values studied for the thickness of the Au layer, at the thickness of 80 nm, the highest improvement was obtained compared to the bare structure. According to the results, the use of Au layer at the bottom of the NWs prevented the transmission of light into the substrate and improved the absorption at long wavelengths, especially around the band gap wavelength, where light absorption was usually poor. The light trapping due to the multiple reflections of the Au layer increased the optical path length of the excited modes in the NWs. Enhancing the absorption of light also increased the photogenerated carriers and ultimately increased the J_{sc}. The conical-shaped parabola of the top part of the NW also reduced light reflection from the upper part of the structure and improved cell performance. The proposed structure increased J_{sc} from 24.1 mA/cm^2 for the bare cell to 27.64 mA/cm^2 for the proposed cell with ohmic contacts and from 24.71 mA/cm^2 to 28.2 mA/cm^2 for the carrier selective contacts, which improved to 14.69% and 14.12%, respectively.

Author Contributions: F.A. designed and performed simulations and analyzed data, S.O. edited and prepared the final draft of the manuscript. All authors have read and agreed to the published version of the manuscript.

Funding: This research received no external funding.

Data Availability Statement: Not applicable.

Acknowledgments: This work has been done in Nano-photonics and Optoelectronics Research Laboratory (NORLab), Shahid Rajaee University.

Conflicts of Interest: The authors declare that they have no known competing financial interests or personal relationships that could have appeared to influence the work reported in this paper.

References

1. Mauthe, S.; Baumgartner, Y.; Sousa, M.; Ding, Q.; Rossell, M.D.; Schenk, A.; Czornomaz, L.; Moselund, K.E. High-speed III-V nanowire photodetector monolithically integrated on Si. *Nat. Commun.* **2020**, *11*, 1–7. [CrossRef]
2. Shen, L.; Pun, E.Y.; Ho, J.C. Recent developments in III–V semiconducting nanowires for high-performance photodetectors. *Mater. Chem. Front.* **2017**, *1*, 630–645. [CrossRef]
3. Ren, D.; Azizur-Rahman, K.M.; Rong, Z.; Juang, B.-C.; Somasundaram, S.; Shahili, M.; Farrell, A.C.; Williams, B.S.; Huffaker, D.L. Room-temperature midwavelength infrared InAsSb nanowire photodetector arrays with Al$_2$O$_3$ passivation. *Nano Lett.* **2019**, *19*, 2793–2802. [CrossRef] [PubMed]
4. Berg, A.; Yazdi, S.; Nowzari, A.; Storm, K.; Jain, V.; Vainorius, N.; Samuelson, L.; Wagner, J.B.; Borgström, M.T. Radial Nanowire Light-Emitting Diodes in the (Al$_x$Ga$_{1-x}$)$_y$In$_{1-y}$P Material System. *Nano Lett.* **2016**, *16*, 656–662. [CrossRef]
5. Gagliano, L.; Kruijsse, M.; Schefold, J.D.; Belabbes, A.; Verheijen, M.A.; Meuret, S.; Koelling, S.; Polman, A.; Bechstedt, F.; Haverkort, J.E. Efficient Green Emission from Wurtzite Al x In1–x P Nanowires. *Nano Lett.* **2018**, *18*, 3543–3549. [CrossRef] [PubMed]
6. Bui, H.Q.T.; Velpula, R.T.; Jain, B.; Aref, O.H.; Nguyen, H.-D.; Lenka, T.R.; Nguyen, H.P.T. Full-color InGaN/AlGaN nanowire micro light-emitting diodes grown by molecular beam epitaxy: A promising candidate for next generation micro displays. *Micromachines* **2019**, *10*, 492. [CrossRef]
7. Kim, H. *III-V Semiconductor Nanowire Lasers on Silicon*; UCLA: Los Angeles, CA, USA, 2018.
8. Zhang, G.; Takiguchi, M.; Tateno, K.; Tawara, T.; Notomi, M.; Gotoh, H. Telecom-band lasing in single InP/InAs heterostructure nanowires at room temperature. *Sci. Adv.* **2019**, *5*, eaat8896. [CrossRef]
9. Gao, Y.; Murai, S.; Zhang, F.; Tamura, S.; Tomita, K.; Tanaka, K. Enhancing upconversion photoluminescence by plasmonic-photonic hybrid mode. *Opt. Express* **2020**, *28*, 886–897. [CrossRef] [PubMed]
10. Singh, B.; Shabat, M.M.; Schaadt, D.M. Wide angle antireflection in metal nanoparticles embedded in a dielectric matrix for plasmonic solar cells. *Prog. Photovolt. Res. Appl.* **2020**, *28*, 682–690. [CrossRef]
11. Samajdar, D. Performance enhancement of Nanopyramid based Si hybrid solar cells utilizing the plasmonic properties of oxide coated Metal Nanoparticles. *Opt. Mater.* **2020**, *107*, 110166.
12. Ghahremanirad, E.; Olyaee, S.; Hedayati, M. The Influence of Embedded Plasmonic Nanostructures on the Optical Absorption of Perovskite Solar Cells. *Photonics* **2019**, *6*, 37. [CrossRef]

13. Liu, B.; Chen, S.; Zhang, J.; Yao, X.; Zhong, J.; Lin, H.; Huang, T.; Yang, Z.; Zhu, J.; Liu, S. A plasmonic sensor array with ultrahigh figures of merit and resonance linewidths down to 3 nm. *Adv. Mater.* **2018**, *30*, 1706031. [CrossRef]

14. Pathak, N.K.; Chander, N.; Komarala, V.K.; Sharma, R. Plasmonic perovskite solar cells utilizing Au@SiO 2 core-shell nanoparticles. *Plasmonics* **2017**, *12*, 237–244. [CrossRef]

15. Jangjoy, A.; Bahador, H.; Heidarzadeh, H. A Comparative Study of a Novel Anti-reflective Layer to Improve the Performance of a Thin-Film GaAs Solar Cell by Embedding Plasmonic Nanoparticles. *Plasmonics* **2021**, *16*, 395–401. [CrossRef]

16. Ferry, V.E.; Verschuuren, M.A.; Lare, M.C.v.; Schropp, R.E.; Atwater, H.A.; Polman, A. Optimized spatial correlations for broadband light trapping nanopatterns in high efficiency ultrathin film a-Si: H solar cells. *Nano Lett.* **2011**, *11*, 4239–4245. [CrossRef] [PubMed]

17. Deka, N.; Islam, M.; Sarswat, P.K.; Kumar, G. Enhancing solar cell efficiency with plasmonic behavior of double metal nanoparticle system. *Vacuum* **2018**, *152*, 285–290. [CrossRef]

18. Diukman, I.; Orenstein, M. How front side plasmonic nanostructures enhance solar cell efficiency. *Sol. Energy Mater. Sol. Cells* **2011**, *95*, 2628–2631. [CrossRef]

19. Cao, S.; Yu, D.; Lin, Y.; Zhang, C.; Lu, L.; Yin, M.; Zhu, X.; Chen, X.; Li, D. Light Propagation in Flexible Thin-film Amorphous Silicon Solar Cells with Nanotextured Metal Back Reflectors. *ACS Appl. Mater. Interfaces* **2020**, *12*, 26184–26192. [CrossRef]

20. Sun, T.; Shi, H.; Cao, L.; Liu, Y.; Tu, J.; Lu, M.; Li, H.; Zhao, W.; Li, Q.; Fu, T. Double grating high efficiency nanostructured silicon-based ultra-thin solar cells. *Results Phys.* **2020**, *19*, 103442. [CrossRef]

21. Hungerford, C.D.; Fauchet, P.M. Design of a plasmonic back reflector using Ag nanoparticles with a mirror support for an a-Si: H solar cell. *AIP Adv.* **2017**, *7*, 75004. [CrossRef]

22. Liu, K.; Zhang, R.; Liu, Y.; Chen, X.; Li, K.; Pickwell-Macpherson, E. Gold nanoparticle enhanced detection of EGFR with a terahertz metamaterial biosensor. *Biomed. Opt. Express* **2021**, *12*, 1559–1567. [CrossRef]

23. Devine, E.P. Mid-infrared photodetector based on 2D material metamaterial with negative index properties for a wide range of angles near vertical illumination. *Appl. Phys. A* **2021**, *127*, 1–5.

24. Jiang, X.; Wang, T.; Zhong, Q.; Yan, R.; Huang, X. A near-ideal solar selective absorber with strong broadband optical absorption from UV to NIR. *Nanotechnology* **2020**, *31*, 315202. [CrossRef]

25. Jiang, X.; Wang, T.; Zhong, Q.; Yan, R.; Huang, X. Ultrabroadband light absorption based on photonic topological transitions in hyperbolic metamaterials. *Opt. Express* **2020**, *28*, 705–714. [CrossRef] [PubMed]

26. Chang, S.; Lee, G.J.; Song, Y.M. Recent Advances in Vertically Aligned Nanowires for Photonics Applications. *Micromachines* **2020**, *11*, 726. [CrossRef] [PubMed]

27. Lu, W.; Lieber, C.M. Semiconductor nanowires. *J. Phys. D Appl. Phys.* **2006**, *39*, R387. [CrossRef]

28. Palik, E.D. *Handbook of Optical Constants of Solids*; Academic Press: Cambridge, MA, USA, 1998; Volume 3.

29. Darling, R.B. Defect-state occupation, Fermi-level pinning, and illumination effects on free semiconductor surfaces. *Phys. Rev.* **1991**, *43*, 4071. [CrossRef]

30. Adibzadeh, F.; Olyaee, S. Performance improvement of InP nanowire array solar cells by decorated nanowires and using back reflector. *Opt. Mater.* **2020**, *109*, 110397. [CrossRef]

31. Rosenwaks, Y.; Shapira, Y.; Huppert, D. Metal reactivity effects on the surface recombination velocity at InP interfaces. *Appl. Phys. Lett.* **1990**, *57*, 2552–2554. [CrossRef]

32. Vurgaftman, I.; Meyer, J.Á.; Ram-Mohan, L.Á. Band parameters for III–V compound semiconductors and their alloys. *J. Appl. Phys.* **2001**, *89*, 5815–5875. [CrossRef]

33. Shur, M.S. *Handbook Series on Semiconductor Parameters*; World Scientific: Singapore, 1996; Volume 1.

34. Anttu, N.; Xu, H. Efficient light management in vertical nanowire arrays for photovoltaics. *Opt. Express* **2013**, *21*, A558–A575. [CrossRef]

35. Aghaeipour, M.; Anttu, N.; Nylund, G.; Samuelson, L.; Lehmann, S.; Pistol, M.-E. Tunable absorption resonances in the ultraviolet for InP nanowire arrays. *Opt. Express* **2014**, *22*, 29204–29212. [CrossRef] [PubMed]

36. Diedenhofen, S.L.; Janssen, O.T.; Grzela, G.; Bakkers, E.P.; Gómez Rivas, J. Strong geometrical dependence of the absorption of light in arrays of semiconductor nanowires. *ACS Nano* **2011**, *5*, 2316–2323. [CrossRef] [PubMed]

37. Seo, K.; Wober, M.; Steinvurzel, P.; Schonbrun, E.; Dan, Y.; Ellenbogen, T.; Crozier, K.B. Multicolored vertical silicon nanowires. *Nano Lett.* **2011**, *11*, 1851–1856. [CrossRef]

38. Li, J.; Yu, H.; Li, Y. Solar energy harnessing in hexagonally arranged Si nanowire arrays and effects of array symmetry on optical characteristics. *Nanotechnology* **2012**, *23*, 194010. [CrossRef] [PubMed]

39. Wallentin, J.; Anttu, N.; Asoli, D.; Huffman, M.; Åberg, I.; Magnusson, M.H.; Siefer, G.; Fuss-Kailuweit, P.; Dimroth, F.; Witzigmann, B.; et al. InP nanowire array solar cells achieving 13.8% efficiency by exceeding the ray optics limit. *Science* **2013**, *339*, 1057–1060. [CrossRef] [PubMed]

40. Wang, B.; Leu, P.W. Tunable and selective resonant absorption in vertical nanowires. *Opt. Lett.* **2012**, *37*, 3756–3758. [CrossRef]

41. Anttu, N. Physics and design for 20% and 25% efficiency nanowire array solar cells. *Nanotechnology* **2018**, *30*, 074002. [CrossRef]

Article

Interband, Surface Plasmon and Fano Resonances in Titanium Carbide (MXene) Nanoparticles in the Visible to Infrared Range

Manuel Gonçalves [1,*], **Armen Melikyan** [2], **Hayk Minassian** [3], **Taron Makaryan** [4], **Petros Petrosyan** [5] **and Tigran Sargsian** [6]

[1] Institute of Experimental Physics, Ulm University, Albert-Einstein-Allee 11, 89069 Ulm, Germany
[2] Institute of Engineering and Physics, Russian-Armenian (Slavonic) State University, Yerevan 0051, Armenia; armen.melikyan@rau.am
[3] A. Alikhanian National Science Laboratory, Yerevan 0036, Armenia; hminassian@yerphi.am
[4] Murata Manufacturing Co., Ltd., 1-10-1 Higashikotari, Nagaokakyo-shi, Kyoto 617-8555, Japan; taron.makaryan@murata.com
[5] Department of Physics, Yerevan State University, Yerevan 0025, Armenia; petros.petrosyan@ysumail.am
[6] Department of General Physics and Quantum Nanostructures, Russian-Armenian (Slavonic) State University, Yerevan 0051, Armenia; tigran.sargsian@rau.am
* Correspondence: manuel.goncalves@uni-ulm.de

Abstract: Since the discovery of the optical properties of two-dimensional (2D) titanium carbide (MXene) conductive material, an ever increasing interest has been devoted towards understanding it as a plasmonic substrate or nanoparticle. This noble metal-free alternative holds promise not only due to its lower cost but also its 2D nature, hydrophilicity and apparent bio-compatibility. Herein, the optical properties of the most widely studied $Ti_3C_2T_x$ MXene nanosheets are theoretically analyzed and absorption cross-sections are calculated exploiting available experimental data on its dielectric function. The occurrence of quadrupole surface plasmon mode in the optical absorption spectra of large MXene nanoparticles is demonstrated for the first time. The resonance wavelengths corresponding to interband transitions, longitudinal and transversal dipole oscillations and quadrupole longitudinal surface plasmon mode are identified for single and coupled nanoparticles by modeling their shapes as ellipsoids, disks and cylinders. A new mechanism of excitation of longwave transversal surface plasmon oscillations by an external electric field perpendicular to the direction of charge oscillations is presented. Excitingly enough, a new effect in coupled MXene nanoparticles—Fano resonance—is unveiled. The results of calculations are compared to known experimental data on electron absorption spectroscopy, and good agreement is demonstrated.

Keywords: two-dimensional material; titanium carbide MXene; near-field enhancement; plasmonic material

Citation: Gonçalves, M.; Melikyan, A.; Minassian, H.; Makaryan, T.; Petrosyan, P.; Sargsian, T. Interband, Surface Plasmon and Fano Resonances in Titanium Carbide (MXene) Nanoparticles in the Visible to Infrared Range. *Photonics* **2021**, *8*, 36. https://doi.org/10.3390/photonics8020036

Received: 26 December 2020
Accepted: 28 January 2021
Published: 1 February 2021

Publisher's Note: MDPI stays neutral with regard to jurisdictional clai-ms in published maps and institutio-nal affiliations.

1. Introduction

A recently discovered family of 2D materials—transition metal carbides/nitrides called MXenes—has become the subject of versatile intensive studies due to the materials' unique optoelectronic properties as well as hydrophilicity, flexibility and metal-like electronic conductivity [1–10]. This interest in MXenes is fueled by the possibility of applications as supercapacitors [11] and in electromagnetic interference shielding [5], in optical sensing and light detection [7], and even in communication and biology [9]. The study of the optical properties of MXenes showed the possibility of creating transparent conductive electrodes [9] and saturable absorbers for femtosecond mode-locked lasers [12], for photonic devices [4] and for plasmonic applications at near-infrared wavelengths [13].

MXenes are mostly synthesized by etching out the A layers from parent MAX phases [11,14,15], which have a composition generally denoted as $M_{n+1}AX_n$, where M is an early transition metal, A is mainly a group IIIA or IVA element, X is C and/or N, and n = 1, 2, 3 or 4. In most cases the etching process with acids or alkalis results in the fabricated

MXene, $M_{n+1}X_nT_x$, that is terminated by some surface functional groups—T_x, such as OH, O, Cl or F (depending on the synthesis method). The synthesis method is a crucial aspect for controlling the surface functionality of MXenes and hence the underlying properties.

Much work has been performed for exploring the optical properties of MXenes. For instance, the plasmonic properties of MXenes were investigated by electron energy loss spectroscopy (EELS) [16,17] and by optical measurements in the IR, visible and UV ranges [13,18,19]. In these studies, both surface plasmons (SPs) as well as bulk plasmons and interband transitions (IBTs) were identified. Optical forces on MXene nanoparticles in the near-infrared have been calculated using the finite-difference time-domain method [20]. Non-linear optical properties of MXenes have also been studied [21]. The exploitation of optical properties of MXenes is not limited to the possibility of creating new photonics devices. Another promising area of their application is molecular sensing by surface enhanced Raman scattering (SERS). For the most widely studied MXene, $Ti_3C_2T_x$, the first experimental result demonstrated a SERS enhancement factor in the order of 10^6 [19]. Afterwards, the following studies included other types of MXenes as well, such as nitride Ti_2N, which yielded a notable enhancement factor of 10^{12} [22]. Recently, more publications appeared on thickness dependence of SERS in $Ti_3C_2T_x$ nanosheets [23], or for using it as a SERS active substrate for reliable and sensitive detection of organic pollutants [24]. A systematic theoretical understanding of the optical properties of MXenes is not developed yet, which makes difficult the interpretation of all the experimental data.

In this paper we analyze the experimental data obtained in $Ti_3C_2T_x$ nanoparticles by both optical absorption and EELS and present the results of calculations for the absorption cross-section in isolated and interacting nanoparticles composed of MXene sheets. To effectively use the simulation software, an approximate analytical method is applied to identify the ranges of wavelengths corresponding to the peculiarities of spectra. The presented approach allows interpretation of experimental results on EELS in MXene in the vis-infrared range. Moreover, an interesting phenomenon related to the interaction between longitudinal quadrupole and transversal dipole modes suggests the appearance of a Fano resonance.

2. Analysis of Experimental Data of EELS and Optical Absorption Spectroscopy of $Ti_3C_2T_x$

To assess the physical properties of MXenes it is crucial to reveal the peculiarities of the energy spectrum of free and bounded carriers. EELS experiments in $Ti_3C_2T_x$, covering a fairly wide range of energies up to 30 eV, showed that the maxima at lower energy losses between 0.2 and 0.7 eV in stacks of MXenes occur due to dipole oscillations [16]. Furthermore, in [16], for the energy losses at higher energies up to 1.5 eV a weaker maximum was also found, which was independently revealed in optical spectra of $Ti_3C_2T_x$ as well in [18,19]. Detected by different methods, this peculiarity was prescribed to low-energy IBTs (see Figure 2 and Figure S6 in [19]). At energies over 5 eV, rather strong IBTs were detected, after which bulk plasmons occured [16]. It was also shown that dipole plasmonic frequencies can be tuned in the mid-infrared range by controlling the sheet geometries and terminations. Moreover, the experiment of [16] revealed that in multilayered (ML) MXene sheets the individual 2D flakes interact weakly, apparently reducing the intensity of bulk excitations and leaving longitudinal surface plasmons (LSPs) as a source of dominant screening mechanism. In [17], a strong IBT centered around 4 eV was revealed by exploring the role of the localization of termination groups on the optical properties of ML $Ti_3C_2T_x$. It was shown that the T_x functional group localization on the surface has a prominent effect on the optical properties of the ML MXene, leading to 40% variations in the optical conductivity in the middle of the visible spectrum.

For a complete understanding of the optical properties of ML MXenes, it is necessary to know the permittivity tensor of the material. For the first time, the $Ti_3C_2T_x$ permittivity was measured by the ellipsometric method at 45° polarized light in [13]. We recall that setting the real part of the transversal component of a dielectric function equal to 0 ($\Re[\epsilon(\omega)] = 0$) of a thin film is the condition for the realization of transversal surface plasmons (TSPs).

According to the measurements of [13], this condition for the films of thicknesses from 14 to 75 nm is satisfied for the wavelength range $\lambda = 1130 - 1260$ nm. Thus, in ML MXene one can expect the appearance of TSP peaks in the EELS and optical absorption spectra with a photon energy of less than 1.1 eV. This means that the absorption in ML $Ti_3C_2T_x$ in the photon energy region above the values corresponding to TSP can be attributed to IBT except if higher plasmonic multipoles are excited.

In [25], using scanning transmission electron microscopy (STEM) combined with ultra-high resolution EELS, the spatial and energy distribution of electron excitations in $Ti_3C_2T_x$ were obtained. It was shown that the longitudinal multipolar SP modes strongly depend on the geometry of the sample. An interesting result related to the similarity in the behavior of polarizability of monolayer (1L) and ML samples attributed to unusual weak interlayer coupling was obtained. The origin of the peak at energy of 1.7 eV in EELS spectra [25] will be discussed in the next section.

Recently, $Ti_3C_2T_x$ MXene disk arrays on glass and on Au/alumina were fabricated, and it was shown that the new metamaterial exhibits strong broadband absorption in a wide range of wavelengths, $\lambda = 500 - 1500$ nm [26]. The origin of this effect is conditioned by the appearance of gap plasmons, localized SPs and plasmon-polaritons. Full wave three-dimensional finite element method (FEM) simulations were performed using the COMSOL package, and a good agreement with the experiment was demonstrated. The broadband absorption, obviously being the result of hybridization of different plasmonic modes via interparticle interaction in the array, causes splitting of the absorption frequency. While in a system of ν noninteracting resonators the resonance frequency is ν-fold degenerated, in the case of interacting nanoparticles within an array ν spectral lines appear in the absorption band because of splitting. The absorption spectrum of a single disk for both TE and TM polarizations will be considered in the next section, and it will be demonstrated that a strong broadening takes place already in the isolated particle.

3. Calculation of Optical Spectra of Single $Ti_3C_2T_x$ Nanoparticles

In this section, we consider the excitation of the quadrupole SP mode in large enough MXene nanoparticles. Moreover, we show that hybridization of transversal modes in large coupled particles initiates Fano resonance. Theoretical study of optical properties such as wavelength dependence of dielectric function, reflection and absorption for 1Ls of pristine MXenes is performed by DFT calculations of density of states (DOS) and band structures [27]. In [28,29], analogous DFT calculations were performed and substantial differences were revealed in the absorption spectra of 1L MXenes with various terminations particularly in the lower photon energy regime. Furthermore, in case of ML MXene sheets with thicknesses ranging from 5 to 45 nm, ab initio calculations revealed LSPs [16]. It was also shown that the thinner the sheets are, the stronger are the contribution of LSP modes in the total cross section, and with the increase of the thickness the role of surface terminations becomes negligible. In [17] ab initio calculations were carried out to investigate the role of surface functionalization on dielectric properties of ML $Ti_3C_2T_x$ for the termination cases of either OH or F, and intense IBT at 4 eV and higher energies was identified. However, the lower energy region was not investigated and consequently SPs were not studied.

Below, we present our numerical simulations on optical spectra of MXene particles of sizes up to 2000 nm, exploiting analytically a priori identified ranges of resonance frequencies for various configurations and shapes (see Supplementary Information (SI)—S1 Analytical approach). For that purpose, we applied the COMSOL Multiphysics® package, modeling the shapes and sizes of sheets as close as possible to the synthesized samples in the above-mentioned experiments with a known dielectric function [13,26]. It is important that scanning electron microscope (SEM) and atomic force microscope (AFM) images show a continuous coverage of $Ti_3C_2T_x$ flakes predominantly oriented parallel to the substrate. MXenes flakes parallel to Si-SiO2 substrates, as shown in [13], take place even for thicknesses of less than 10 nm, corresponding to just a few MXene layers. A more detailed description of the modeling method is provided in Section S2 of the SI. The resulting plots are accompanied with corresponding intensity maps (spatial distribution of electric field normalized to the amplitude of the incident field). Throughout the paper the scale bars on the right side of each inset correspond to the absolute value of the ratio between the amplitude of the total electric field and the amplitude of the incident field. The direction of wave incidence coincides with the Z-axis in the figures, and the electric field oscillates along the Y-axis. In all of the calculations, MXene particles are embedded in air, i.e., the effect of the substrate that could be important in certain experiments is not considered here. We also compare our results with experiments performed with MXene films on substrates. This approach is justified by the very small difference of the reflectance between common substrates and MXene [13]. On the other hand, neglecting the role of the substrate of MXene films is also justified by the fact that our results, which are based on experimental values of the dielectric function from [26], match well with measured absorption data.

Calculation of the absorption in $Ti_3C_2T_x$ in the range of 400 to 1600 nm, when the single sheets are modeled as small oblate spheroids with a shape approximation to a disk, are presented in the Supplementary Information of [19]. As shown for small MXene particles of different thickness, in case of an electric field directed along the long axis the IBT occurs at a wavelength of about 760 nm. This value is formed under the influence of strong LSP absorption at longer wavelengths—the effect of "attraction" of peaks (see Figure S7a in [19], and S1 Analytical approach of this paper). The calculations were also performed with different aspect ratios (η) of spheroids in the case of light polarization along the short axis and an IBT at 600 nm was identified. It was shown that for small nanoparticles, the absorption of light polarized along the short axis is relatively weak and the "attraction" by the TSP peak is negligible. Thus, for the polarization considered, the redshift of IBT peaks does not appear, independently of η and in accordance with the experiment (see Figure S7b in [18]).

In [26], for the wavelength range of 600 to 2100 nm the absorption spectrum of unpatterned nanometer-thin MXene film was measured for TE polarization of incident light. On the absorption curve, two maxima decreasing with the increase of wavelength were observed at around 600 and 1000 nm (see red curve corresponding to unpatterned MXene film in Figure 4c of [26]). To assess the thickness dependence of absorption cross-section in unpatterned film, we performed COMSOL simulations for different heights of single MXene disks. The result of calculations for the absorption in a $Ti_3C_2T_x$ disk of 500 nm diameter with a thickness of 14 nm using $\epsilon(\omega)$ from [13] is presented in Figure 1. The maxima in the short wavelength range of 640 nm, also observed in [26], corresponds to IBT. Indeed, our calculations give the same value of 640 nm for the left-side maxima regardless of nanoparticle shapes (see Figure 2, Figure S3), if the electric field polarization allows excitation of only TSP. This behavior is expected since the absorption peak corresponding to IBT always essentially dominates that of TSPs located nearby and therefore the IBT resonance wavelength is not shifted. The second peak for the chosen polarization located at 1200 nm obviously corresponds to TSP resonance for a disk of a large aspect ratio [30]. Note that this wavelength is very close to $\lambda = 1260$ nm, satisfying the condition $\Re[\epsilon(\lambda)] = 0$ for the $\epsilon(\lambda)$ measured in [13]. The minor shift here is conditioned by the contribution of the imaginary part of $\epsilon(\omega)$ into the polarizability. We also calculated the absorption cross-section for the disk with a diameter of 1000 nm (see Figure S3) and found that the resonances are located at the same wavelengths as for that with a diameter of 500 nm. This is not surprising since also our analytic approach shows (see SI) that the peak position of TSP resonance in a small particle does not depend on the diameter at large aspect ratios. On the other hand, the peak values of absorption for a larger disk are four times higher, which is a consequence of the four-fold increase of volume of the particle.

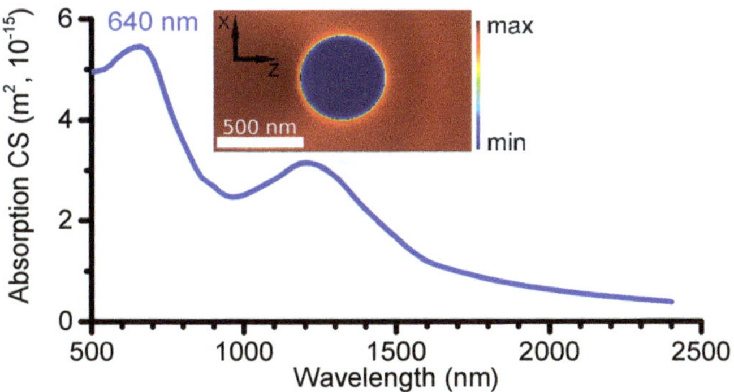

Figure 1. Simulated absorption cross-section (CS) of $Ti_3C_2T_x$ disk of diameter 500 nm with 14 nm height. Inset: electric field intensity map at IBT resonance—640 nm.

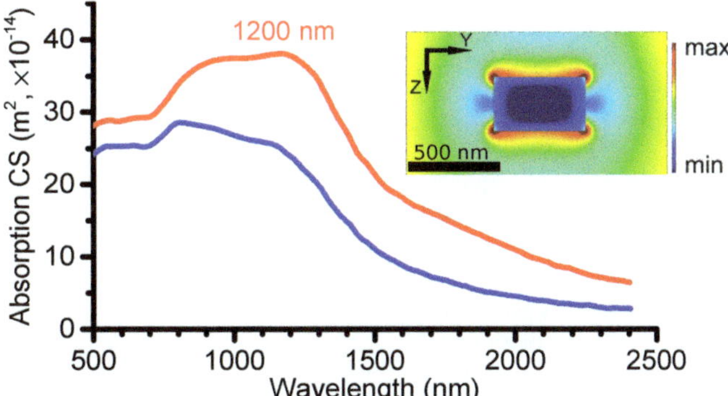

Figure 2. Simulated absorption CS of $Ti_3C_2T_x$ cylinder of diameter 500 nm with heights 200 nm (blue) and 300 nm (red). Inset: electric field intensity map of cylinder of diameter 500 nm with height 300 nm at the wavelength of TSP resonance—1200 nm.

To counter for the effect of particle thickness, we calculated the absorption cross-section of a cylinder of 1000 nm diameter and 300 nm height using the $\epsilon(\omega)$ measured in [26], where the nanoparticles of these sizes were experimentally studied. Herein, two directions of electric field are applied—parallel and perpendicular to the base of the disk. Interestingly, for parallel polarization, a new peak appeared next to IBT one (600 nm) at around 1000 nm, which we attribute to quadrupole SP (QSP) absorption (see Figure 3). Indeed, the penetration depth d of an incident wave defined by $d^{-1} = (2\pi/\lambda)\,\Im[\tilde{n}(\lambda)]$ (where λ is the vacuum wavelength and $\tilde{n}$ is the complex index of refraction) at $\lambda = 1000$ nm takes a value of 100 nm. Obviously, the decrease and vanishing of the electric field amplitude over the penetration depth and consequent fading of the dipolar polarization of MXene along with propagation of light leads to an increase in the symmetry of charge distribution. Indeed, the loss of axial symmetry leads to full isotropy of charge distribution allowing the excitation of higher multipoles. This creates favorable conditions (full symmetry and appropriate value of QSP wavelength in the material) for inhomogeneity of charge distribution. This causes the appearance of a quadrupole mode due to fluctuations of charge distribution.

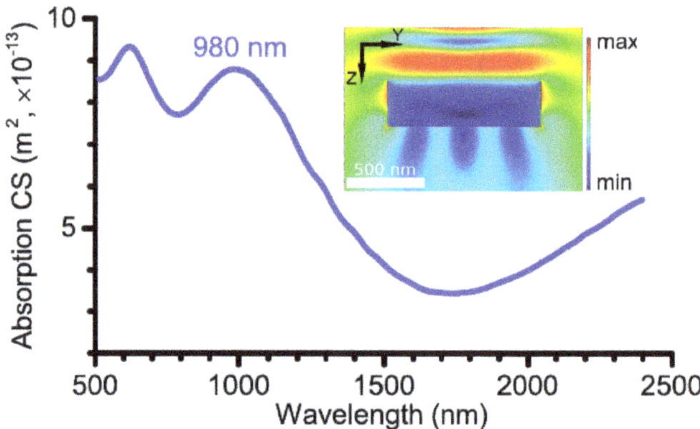

Figure 3. Simulated absorption CS of $Ti_3C_2T_x$ cylinder of diameter 1000 nm with height 300 nm. Inset: electric field intensity map at the wavelength of QSP resonance—980 nm.

The wavelength λ_m denoting the longitudinal oscillation induced by the pump wave at 980 nm in the medium is determined by the formula $\lambda_m = \lambda / \sqrt{\Re[\epsilon(\lambda)]}$ and gives for $\lambda_m = 363$ nm. Such a value of λ_m obviously allows the excitation of longitudinal quadrupole mode in a cylinder with a diameter of 1000 nm. The excitation of the QSP mode at a shorter wavelength than the dipole SP mode is well-known in the quasistatic approximation [30]. The absorption cross-section of TE polarized light in unpatterned MXene film was measured and two absorption peaks at around 600 nm and 1000 nm were revealed (see Figure 4c of [26]). Our simulation for the same polarization of incident light for the cylinder, as was mentioned above (see Figure 3), gives exactly the same maxima, which we identify as IBT and QSP resonances, respectively. The close values of resonance wavelengths of QSP for the unpatterned film and cylinder are conditioned by the fact that the electric field strength of quadrupole charge distribution decreases very fast, in the order of R^{-4}, where R is the distance from the center of charge distribution to the point of observation. This behavior of the field is associated with a weak dependence of the sensitivity of QSP resonance on the shape of the nanoparticle. As can be seen from the inset of Figure 3, for the considered polarization there should be a peak corresponding to LSP resonance at much longer wavelengths. However, to demonstrate it, we need the experimental dielectric function of MXene in that region, which is not yet available.

The increase in the absorption cross-section at longer wavelengths ($\lambda > 2400$ nm) is conditioned by the broadened weak LSP resonance. We interpret this behavior as a consequence of the fact that the diameter of the cylinder is much larger than the mean free path of charge carriers. In the inset of Figure 3 the presented electric field intensity map corresponding to the peak wavelength of QSP absorption at 980 nm for the cylinder of 300 nm height shows inhomogeneous field distribution around the particle. This asymmetry is caused by the extinction of the electric field along the propagation direction. The occurrence of four maxima beneath the cylinder shown in the inset of Figure 3 (see also Figure S7) indicates the excitation of a quadrupole plasmonic mode. We recall that the QSP absorption is proportional to the fourth power of the resonator size, whereas the dipole absorption of LSP is proportional to the square of resonator size. Our calculation shows that due to that reason the QSP resonance is not excited in smaller cylinders. Particularly, the simulation of the absorption cross-section for a cylinder with a diameter of 500 nm and height of 200 nm does not indicate a QSP resonance (see Figure S4). However, for a larger height of 300 nm, there appears a broad peak caused by a joint contribution of IBT resonance and QSP oscillations.

For the perpendicular polarization (electric field is directed along the cylinder axis) the change of cylinder diameters from 500 to 1000 nm and heights from 200 to 400 nm does not cause qualitative differences in absorption cross-sections. In Figure 2 we present the results of calculations of absorption cross-section spectra of a $Ti_3C_2T_x$ cylinder of diameter 500 nm with heights 200 nm (blue curve) and 300 nm (red curve) with an electric field that is perpendicular to the disk base. Particularly, one can see a broadened absorption band due to the joint contribution of three processes—the IBT, QSP and TSP modes. In addition, the described behavior of the cross-section takes place for a larger particle (cylinder diameter 1000 nm) as well (see Figure S5).

To verify the possibility of the appearance of a QSP resonance in MXene, we also calculated the absorption cross-section of a large spheroidal particle (with semiaxes $c = 500$ nm, $a = b = 200$ nm). As expected, a peak at 980 nm was revealed, indicating the appearance of a QSP resonance (see Figure 4) in the spheroidal nanoparticle as well. The symmetry braking of the field distribution with respect to the XY-plane typical for the QSP mode can be observed in the inset of Figure 4. Note that the exhibited sharp rise of the cross-section with wavelengths is due to LSP resonance for $\lambda > 2400$ nm, which is typical for sub-micrometer spheroidal nanoparticles.

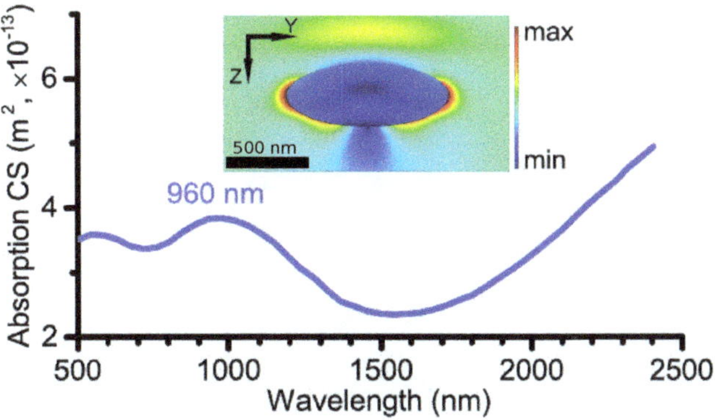

Figure 4. Simulated absorption CS of large spheroidal particle with semiaxes of 500, 200 and 200 nm. Inset: electric field intensity map at quadrupole resonance wavelength, 960 nm.

Thus, the resonance wavelengths corresponding to QSP modes for a single cylinder (Figure 3) and a spheroid (Figure 4) are very close to each other, which is a consequence of insensitivity of QSP resonance on the shape of such a large nanoparticle.

Next, we apply the measured dielectric function of [13] to analyze the experimental data of [25], where in the vicinity of 1.7 eV (730 nm) in the EELS spectrum of a triangular $Ti_3C_2T_x$ film a peak was ascribed to TSPs. To analyze this experiment, we look for the positions of SP and IBT modes of a triangular 7.5 nm-thick MXene (using values of $\epsilon(\omega)$ from [13] for the thinnest available sample, i.e., 14 nm-thick film) by carrying out COMSOL calculations. We found the absorption spectra for two geometries, corresponding to the excitation of LSP and TSP resonances for the triangular sheet of sides of 300, 400 and 500 nm. In Figure 5 we present the spectra for the cases when the incident electric field is directed perpendicular to the plane of the film to excite TSP (red curve), and when the electric field is directed parallel to the plane of film (perpendicular to the longest side of the triangle) to excite LSP (blue curve). We see that for the geometry close to the experimentally studied one (by EELS), there are two peaks in each spectrum.

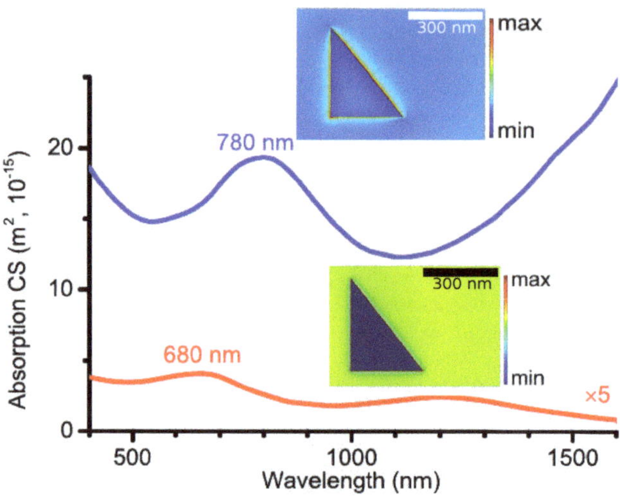

Figure 5. Simulated absorption CS of 7.5 nm thick triangular $Ti_3C_2T_x$ sheet with side lengths 300, 400 and 500 nm. For the ease of modeling, the triangles are capped at corners by arcs of 2 nm curvature radius. Incident electric field is: perpendicular (TM polarization, red curve) and parallel (TE polarization, blue curve) to the plane of sheet being perpendicular to the hypotenuse. Insets: electric field intensity maps placed above each curve, at the excitation wavelengths of 680 nm for the red one, and 780 nm for the blue.

The maxima at longer wavelengths correspond to TSP, at 1215 nm (red curve) and LSP, exceeding 2300 nm (blue curve). As expected, the LSP peak is much stronger than the TSP. We interpret the peaks at shorter wavelengths on both curves in the range of 650 to 800 nm in Figure 5 as IBT, based on the following reasoning. First, when the electric field is directed perpendicularly to the film plane our calculations for the triangular sheets of differing thickness from 7.5 to 30 nm give the same peak at the wavelength of 600 nm. Obviously, this is because the IBT peak values, which depend solely on material properties, are stronger and cannot be shifted by the weaker TSP resonance peaks located at higher wavelengths. Note that the maxima of IBT peaks for two polarizations in Figure 5 do not match since the higher peak of LSP, unlike TSP, causes a redshift of IBT maxima by 140 nm (effect of "attraction" of peaks, see SI—Analytical approach). Thus, we can state that the detected peak at 1.7 eV in EELS spectra of $Ti_3C_2T_x$ [25] is due to IBTs. It can also be seen from Figure 5 (blue curve) that the values of the LSP resonance absorption cross-section are much higher (25 times) than the TSP peak around 1200 nm. This could be the reason why the TSP resonance was suppressed in the EELS spectrum of [25]. Note that in case of excitation of SP, higher multipoles cannot be optically excited because of the shortness of the propagation length (7.5 nm) compared to penetration depth. The peak at 0.66 eV in the EELS spectrum prescribed to the QSP mode (see Figure 1 in [25]) arises due to the strong inhomogeneity of the field created by the electron beam. Dissimilar to that, the QSP mode of the cylinder revealed in this study (Figure 3) appears because of light intensity decay during prorogation in the nanoparticle. Interestingly, these two physical mechanisms of inhomogeneity excite QSPs at very different wavelengths. The direction of light propagation for the case of TM polarization is parallel to the plane of the triangle and, consequently, the absorbing area is much smaller than for the TE polarization where the light incidence is normal to the triangle. Thus, the electric field intensity in the vicinity of the triangle in TE case is much stronger than in the TM case, as demonstrated in the insets of Figure 5.

Thus, the interpretation of the peculiarities of spectra of ML MXene sheets exploiting the measured dielectric function provides good agreement with a wide range of wavelengths, namely 500 to 2500 nm, and predicts the appearance of QSP in large nanoparticles.

4. Calculation of Optical Spectra of Coupled $Ti_3C_2T_x$ Nanoparticles

It is known that large and spatially confined electromagnetic enhancement effects can be reached only for the case of strongly coupled nanoparticles such as dimer configurations [31]. Using experimental data for the dielectric function of $Ti_3C_2T_x$ measured in [26] we also calculate the optical absorption in coupled particles that can present interest for reaching large values of enhancement factors of SERS. This type of physical enhancement mechanism commonly termed as hot-spots is important to consider for material design in practical applications.

To reveal the peculiarities of LSP resonances in coupled MXene nanoparticles, we calculated the absorption spectra of two identical interacting spheroidal $Ti_3C_2T_x$ particles in an end-to-end configuration at separations of $D = 0.8$ nm, $D = 1.0$ nm and $D = 1.5$ nm. The corresponding semiaxes of the particles are: (a) (20 nm, 5 nm, 5 nm), (b) (25 nm, 5 nm, 5 nm) and (c) (30 nm, 5 nm, 5 nm). The electric field of the light is again directed along the symmetry axis of the system. The results of the COMSOL simulation are presented in Figure 6. The positions of the IBT peaks around 760–780 nm did not change significantly when varying the aspect ratios of spheroids and the interparticle separations, since the IBT and LSP peaks' separation is large.

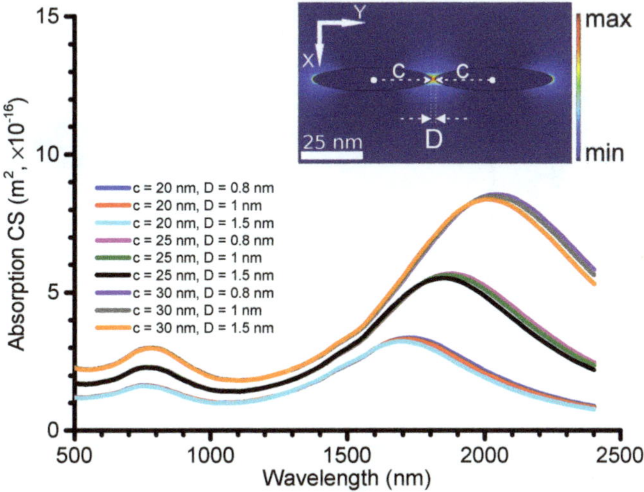

Figure 6. Simulated absorption CS of coupled MXene nanospheroids in end-to-end configuration with equal short semiaxes ($a = b = 5$ nm) and varying long semiaxis ($c = 20$, 25 or 30 nm). Inset: electric field intensity map in the case of coupled spheroids with semiaxis $c = 25$ nm and $a = b = 5$ nm, irradiated by parallel electric field at 760 nm.

However, contrary to IBT, the LSP peaks were drastically redshifted. It is interesting that the proposed simple model consisting of coupled $Ti_3C_2T_x$ ellipsoids for the SP resonance energy of 0.59–0.73 eV is in good agreement with the results observed in EELS [16] and optical absorption experiments [19]. The corresponding electric field intensity map at 760 nm, presented in the inset of Figure 6, clearly demonstrates the hotspot induced by IBT polarization.

Note that when aspect ratios of spheroids increase from $\eta = 4$ to $\eta = 6$ (the semiaxis c varies from 20 to 30 nm), peak positions of LSP resonances shift by 100 nm for all interparticle separations. Furthermore, the peak positions of the absorption in the range of

LSP resonances depend weakly on the interparticle separation. Thus, the corresponding shift for different separations makes no more than 50 nm for all three values of $\eta = 4, 5, 6$. Consequently, the main peculiarities of the SP absorption spectrum can be prescribed to the geometry of a single nanoparticle of MXene rather than the random spatial deposition of their plurality on the substrate. Hence, for small nanoparticles wherein the spatial dependence of the EM field in the ignored, the simple model allows to describe the optical properties of individual MXene sheets. Note that the spectral shift of coupled SPs in the MXene dimer (50 nm in considered case) is smaller than that of the coupled noble metallic nanoparticles [32]. This difference is due to a higher density of free charge carriers in noble metals as compared to MXene.

To see how the peculiarities revealed above can change in the case of relatively large particles, we further carried out simulations of the absorption spectra for the identical coupled MXene ellipsoids of semiaxes $c = 100$ nm, $a = 40$ nm and $b = 10$ nm. The calculation results presented in Figure 7 show that the absorption spectrum of medium-sized coupled ellipsoidal particles still does not contain QSP resonance.

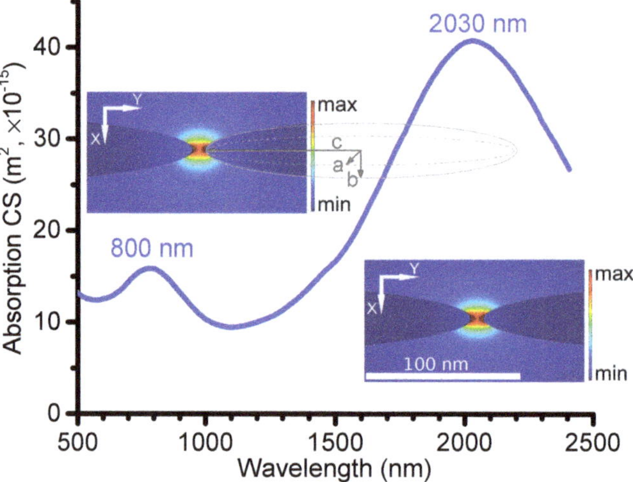

Figure 7. Simulated absorption CS of coupled identical MXene ellipsoids with semiaxes $c = 100$ nm, $a = 40$ nm and $b = 10$ nm. Insets: electric field intensity maps at IBT maxima—800 nm and LSP resonance at 2030 nm.

The optical spectrum of a single ellipsoid of the same size, presented in Figure S6 of the SI, confirms that the IBT resonance wavelength is the same as that in Figure 7 while the LSP resonance appears redshifted by 200 nm. This consistent IBT peak in the range of 700–800 nm (irrespective of particle size or colloidal solution concentration) is in accord with the UV-vis absorption results of $Ti_3C_2T_x$ often reported in the literature.

In order to assess the possibility of the appearance of QSP resonance in interacting MXene nanoparticles, we also simulated the absorption in notably larger coupled spheroids with semiaxes: 500, 200 and 200 nm. The results for the end-to-end configuration of spheroids with an inter-particle gap of 1.5 nm, when the incident electric field is directed along the long axis, are presented in Figure 8. They clearly show the existence of a QSP mode at 980 nm, analogous to those revealed for the single cylinder and spheroid (see Figures 3 and 4).

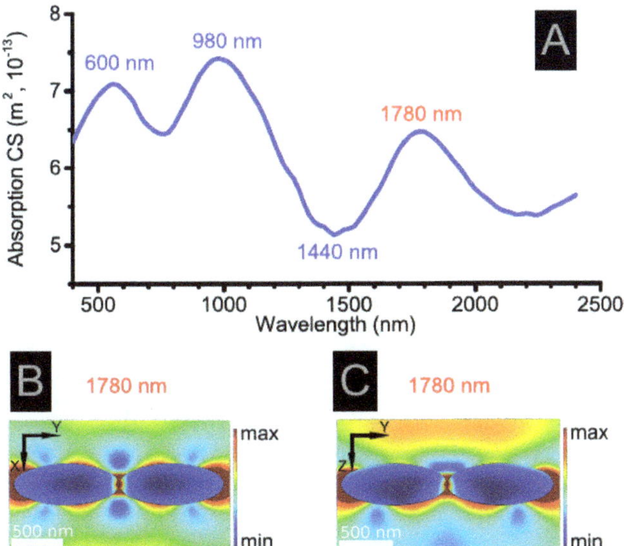

Figure 8. (**A**) Coupled spheroids with semiaxes of 500, 200 and 200 nm in end-to-end configuration. Electric field intensity maps at TSP maximum—1780 nm in the XY-plane (**B**) and YZ-plane (**C**) are presented.

Interestingly, as can be seen from Figure 8A, in addition to LSP at resonance wavelength $\lambda > 2300$ nm (the highest field strength at the far apexes of spheroids) and the quadrupole resonance detected at 980 nm, there is also a new resonance at 1780 nm. We attribute this resonance to longwave transversal plasmon oscillations, the mechanism of occurrence of which differs significantly from the classical one. The point is that typically the directions of the external field and plasmon oscillations coincide and correspondingly the transversal plasmon oscillations should not occur when the electric field of light oscillates along the long axis of coupled spheroids. The physical mechanism of excitation of TSP when the directions of incident electric field and SP oscillations are orthogonal is the following. In the center of the symmetry of the quadrupole charge distribution a repulsive alternating force appears orthogonal to the incident electric field (see schematics in SI 2). It is this force that causes dipole transversal plasmon oscillations in both spheroids. It is easy to see from the electric field intensity maps of Figure 8B,C that there is an attraction between the transversal plasmons, which is also demonstrated in SI 2. Thus, the QSP–TSP coupling in each spheroid and TSP–TSP hybridization in interacting spheroids lead to a strong redshift of the new TSP resonance wavelength.

Figures 8B,C apparently demonstrate the occurrence of the TSP mode in the high field intensity regions close to middle parts of spheroids. A small shift of bright regions from the centers of spheroids is a consequence of attraction between counter-phase oscillating transversal dipoles. Moreover, Figure 8C clearly shows the shift of all induced dipoles in the incident wave propagation direction.

Furthermore, in Figure 9 the electric field intensity maps at QSP maximum (980 nm) in the XY-plane (A) and YZ-plane (B) are presented. Note that the resonance wavelength of QSP remains the same for single and coupled nanoparticles.

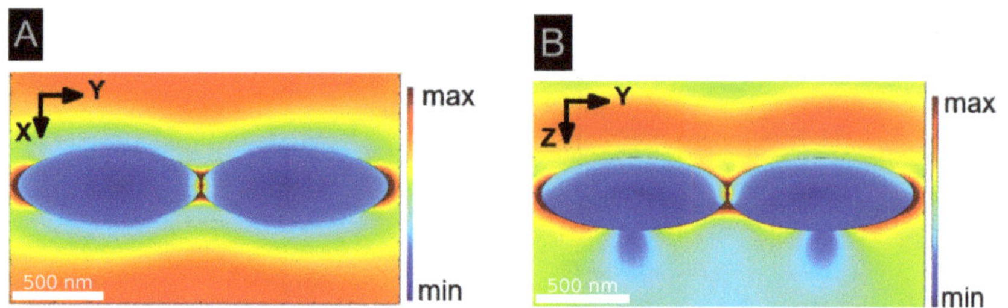

Figure 9. Electric field intensity maps at QSP maximum—980 nm in the XY-plane (**A**) and YZ-plane (**B**).

It is important to mention that as a result of strong coupling of QSP and induced TSP modes, as follows from Figure 8A, there appears a Fano dip at 1440 nm. The lowest minimum at 1440 nm in strongly interacting nanoparticles (Figure 8) is a result of destructive interference between QSP and TSP modes, which is a manifestation of Fano resonance [33]. Moreover, the attraction between dipoles induced by TSP oscillations in two spheroids is clearly demonstrated. In Figure 8C the symmetry of intensity distribution with respect to incident electric field is violated, obviously demonstrating the appearance of a QSP resonance. Furthermore, in Figure 9 the electric field intensity maps at QSP maximum (980 nm) in the XY-plane (A) and YZ-plane (B) are presented. Note that the resonance wavelength of QSP remains the same for single and coupled nanoparticles.

It can be seen from Figure 10A that the dipole and QSP modes are suppressed ("dark" mode in Fano resonance), whereas Figure 10B shows that the energy is cumulated in the TSP mode ("bright" mode). Unlike the 1780 nm wavelength, the field distribution around the hotspot at the Fano resonance (1440 nm) is negligibly small, since the transversal dipoles oscillate counter-phase and their cumulative superimposed field vanishes. Our result is different from conventional Au nanoparticle systems where the Fano resonance appears only in the case of non-identical small nanoparticles (with sizes less than the wavelength of the incident wave) [34]. It is clear, however, that in a realistic case of non-identical large interacting MXene flakes a Fano resonance can take place as well. The small asymmetry in the Fano deep, is a consequence of the moderate difference between the damping constants of the two modes considered.

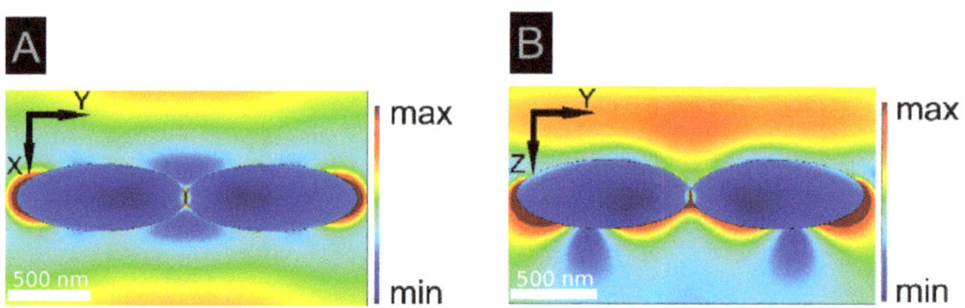

Figure 10. Electric field intensity maps at Fano resonance—1440 nm in the XY-plane (**A**) and YZ-plane (**B**).

Summarizing, all optical phenomena taking place in the visible and infrared ranges for arbitrary geometries of ML MXene single and coupled particles originate via strong polarization induced by IBT, QSP and TSP resonances. We stress that the QSP resonance in the absorption spectrum yields stronger peaks than the TSP. Moreover, the plasmonic behavior of this MXene starts not at the TSP resonance but at a shorter wavelength (980 nm).

5. Discussion

Experimental MXene 2D flakes present a large shape variety, depending on the fabrication conditions. It is therefore not easy to define a general shape for samples appropriate for all experiments. Hence we tried to model several flake shapes to see what kind of shape-dependent peculiar plasmonic effects are also evidenced in experiments. We studied the absorption spectra of MXene nanoparticles with shapes and sizes matching those studied experimentally. Thus, we considered both thin and thick samples.

The substrate was not taken into account in our simulations. Although we have used a homogeneous medium with a refractive index of $n = 1$ (air), this situation is similar to MXene flakes suspended in aqueous solution (the refractive index of water in the green region of the visible spectrum is $n \sim 1.33$). On the other hand, the MXene films in [13,26] were deposited on quartz, glass, and SiO_2/Si. As was shown experimentally, the difference in the reflectance between $Ti_3C_2T_x$ and quartz at $\lambda = 550$ nm is only 4.7% (see Figure S5 in the SI of [13]). Not taking into account the role of substrate of MXene film in the interpretation of absorption spectra can be largely justified by the fact that our results, which are based on experimental values of the dielectric function from [26], match well with measured absorption data.

In the analysis of EELS spectra of MXenes it is important to stress that the energy loss experiments allow revealing all resonances of the sample. However, in some parts of experimental EELS spectra of ML MXenes, it is difficult to distinguish closely located maxima, such as resonances corresponding to IBT and TSP. On the other hand, optical absorption spectra expose the resonances selectively, depending on the polarization of incident light. Thus, for exciting TSP in optical experiments the electric field of the light wave should have a component perpendicular to the surface of the flake. Accordingly, changing the polarization of light will allow the identification of the TST mode revealed in EELS experiments. As for the peak frequency of IBT, it is independent on the geometry of the sample and is determined only by the dielectric properties of the material. Consequently, this resonance appears in both experiments.

To study the interaction of MXene sheets closely placed on the substrate, we considered coupled large nanospheroids with semiaxes of $c = 500$ nm, $b = 200$ nm and $a = 200$ nm in the end-to-end configuration under plane wave irradiation, as depicted in Figure 7. The polarization is along the Y-axis and the propagation is along the Z-axis. Since the particle lengths are large enough compared to the penetration depth, higher multipole plasmon oscillations can be excited together with dipole LSP. In the case considered, quadrupole oscillations occur, since the QSP wavelength $\lambda_m = \lambda / \Re[n(\lambda)] = 366$ nm for $\lambda = 980$ nm is less than half the long axis of the spheroid. Here n is the refractive index of the MXene. The configuration of dipole and quadrupole LSP in each spheroid is presented in the Figure 8A,B. Furthermore, the TSP mode in the system is excited due to repulsion forces acting on the free carriers in the center of symmetry of the quadrupole charge distribution. Thus, a new mechanism of excitation of the longwave TSP at $\lambda = 1780$ nm by an external electric field perpendicular to the direction of charge oscillations is presented. We note that this mechanism strongly differs from that inducing classical transversal oscillations.

In order to observe the typical asymmetric Fano dip characteristic of atomic ionization resonances, as well as in very small interacting nanoparticles, one needs narrow resonance and a continuum, which leads to the strong asymmetry. Obviously, in identical and very small interacting nanoparticles, when the damping of both rates is the same, a Fano resonance does not arise. However, for identical and large enough nanoparticles a Fano dip with a small asymmetry appears, when plasmonic modes (in our case QPS and induced TSP) are excited. The small asymmetry in our results is due to the small difference between the damping constants of two modes. Contrarily to atomic systems, in experiments and for some configurations of interacting metallic nanoparticles it is possible to detect the Fano dip with a small asymmetry, which was demonstrated theoretically is our study. In sensing applications of MXenes it is interesting to analyze the quality factor of resonances. It can be calculated using the method presented in [35].

6. Conclusions

Incorporating the experimentally measured dielectric function of $Ti_3C_2T_x$ MXene, we performed simulations of optical absorption spectra of 2D nanosheets and showed that the main peculiarities of recently reported electro-optical experiments in the vis-infrared range can be reproduced. Moreover, some details of electron energy loss spectra in a the range of low eV energy, particularly concerning the inter-band transition wavelength range, were clarified. The resonance wavelengths of dipole longitudinal and transversal, as well as quadrupole surface plasmon oscillations in MXene nanosheets of different geometries, were identified. It was shown that the resonance wavelength of the dipole SP mode for large nanoparticles can be tuned up to the 1700 nm range, and that of a rather strong quadrupole mode can be as short as 980 nm. We demonstrated that the plasmonic phenomena of the $Ti_3C_2T_x$ MXene occurs at the quadrupole SP resonance wavelength, that is, at a shorter wavelength than that of the transversal one. A new mechanism of excitation of TSP oscillations by an orthogonal external electric field was presented. Owing to a strong quadrupole-transversal SP interaction, the possibility of realization of Fano resonance in coupled large spheroidal nanoparticles was demonstrated for the first time, indicating a new direction for exploring additional optical resonance effects in MXenes.

Supplementary Materials: The following are available online at www.mdpi.com/xxx/s1: Supplementary information including figures is in a separate file.

Author Contributions: Conceptualization, M.G., A.M., and H.M.; methodology, A.M., H.M., T.S., P.P.; software, M.G., P.P., T.S.; writing—original draft preparation, A.M., H.M.; writing—review and editing, M.G., A.M., H.M. and T.M.; visualization, P.P., T.S.; All authors have read and agreed to the published version of the manuscript.

Funding: A.M., H.M. and T.S. acknowledge the Science Committee of the Ministry of Education, Science, Culture and Sport of the Republic of Armenia (Grant 18T-1C222) for financial support provided over the last two years (http://www.scs.am). This work did not receive any funding from the Murata Manufacturing Co., Ltd.

Data Availability Statement: The data that support the findings of this study are available from the corresponding author upon reasonable request.

Acknowledgments: The authors acknowledge N. Minasyan for technical assistance.

Conflicts of Interest: The authors declare no conflict of interest.

References

1. Naguib, M.; Kurtoglu, M.; Presser, V.; Lu, J.; Niu, J.; Heon, M.; Hultman, L.; Gogotsi, Y.; Barsoum, M.W. Two-Dimensional Nanocrystals Produced by Exfoliation of Ti_3AlC_2. *Adv. Mater.* **2011**, *23*, 4248–4253. [CrossRef] [PubMed]
2. Xie, Y.; Kent, P.R.C. Hybrid density functional study of structural and electronic properties of functionalized $Ti_{n+1}X_n$ ($X = C, N$) monolayers. *Phys. Rev. B* **2013**, *87*, 235441. [CrossRef]
3. Hantanasirisakul, K.; Gogotsi, Y. Electronic and Optical Properties of 2D Transition Metal Carbides and Nitrides (MXenes). *Adv. Mater.* **2018**, *30*, 1804779. [CrossRef] [PubMed]
4. Dong, Y.; Chertopalov, S.; Maleski, K.; Anasori, B.; Hu, L.; Bhattacharya, S.; Rao, A.M.; Gogotsi, Y.; Mochalin, V.N.; Podila, R. Saturable Absorption in 2D Ti_3C_2 MXene Thin Films for Passive Photonic Diodes. *Adv. Mater.* **2018**, *30*, 1705714. [CrossRef]
5. Shahzad, F.; Alhabeb, M.; Hatter, C.B.; Anasori, B.; Hong, S.M.; Koo, C.M.; Gogotsi, Y. Electromagnetic interference shielding with 2D transition metal carbides (MXenes). *Science* **2016**, *353*, 1137–1140. [CrossRef] [PubMed]
6. Zhu, Z.; Zou, Y.; Hu, W.; Li, Y.; Gu, Y.; Cao, B.; Guo, N.; Wang, L.; Song, J.; Zhang, S.; Gu, H.; Zeng, H. Near-Infrared Plasmonic 2D Semimetals for Applications in Communication and Biology. *Adv. Funct. Mater.* **2016**, *26*, 1793–1802. [CrossRef]
7. Montazeri, K.; Currie, M.; Verger, L.; Dianat, P.; Barsoum, M.W.; Nabet, B. Beyond Gold: Spin-Coated Ti_3C_2-Based MXene Photodetectors. *Adv. Mater.* **2019**, *31*, 1903271. [CrossRef]
8. Hantanasirisakul, K.; Zhao, M.Q.; Urbankowski, P.; Halim, J.; Anasori, B.; Kota, S.; Ren, C.E.; Barsoum, M.W.; Gogotsi, Y. Fabrication of $Ti_3C_2T_x$ MXene Transparent Thin Films with Tunable Optoelectronic Properties. *Adv. Electron. Mater.* **2016**, *2*, 1600050. [CrossRef]
9. Naguib, M.; Mochalin, V.N.; Barsoum, M.W.; Gogotsi, Y. 25th Anniversary Article: MXenes: A New Family of Two-Dimensional Materials. *Adv. Mater.* **2013**, *26*, 992–1005. [CrossRef]
10. Jiang, X.; Kuklin, A.V.; Baev, A.; Ge, Y.; Ågren, H.; Zhang, H.; Prasad, P.N. Two-dimensional MXenes: From morphological to optical, electric, and magnetic properties and applications. *Phys. Rep.* **2020**, *848*, 1–58. [CrossRef]

11. Ghidiu, M.; Lukatskaya, M.R.; Zhao, M.Q.; Gogotsi, Y.; Barsoum, M.W. Conductive two-dimensional titanium carbide 'clay' with high volumetric capacitance. *Nature* **2014**, *516*, 78–81. [CrossRef] [PubMed]

12. Jhon, Y.I.; Koo, J.; Anasori, B.; Seo, M.; Lee, J.H.; Gogotsi, Y.; Jhon, Y.M. Metallic MXene Saturable Absorber for Femtosecond Mode-Locked Lasers. *Adv. Mater.* **2017**, *29*, 1702496. [CrossRef] [PubMed]

13. Dillon, A.D.; Ghidiu, M.J.; Krick, A.L.; Griggs, J.; May, S.J.; Gogotsi, Y.; Barsoum, M.W.; Fafarman, A.T. Highly Conductive Optical Quality Solution-Processed Films of 2D Titanium Carbide. *Adv. Funct. Mater.* **2016**, *26*, 4162–4168. [CrossRef]

14. Naguib, M.; Mashtalir, O.; Carle, J.; Presser, V.; Lu, J.; Hultman, L.; Gogotsi, Y.; Barsoum, M.W. Two-Dimensional Transition Metal Carbides. *ACS Nano* **2012**, *6*, 1322–1331. [CrossRef] [PubMed]

15. Hantanasirisakul, K.; Alhabeb, M.; Lipatov, A.; Maleski, K.; Anasori, B.; Salles, P.; Ieosakulrat, C.; Pakawatpanurut, P.; Sinitskii, A.; May, S.J.; et al. Effects of Synthesis and Processing on Optoelectronic Properties of Titanium Carbonitride MXene. *Chem. Mater.* **2019**, *31*, 2941–2951. [CrossRef]

16. Mauchamp, V.; Bugnet, M.; Bellido, E.P.; Botton, G.A.; Moreau, P.; Magne, D.; Naguib, M.; Cabioc'h, T.; Barsoum, M.W. Enhanced and tunable surface plasmons in two-dimensional Ti_3C_2 stacks: Electronic structure versus boundary effects. *Phys. Rev. B* **2014**, *89*, 235428. [CrossRef]

17. Magne, D.; Mauchamp, V.; Célérier, S.; Chartier, P.; Cabioc'h, T. Spectroscopic evidence in the visible-ultraviolet energy range of surface functionalization sites in the multilayer Ti_3C_2 MXene. *Phys. Rev. B* **2015**, *91*, 201409. [CrossRef]

18. Satheeshkumar, E.; Makaryan, T.; Melikyan, A.; Minassian, H.; Gogotsi, Y.; Yoshimura, M. One-step Solution Processing of Ag, Au and Pd@MXene Hybrids for SERS. *Sci. Rep.* **2016**, *6*, 32049. [CrossRef]

19. Sarycheva, A.; Makaryan, T.; Maleski, K.; Satheeshkumar, E.; Melikyan, A.; Minassian, H.; Yoshimura, M.; Gogotsi, Y. Two-Dimensional Titanium Carbide (MXene) as Surface-Enhanced Raman Scattering Substrate. *J. Phys. Chem. C* **2017**, *121*, 19983–19988. [CrossRef]

20. Spector, M.; Ang, A.S.; Minin, O.V.; Minin, I.V.; Karabchevsky, A. Photonic hook formation in near-infrared with MXene Ti3C2 nanoparticles. *Nanoscale Adv.* **2020**, *2*, 5312–5318. [CrossRef]

21. Song, Y.; Chen, Y.; Jiang, X.; Ge, Y.; Wang, Y.; You, K.; Wang, K.; Zheng, J.; Ji, J.; Zhang, Y.; et al. Nonlinear Few-Layer MXene-Assisted All-Optical Wavelength Conversion at Telecommunication Band. *Adv. Opt. Mater.* **2019**, *7*, 1801777. [CrossRef]

22. Soundiraraju, B.; George, B.K. Two-Dimensional Titanium Nitride (Ti_2N) MXene: Synthesis, Characterization, and Potential Application as Surface-Enhanced Raman Scattering Substrate. *ACS Nano* **2017**, *11*, 8892–8900. [CrossRef] [PubMed]

23. Limbu, T.B.; Chitara, B.; Garcia Cervantes, M.Y.; Zhou, Y.; Huang, S.; Tang, Y.; Yan, F. Unravelling the Thickness Dependence and Mechanism of Surface-Enhanced Raman Scattering on $Ti_3C_2T_x$ MXene Nanosheets. *J. Phys. Chem. C* **2020**, *124*, 17772–17782. [CrossRef]

24. Liu, R.; Jiang, L.; Lu, C.; Yu, Z.; Li, F.; Jing, X.; Xu, R.; Zhou, W.; Jin, S. Large-scale two-dimensional titanium carbide MXene as SERS-active substrate for reliable and sensitive detection of organic pollutants. *Spectrochim. Acta Part A Mol. Biomol. Spectrosc.* **2020**, *236*, 118336. [CrossRef]

25. El-Demellawi, J.K.; Lopatin, S.; Yin, J.; Mohammed, O.F.; Alshareef, H.N. Tunable Multipolar Surface Plasmons in 2D $Ti_3C_2T_x$ MXene Flakes. *ACS Nano* **2018**, *12*, 8485–8493. [CrossRef]

26. Chaudhuri, K.; Alhabeb, M.; Wang, Z.; Shalaev, V.M.; Gogotsi, Y.; Boltasseva, A. Highly Broadband Absorber Using Plasmonic Titanium Carbide (MXene). *ACS Photonics* **2018**, *5*, 1115–1122. [CrossRef]

27. Lashgari, H.; Abolhassani, M.R.; Boochani, A.; Elahi, S.M.; Khodadadi, J. Electronic and optical properties of 2D graphene-like compounds titanium carbides and nitrides: DFT calculations. *Solid State Commun.* **2014**, *195*, 61–69. [CrossRef]

28. Berdiyorov, G.R. Optical properties of functionalized $Ti_3C_2Ti_2$ (T = F, O, OH) MXene: First-principles calculations. *AIP Adv.* **2016**, *6*, 055105. [CrossRef]

29. Bai, Y.; Zhou, K.; Srikanth, N.; Pang, J.H.L.; He, X.; Wang, R. Dependence of elastic and optical properties on surface terminated groups in two-dimensional MXene monolayers: a first-principles study. *RSC Adv.* **2016**, *6*, 35731–35739. [CrossRef]

30. Bohren, C.F.; Huffman, D.R. *Absorption and Scattering of Light by Small Particles*; Wiley: Weinheim, Germany, 1998. [CrossRef]

31. Xu, H.; Aizpurua, J.; Käll, M.; Apell, P. Electromagnetic contributions to single-molecule sensitivity in surface-enhanced Raman scattering. *Phys. Rev. E* **2000**, *62*, 4318–4324. [CrossRef]

32. Jain, P.K.; El-Sayed, M.A. Surface Plasmon Coupling and Its Universal Size Scaling in Metal Nanostructures of Complex Geometry: Elongated Particle Pairs and Nanosphere Trimers. *J. Phys. Chem. C* **2008**, *112*, 4954–4960. [CrossRef]

33. Luk'yanchuk, B.; Zheludev, N.I.; Maier, S.A.; Halas, N.J.; Nordlander, P.; Giessen, H.; Chong, C.T. The Fano resonance in plasmonic nanostructures and metamaterials. *Nat. Mater.* **2010**, *9*, 707–715. [CrossRef] [PubMed]

34. Gonçalves, M.R.; Melikyan, A.; Minassian, H.; Makaryan, T.; Marti, O. Strong dipole-quadrupole coupling and Fano resonance in H-like metallic nanostructures. *Opt. Express* **2014**, *22*, 24516. [CrossRef] [PubMed]

35. Christopoulos, T.; Tsilipakos, O.; Sinatkas, G.; Kriezis, E.E. On the calculation of the quality factor in contemporary photonic resonant structures. *Opt. Express* **2019**, *27*, 14505. [CrossRef] [PubMed]

Article

Second-Order Dispersion Sensor Based on Multi-Plasmonic Surface Resonances in D-Shaped Photonic Crystal Fibers

Markos P. Cardoso [1,*], **Anderson O. Silva** [2], **Amanda F. Romeiro** [1], **M. Thereza R. Giraldi** [3], **João C. W. A. Costa** [1], **José L. Santos** [4,5], **José M. Baptista** [4,6] and **Ariel Guerreiro** [4,5]

1   Applied Electromagnetism Laboratory, Federal University of Pará, Belém 66075-110, Brazil; amanda.romeiro@itec.ufpa.br (A.F.R.); jweyl@ufpa.br (J.C.W.A.C.)
2   Federal Center for Technological Education Celso Suckow da Fonseca, Rio de Janeiro 20271-110, Brazil; anderson.silva@cefet-rj.br
3   Military Engineering Institute, Laboratory of Photonics, Rio de Janeiro 22290-270, Brazil; mtmrocco@ime.eb.br
4   INESC TEC, 4200-465 Porto, Portugal; josantos@fc.up.pt (J.L.S.); jmb@staff.uma.pt (J.M.B.); ariel@fc.up.pt (A.G.)
5   Faculty of Sciences, University of Porto, 4169-007 Porto, Portugal
6   Faculty of Exact Sciences and Engineering, University of Madeira, 9020-105 Funchal, Portugal
*   Correspondence: markosdenardi@gmail.com or markos.cardoso@itec.ufpa.br

Abstract: This paper proposes a scheme to determine the optical dispersion properties of a medium using multiple localized surface plasmon resonances (SPR) in a D-shaped photonic crystal fiber (PCF) whose flat surface is covered by three adjacent gold layers of different thicknesses. Using computational simulations, we show how to customize plasmon resonances at different wavelengths, thus allowing for obtaining the second-order dispersion. The central aspect of this sensing configuration is to balance miniaturization with low coupling between the different localized plasmon modes in adjacent metallic nanostructures. The determination of the optical dispersion over a large spectral range provides information on the concentration of different constituents of a medium, which is of paramount importance when monitoring media with time-varying concentrations, such as fluidic media.

Keywords: surface plasmon resonance; photonic crystal D-shaped fiber; refractive index sensor; dispersion sensor; second-order dispersion sensor

Citation: Cardoso, M.P.; Silva, A.O.; Romeiro, A.F.; Giraldi, M.T.R.; Costa, J.C.W.A.; Santos, J.L.; Baptista, J.M.; Guerreiro, A. Second-Order Dispersion Sensor Based on Multi-Plasmonic Surface Resonances in D-Shaped Photonic Crystal Fibers. *Photonics* **2021**, *8*, 181. https://doi.org/10.3390/photonics8060181

Received: 27 April 2021
Accepted: 21 May 2021
Published: 24 May 2021

Publisher's Note: MDPI stays neutral with regard to jurisdictional claims in published maps and institutional affiliations.

1. Introduction

Surface plasmon polaritons are electromagnetic modes that arise from the coupling between photons and free-electron oscillations at a conducting surface [1]. The practical excitation of a surface plasmon resonance (SPR) promotes the confinement of optical power at subwavelength dimensions and represents a milestone in the development of nano-optical sensors. Since surface plasmon modes are quite dependent on the refractive index of the surrounding medium, higher levels of sensitivity and resolution in a broad spectral range can be reached, which is the central reason for the huge effort dispended in related studies in the two last decades [2]. As a result, the losses variations and mode phase shifts can be used to describe medium properties in terms of its refractive index [3].

A conventional SPR prism-based sensor, although rather efficient to excite surface plasmon modes [4,5], suffers drawbacks due to its bulky size, which hampers remote sensing applications, and high-cost fabrication process. As an alternative, a common configuration consists of an optical fiber with partial cladding removed for the deposition of a thin metallic layer. The plasmonic resonance is achieved when the fiber core guided mode phase is equal to the surface plasmon mode phase at the conducting surface. Multiple sensing devices are designed to reach this condition as tapered fibers, grating-based fiber sensors and D-shaped fibers. In [6], a tapered optical fiber covered by gold nanoparticles is applied for biomolecular sensing. By its turn, the deposition of gold nanoparticles over a grating fiber sensor provides a highly sensitive and selective platform, which is quite useful for chemical applications [7,8]. In [9], a surface plasmon resonance sensor operating at the near-infrared band is constructed by coating the flat face of a D-shaped optical fiber with a graphene-based metal oxide layer.

Photonic crystal fibers (PCF) are highly suitable to excite and enhance surface plasmon resonances due to their unique characteristics, such as fine control of the evanescent field penetration into the conducting medium and high mode confinement with a large mode area [10–12]. A PCF has an arrangement of air holes periodically distributed over its cross-sectional area. The light waveguiding is favored by inserting a defect in the set of air holes. The modal properties of the fiber are extremely influenced by the structural parameters, such as the diameter of the air holes and the distance between them. Several designs of SPR sensors based on D-shaped PCF with high levels of sensitivity and resolution have been reported [12–15]. However, to the best of our knowledge, all these sensors are used to interrogate the average refractive index or a specific wavelength of the surrounding medium, with no further information regarding the wavelength dependence of the analyte.

We reported a D-shaped PCF with two gold slabs [16] whose response has a highly linear dependence on the plasmonic resonances with the first-order dispersion. The designed dispersion sensor finds potential applications for the investigation of changes in fluidic media over time, for instance. This is because small variations on the chemical composition produce subtle changes in the dispersion relation of the medium [17,18], which cannot be roughly monitored by conventional refractive-index-based sensors.

In this work, we investigate theoretically the viability of applying an SPR sensor based on a D-shaped PCF to monitor the second-order dispersion of an optical medium. The design is projected to provide three distinct and independent plasmonic resonances. This is obtained by depositing three gold slabs with different thicknesses on the top of the core region at the flat surface of the PCF. The sensing response is taken by the changes in the amplitude of the resonance peaks and the distance between them for parabolic dispersive profiles. Our final goal is to demonstrate a relationship between the sensor response with the input second-order dispersive medium. In the following sections, we detail the modeling of the SPR sensor based on a D-shaped PCF and discuss the results for second-order dispersion sensing.

2. Designed Structure and Modeling

The cross-section of the proposed sensor is shown in Figure 1. It is composed of a hexagonal arrangement of air holes in silica background. The fiber core (the central region with an absence of air holes) has a diameter of 5 μm, the pitch (distance between two adjacent air holes) is $\Lambda = 2$ μm and the diameter of each air hole is related to the pitch by $d/0.88 = \Lambda$. This type of D-shaped structure can be obtained by the stack-and-draw process [19] followed by a side-polishing or controlled etching technique [20]. Three gold slabs of equal width $w = 1.5$ μm but with different thicknesses ($t_1 = 40$ nm, $t_2 = 15$ nm and $t_3 = 30$ nm) are deposited on the flat surface of the fiber, which can be performed in practice by the CVD (Chemical Vapor Deposition) process [21].

The Finite-Element-Method-based software COMSOL Multiphysics [22] is applied for the numerical modeling. The computational domain comprises the cross-section of Figure 1a with diameter $D = 24$ μm, truncated by a 0.1D thick PML (Perfectly Matched Layer). More specifically, the Wave Optics package in the frequency domain is applied to carry out 2D computational simulations to obtain the eigenvalues of the Helmholtz equation in the angular frequency ω:

$$\nabla_\perp^2 E(r_\perp, \omega) + k_0^2 \left(\varepsilon(\omega) - n_{eff}^2 \right) E(r_\perp, \omega) = 0 \tag{1}$$

where k_0 is the magnitude of the free-space wavenumber and ε is the complex frequency-dependent relative permittivity. $E(r_\perp, \omega)$ is the modal electric field distribution at the position $r_\perp$ perpendicular to the direction of light propagation. The expression for ε

depends on the region at which (1) is solved. The complex effective index n_{eff} is the result of the modal fields that arise from the coupling between the fundamental fiber mode and the plasmonic excitations at the boundaries of the gold layers. While the real part of n_{eff} refers to the mode phase, the imaginary part is related to the losses experienced by the confined mode as the fields are tunneled through the gold slabs.

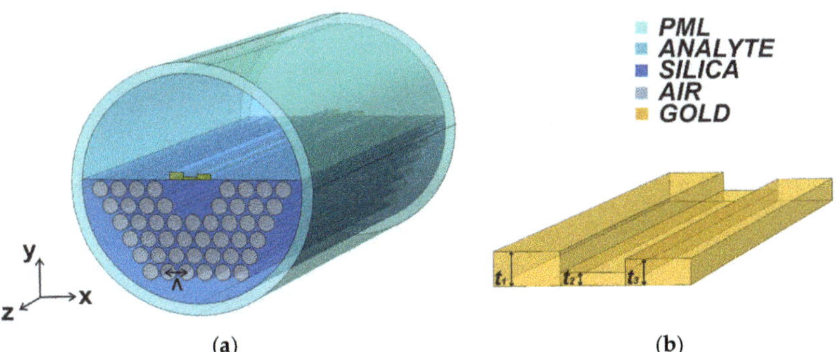

Figure 1. D-shaped photonic crystal fiber with three gold slabs. (**a**) Perspective view of the sensor with diameter D = 24 µm, diameter of air holes d = 1.76 µm and Λ = 2 µm. The outermost layer that encloses the entire domain corresponds to a 0.1D thick PML. (**b**) Highlight of the three gold slabs deposited on the top of the flat face of the PCF. They present an equal 1.5 µm width but with different thicknesses: t_1 = 40 nm, t_2 = 15 nm and t_3 = 30 nm.

For accurate modeling, the dispersive character of all materials in the sensing structure has to be considered. The refractive index of silica is computed from the Sellmeier equation [23] and the air region is modelled by a constant refractive index n_{air} = 1. Drude–Lorentz formalism [24], which is an improvement on the original Drude model, is used to characterize the dispersive character of gold. This model provides a better approach to the corresponding experimental data:

$$\varepsilon_{Au}(\omega) = \varepsilon_\infty - \frac{\omega_p^2}{\omega(\omega + i\gamma)} - \frac{\Delta\epsilon\Omega_L^2}{\left(\omega^2 - \Omega_L^2\right) + i\omega\Gamma_L} \tag{2}$$

The parameters ω_p = 2155.6 THz and γ = 15.92 THz are the plasma frequency and damping factor, respectively, and ε_∞ = 5.9673 stands for the residual polarization of gold at high frequencies. The parameters in the correction factor are $\Delta\epsilon$ = 1.09, Ω_L = 650.07 THz and Γ_L = 104.86 THz [24].

The modeling approach allowed for solving (1) without the requirements of significative computational efforts and extensive time consumption.

3. Results

3.1. Sensitivity Performance Analysis

In the sensing platform schematically depicted in Figure 1, the resonance condition is achieved when there is phase-matching between the fundamental fiber mode and surface plasmon mode. Due to this phenomenon, the spectral confinement loss presents a peak, which is related to the fact that the amount of modal energy penetrating into the gold layers is maximal. When the refractive index of the analyte changes, the wavelength that satisfies the resonance condition also changes. In Figure 2, the imaginary part, expressed in terms of the spectral losses, and the real part of the effective indexes of the fundamental fiber mode and plasmonic modes are shown for an analyte with dispersive parabolic refractive index $Ri_{analyte} = 7.41 \times 10^{-6}\lambda^2 - 0.0083\lambda + 3.65$, with the wavelength λ measured in nanometers. The effective indexes were computed from the numerical modeling described previously applied to the solution of (1) over the entire sensing structure. At the resonance wavelengths, the effective indexes of the surface plasmon mode and fiber mode present the same real part due to phase-matching and, as a consequence, the spectral losses of the fiber mode reach a peak. These modes are uncoupled at non-resonance wavelengths, and the corresponding effective indexes are completely distinct. By adding metallic layers of different thicknesses, several resonant peaks are expected in the spectral losses. In Figure 2, the real part of the effective index of the fundamental mode intercepts SPP modes at the wavelengths 568, 586 and 632 nm, respectively. Even though the fundamental mode of the PCF is degenerate for two orthogonal polarizations, only the polarization perpendicular to the gold interfaces allows for the excitation of plasmonic modes.

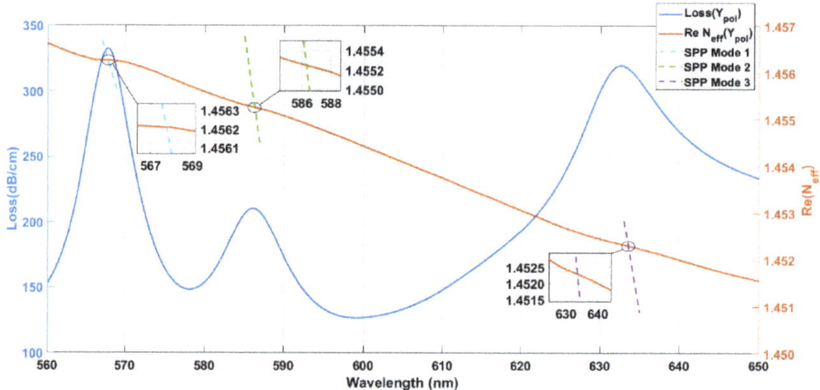

Figure 2. Dispersion curve of the fundamental fiber mode and SPP modes for a variable refractive index with parabolic profile analyte $Ri_{analyte} = 7.41 \times 10^{-6}\lambda^2 - 0.0083\lambda + 3.65$. Y_{pol} is the perpendicular polarization to the gold slabs. The insets show the intersections between the dispersion curve of the fundamental Y_{pol} fiber mode and the plasmonic mode at the gold interfaces.

In [16], we have explored the dependence of SPR spectra on the first-order dispersion profile of the refractive index of the analyte. In the present work, we focus on the relation of the second-order dispersion with the spectral response of the sensor. For that purpose, we carried out computations of the SPR spectral losses for three dispersive characters: constant refractive index, normal dispersion and anomalous dispersion. The spectral losses of the SPR D-shaped PCF with three gold slabs and their dispersion regimes are shown in Figure 3. For a constant refractive index equal to 1.36 (solid red line), three distinct regions of maximum absorption appear (solid blue line) due to the nonuniformity in the thickness of the three gold slabs at the flat face of the sensor.

Moreover, we consider two arbitrary media with parabolic dispersion profiles of opposite concavities. In Figure 3, the dotted blue line represents the spectral losses due to the anomalous dispersion curve (dotted red line), while the dashed blue line is related to the normal dispersion curve (dashed red line). Just as in the case of the constant refractive index, there are three distinct resonance wavelengths for a dispersive index. It is also observed that the magnitude of the losses and the distance between resonance wavelengths is influenced by the dispersion profile.

In a comparative analysis in the case of a constant refractive index, the distances between the wavelengths of the three peaks decrease for the normal dispersion and the amplitude losses increase at 582 and 593 nm. On the other hand, the anomalous dispersion curve leads to a larger difference between resonance wavelengths but to smaller amplitude losses at 568 and 586 nm. For the resonance peak at 632 nm, losses remain close for the different dispersion profiles.

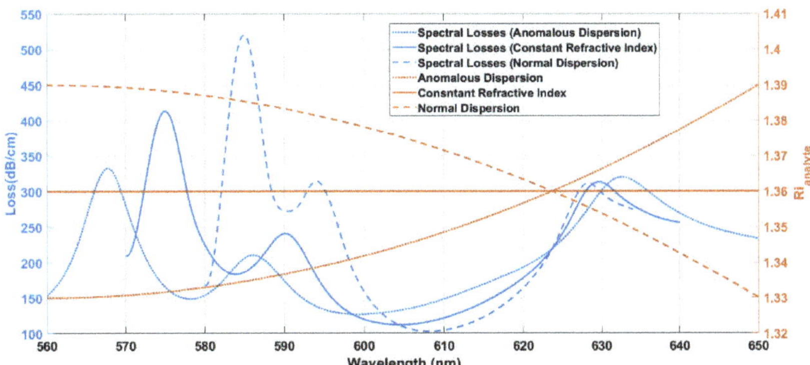

Figure 3. Dispersion profiles and related spectral losses of the SPR D-shaped PCF with three gold slabs. Solid lines represent the case of a constant refractive index of 1.36 and corresponding ing spectral losses. The second-order anomalous and normal dispersion regimes are expressed by $Ri_{analyte} = 7.41 \times 10^{-6}\lambda^2 - 0.0083\lambda + 3.65$ and $Ri_{analyte} = -7.41 \times 10^{-6}\lambda^2 + 0.0083\lambda - 0.93$, respectively. Their related spectral losses are represented by dotted lines and dashed lines, respectively.

3.2. Sensing Response at Second-Order Dispersion

To access the ability of the sensor in retrieving the second-order dispersion of a medium, we applied the multivariate polynomial regression model [25] to obtain the coefficients a, b and c of the general quadratic expression $Ri_{analyte} = a\lambda^2 + b\lambda + c$ from nine parameters obtained from computed SPR spectra, namely: the amplitudes of the three SPR peaks, the corresponding resonance wavelengths and the distances between these wavelengths. The numerical computation of the SPR spectrum was carried out for 270 distinct parabolic dispersion curves in order to achieve a dataset able to ensure an accurate fitting. The multivariate polynomial regression equation to achieve the coefficients β_n can be written in the form as [25]:

$$y = \beta_0 + \beta_1 x_1 + \beta_2 x_2 + \ldots + \beta_{54} x_{54} + \varepsilon \tag{3}$$

where y is the vector containing the parabolic profile parameters (specifically, the $a's$, $b's$ and $c's$), β_n are the multivariate polynomial regression coefficients, $x'_n s$ are the vectors containing the simulated parameters, specifically, the resonance amplitude for the three peaks, the wavelength peak for the three resonances and the product among themselves. The independent term ε is related to the errors.

As an example, Figure 4 shows the points retrieved by applying the multivariate polynomial regression to the SPR spectral parameters for both dispersion regimes.

To characterize quantitatively the performance of the fitting model, we computed the error in estimating the parabolic coefficients a, b and c ($Ri_{analyte} = a\lambda^2 + b\lambda + c$) from the multivariate linear regression model. The histograms depicted in Figure 5 show the Gaussian distribution of the error, which is centered at 0 for the three coefficients. To compare the quality of the regression model for future works, we also calculated the full-width half-maximum for the histograms: 0.1503, 0.0894 and 0.1521 for the parameters a, b and c, respectively. Moreover, the calculated determination coefficients $R^2 = 0.908129$ for a, $R^2 = 0.905276$ for b and $R^2 = 0.901035$ for c demonstrate a reasonable level of accuracy for the estimative of the parabolic dispersion curve from the SPR spectral parameters for the proposed sensing platform.

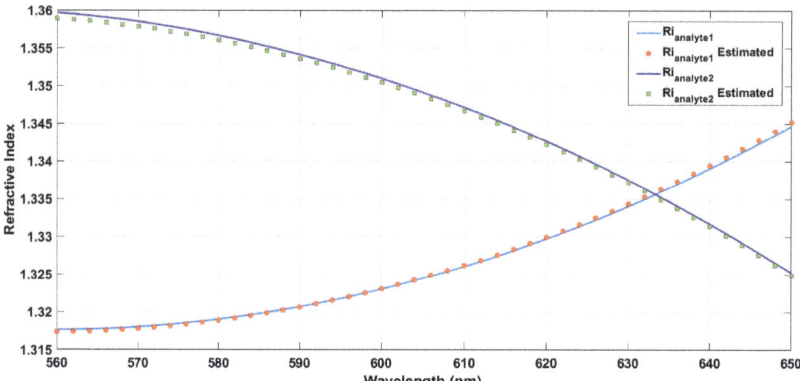

Figure 4. Comparison between two examples of the parabolic curves for normal and anomalous dispersions (continuous lines) and the data estimated using the coefficients calculated by the multivariate linear regression (dotted lines).

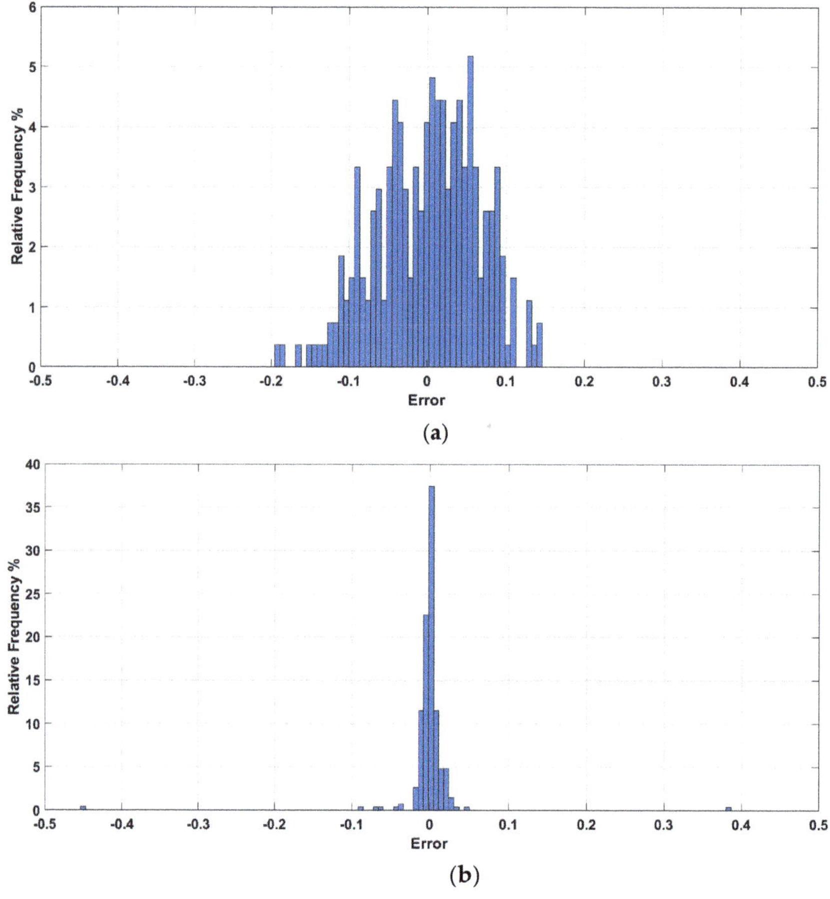

Figure 5. *Cont.*

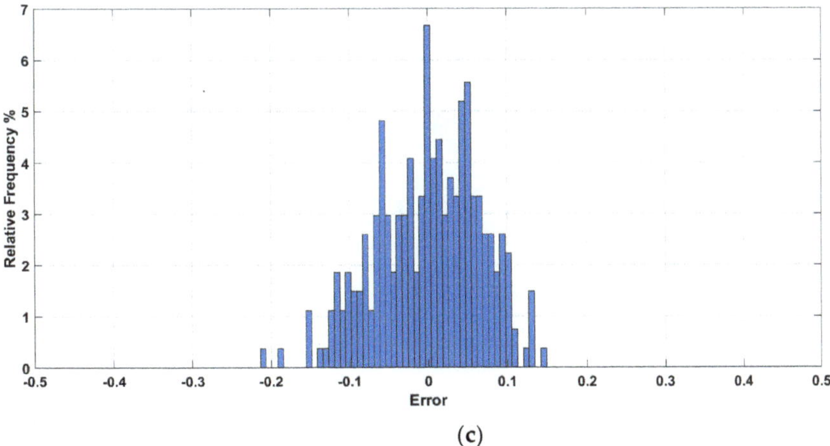

(c)

Figure 5. Distribution of the error in estimating the parabolic coefficients of the refraction index expression $Ri_{analyte} = a\lambda^2 + b\lambda + c$: (**a**) coefficient a, (**b**) coefficient b and (**c**) coefficient c.

4. Conclusions

The characterization of the optical dispersion of the refraction index over a large spectral range can be useful in the determination of the concentration of different constituents of a medium, which is particularly relevant when monitoring media in real time. By adding three gold slabs of different thicknesses at the flat face of the D-shaped photonic crystal fiber, we were able to characterize the second-order dispersion of a medium from the excitation of multiple surface plasmon resonances. This new approach is able to determine the non-local spectral character of the refraction index of realistic media, such as fluids whose composition varies with time. Moreover, since the surface plasmon resonances are excited at distinct spectral channels, the sensing structure can be used to determine simultaneously more than one parameter.

Author Contributions: Theory, conceptualization, review and editing of the manuscript A.G., J.L.S., J.M.B., J.C.W.A.C. and M.T.R.G.; optical design, simulation, investigation M.P.C., A.O.S. and A.F.R.; writing the original draft of the manuscript, A.O.S. and M.P.C. All authors have read and agreed to the published version of the manuscript.

Funding: This study was financed in part by the Coordenação de Aperfeiçoamento de Pessoal de Nível Superior—Brasil (CAPES)—Finance code 001. During the period of elaboration of this work, the authors have obtained support from the Conselho Nacional de Desenvolvimento Cientifico e Tecnologico (CNPq). It was also financed by the ERDF—European Regional Development Fund through the Operational Program for Competitiveness and Internationalization—COMPETE 2020 Program and by National Funds through the Portuguese funding agency, FCT—Fundação para a Ciência e a Tecnologia within project "GreenNanoSensing" POCI-01-0145-FEDER-032257.

Institutional Review Board Statement: Not applicable.

Informed Consent Statement: Not applicable.

Data Availability Statement: The data that support the findings of this study are available from the corresponding author upon reasonable request.

Conflicts of Interest: The authors declare no conflict of interest.

References

1. Barnes, W.L. Surface plasmon–polariton length scales: A route to sub-wavelength optics. *J. Opt. A Pure Appl. Opt.* **2006**, *8*, S87–S93. [CrossRef]
2. Homola, J. Surface Plasmons on Waveguides with a Perturbed Refractive Index Profile. In *Surface Plasmon Polariton Based Sensors*, 1st ed.; Springer: Berlin/Heidelberg, Germany; New York, NY, USA, 2018; Volume 4, pp. 19–26.
3. Prabowo, B.A.; Purwidyantri, A.; Liu, K.C. Surface plasmon resonance optical sensor: A review on light source technology. *Biosensors* **2018**, *8*, 80. [CrossRef]
4. Homola, J.; Koudela, I.; Yee, S.S. Surface plasmon resonance sensors based on diffraction gratings and prims couplers: Sensitivity comparison. *Sens. Actuators B Chem.* **1999**, *54*, 16–24. [CrossRef]
5. Leong, H.-S.; Guo, J.; Lindquist, R.G.; Liu, Q.H. Surface plasmon resonance in nanostructured metal films under the Kretschmann configuration. *J. Appl. Phys.* **2009**, *106*, 124314. [CrossRef]
6. Lin, H.-Y.; Huang, C.-H.; Cheng, G.-L.; Chen, N.K.; Chui, H.C. Tapered optical fiber sensor based on localized surface plasmon resonance. *Opt. Express* **2012**, *20*, 21693–21701. [CrossRef]
7. Heidemann, B.R.; Chiamenti, I.; Oliveira, M.M.; Muller, M.; Fabris, J.L. Functionalized Long Period Grating—Plasmonic Fiber Sensor Applied to the Detection of Glyphosate in Water. *J. Light. Technol.* **2018**, *36*, 863–870. [CrossRef]
8. Si, Y.; Lao, J.; Zhang, X.; Liu, Y.; Cai, S.; Gonzalez-Vila, A.; Li, K.; Huang, Y.; Yuan, Y.; Caucheteur, C.; et al. Electrochemical Plasmonic Fiber-optic Sensors for Ultra-Sensitive Heavy Metal Detection. *J. Light. Technol.* **2019**, *37*, 3495–3502. [CrossRef]
9. Patnaik, A.; Senthilnathan, K.; Jha, R. Graphene-Based Conducting Metal Oxide Coated D-Shaped Optical Fiber SPR Sensor. *IEEE Photonics Technol. Lett.* **2015**, *27*, 2437–2440. [CrossRef]
10. Yu, X.; Zhang, Y.; Pan, S.; Shum, P.; Yan, M.; Leviatan, Y.; Li, C. A selectively coated photonic crystal fiber based surface plasmon resonance sensor. *J. Opt.* **2009**, *12*, 1–4. [CrossRef]
11. Rifat, A.A.; Ahmed, R.; Yetisen, A.K.; Butt, H.; Sabouri, A.; Mahdiraji, G.A.; Yun, S.H.; Adikan, F.R. Photonic crystal fiber based plasmonic sensors. *Sens. Actuators B Chem.* **2017**, *243*, 311–325. [CrossRef]
12. Rifat, A.A.; Haider, F.; Ahmed, R.; Mahdiraji, G.A.; Adikan, F.R.M.; Miroshnichenko, A.E. Highly sensitive selectively coated photonic crystal fiber-based plasmonic sensor. *Opt. Lett.* **2018**, *43*, 891–894. [CrossRef]
13. Santos, D.F.; Guerreiro, A.; Baptista, J.M. SPR Microstructured D-Type Optical Fiber Sensor Configuration for Refractive Index Measurement. *IEEE Sens. J.* **2015**, *15*, 5472–5477. [CrossRef]
14. Liu, Y.; Jing, X.; Li, S.; Zhang, S.; Zhang, Z.; Guo, Y.; Wang, J.; Wang, S. High sensitivity surface plasmon resonance sensor based on D-shaped photonic crystal fiber with circular layout. *Opt. Fiber Technol.* **2018**, *46*, 311–317. [CrossRef]
15. Zhao, L.; Han, H.; Lian, Y.; Luan, N.; Liu, J. Theoretical analysis of all-solid D-type photonic crystal fiber based plasmonic sensor for refractive index and temperature sensing. *Opt. Fiber Technol.* **2019**, *50*, 165–171. [CrossRef]
16. Cardoso, M.P.; Silva, A.O.; Romeiro, A.F.; Giraldi, M.T.R.; Costa, J.C.W.A.; Santos, J.L.; Baptista, J.M.; Guerreiro, A. Multi-plasmonic resonance based sensor for the characterization of optical dispersion using a D-shaped photonic crystal fiber. *IEEE Instrum. Meas. Meg.* accepted for publication.
17. Sai, T.; Saba, M.; Dufresne, E.R.; Steiner, U.; Wilts, B.D. Designing refractive index fluids using the Kramers-Kroing relations. *Faraday Discuss* **2020**, *223*, 136–144. [CrossRef]
18. Daimon, M.; Masumura, A. Measurement of the refractive index of distilled water from the near-infrared region to the ultraviolet region. *Appl. Opt.* **2007**, *46*, 3811–3820. [CrossRef]
19. Mahdiraji, G.A.; Chow, D.M.; Sandoghchi, S.R.; Amirkhan, F.; Dermosesian, E.; Yeo, K.S.; Kakaei, Z.; Ghomeishi, M.; Poh, S.Y.; Gang, S.Y.; et al. Challenges and Solutions in Fabrication of Silica-Based Photonic Crystal Fibers: An Experimental Study. *Fiber Integr. Opt.* **2014**, *33*, 85–104. [CrossRef]
20. Dash, J.N.; Jha, R. Highly sensitive D shaped PCF sensor based on SPR for near IR. *Opt. Quantum Electron.* **2016**, *48*, 1–7. [CrossRef]
21. Sazio, P.J.A.; Amezcua-Correa, A.; Finlayson, C.E.; Hayes, J.R.; Scheidemantel, T.J.; Baril, N.F.; Jackson, B.R.; Won, D.-J.; Zhang, F.; Margine, E.R.; et al. Microstructured Optical Fibers as High-Pressure Microfluidic Reactors. *Science* **2006**, *311*, 1583–1586. [CrossRef]
22. Comsol. Available online: www.br.comsol.com (accessed on 4 February 2021).
23. Sharma, A.K.; Gupta, B.D. On the performance of different bimetallic combinations in surface plasmon resonance based fiber optic sensors. *J. Appl. Phys.* **2007**, *101*, 093111. [CrossRef]
24. Vial, A.; Grimault, A.-S.; Macías, D.; Barchiesi, D.; De La Chapelle, M.L. Improved analytical fit of gold dispersion: Application to the modeling of extinction spectra with a finite-difference time-domain method. *Phys. Rev. B* **2005**, *71*, 085416. [CrossRef]
25. Chapra, S.C. Multiple Linear Regression. In *Applied Numerical Methods with Matlab for Engineers and Scientists*, 4th ed.; McGraw-Hill Education: New York, NY, USA, 2018; pp. 389–391.

Article

Exploration on Structural and Optical Properties of Nanocrystalline Cellulose/Poly(3,4-Ethylenedioxythiophene) Thin Film for Potential Plasmonic Sensing Application

Nur Syahira Md Ramdzan [1], Yap Wing Fen [1,2,*], Josephine Ying Chyi Liew [1], Nur Alia Sheh Omar [1], Nur Ain Asyiqin Anas [3], Wan Mohd Ebtisyam Mustaqim Mohd Daniyal [2] and Nurul Illya Muhamad Fauzi [2]

[1] Department of Physics, Faculty of Science, Universiti Putra Malaysia UPM, Serdang 43400, Selangor, Malaysia; nursyahira.upm@gmail.com (N.S.M.R.); josephine@upm.edu.my (J.Y.C.L.); nuralia_so@upm.edu.my (N.A.S.O.)

[2] Functional Devices Laboratory, Institute of Advanced Technology, Universiti Putra Malaysia UPM, Serdang 43400, Selangor, Malaysia; wanmdsyam@gmail.com (W.M.E.M.M.D.); illyafauzi97@gmail.com (N.I.M.F.)

[3] Physics Unit, Centre of Foundation Studies for Agricultural Science, Universiti Putra Malaysia UPM, Serdang 43400, Selangor, Malaysia; ainasyiqin@upm.edu.my

* Correspondence: yapwingfen@upm.edu.my

Citation: Ramdzan, N.S.M.; Fen, Y.W.; Liew, J.Y.C.; Omar, N.A.S.; Anas, N.A.A.; Daniyal, W.M.E.M.M.; Fauzi, N.I.M. Exploration on Structural and Optical Properties of Nanocrystalline Cellulose/Poly(3,4-Ethylenedioxythiophene) Thin Film for Potential Plasmonic Sensing Application. *Photonics* **2021**, *8*, 419. https://doi.org/10.3390/photonics8100419

Received: 28 August 2021
Accepted: 20 September 2021
Published: 29 September 2021

Publisher's Note: MDPI stays neutral with regard to jurisdictional claims in published maps and institutional affiliations.

Abstract: There are extensive studies on the development of composite solutions involving various types of materials. Therefore, this works aims to incorporate two polymers of nanocrystalline cellulose (NCC) and poly(3,4-ethylenethiophene) (PEDOT) to develop a composite thin film via the spin-coating method. Then, Fourier transform infrared (FTIR) spectroscopy is employed to confirm the functional groups of the NCC/PEDOT thin film. The atomic force microscopy (AFM) results revealed a relatively homogeneous surface with the roughness of the NCC/PEDOT thin film being slightly higher compared with individual thin films. Meanwhile, the ultraviolet/visible (UV/vis) spectrometer evaluated the optical properties of synthesized thin films, where the absorbance peaks can be observed around a wavelength of 220 to 700 nm. An optical band gap of 4.082 eV was obtained for the composite thin film, which is slightly lower as compared with a single material thin film. The NCC/PEDOT thin film was also incorporated into a plasmonic sensor based on the surface plasmon resonance principle to evaluate the potential for sensing mercury ions in an aqueous medium. Resultantly, the NCC/PEDOT thin film shows a positive response in detecting the various concentrations of mercury ions. In conclusion, this work has successfully developed a new sensing layer in fabricating an effective and potential heavy metal ions sensor.

Keywords: nanocrystalline cellulose; poly(3,4-ethylenedioxythiophene); structural properties; optical properties; surface plasmon resonance

1. Introduction

To achieve success in research or experiment, the selection of materials and methods is very significant to ensure the project will be able to accomplish its objectives at the end of the process. Therefore, until recently. researchers have always sought the novelty of materials as a fundamental property to verify the quality of research. Hence, nanocrystalline cellulose (NCC) is one of the novel materials that has been intensively studied in recent years. This biopolymer is a cellulose nanocrystal that has a diameter of 1–5 nm and a length of 150–300 nm [1,2]. As the highlight, it can be synthesized from a variety of natural resources via an acid hydrolysis process [3,4]. Owing to many beneficial properties such as biodegradability, biocompatibility, and low toxicity [5–8], this material has great potential to be applied in various fields including industrial sectors, drug delivery systems, pharmaceutical industries, and sensors [9–16]. However, the potential of NCC can be seen to be further developed through incorporation with other materials to enhance the effectiveness and usefulness in certain fields.

On the other hand, conducting polymers have gained tremendous attention from researchers owing to their unique characteristics. Although there are several examples of conducting polymers, for instance, polyaniline and polypyrrole, poly(3,4-ethylenedioxythiophene) (PEDOT), which is selected for this work [17,18], has additional properties. Compared with other conducting polymers, PEDOT possesses excellent properties thanks to its good conductivity, excellent chemical and electrochemical properties, and high transparency [19–27]. Therefore, in light of these advantages, this polymer has been largely explored for a wide range of applications including batteries, transistors, light-emitting diodes (LEDs), and optical sensors [28–31]. Ravit et al. (2019) had successfully investigated the combination of NCC and PEDOT film using the electrochemical polymerization method to develop a supercapacitor [32]. However, as far as we are aware, there is no published work on this synthesized thin film as a sensing layer for the plasmonic sensor for metal ions' detection. In addition, optical spectroscopy including photoluminescence, inductively coupled plasma mass spectroscopy (ICP-MS), optical metasurfaces-based sensor, and surface-enhanced Raman spectroscopy (SPR) is hampered by some drawbacks [33–36]. Surface plasmon resonance (SPR) is a simple and sensitive plasmonic-based sensor, where it is considered as one of the complementary sensors that offer excellent potential sensing. Furthermore, it has advantageous characteristics such as inexpensive, rapid response time, label-free technique, high sensitivity, and the ability to detect the analyte at very low concentrations. In this plasmonic-based sensor, several configurations have been introduced include grating coupler [37], optical fiber-based [38,39], optical waveguide system [40], and prism coupler [41,42]. The Kretschmann configuration in a prism-based system is commonly used because the metal thin film is attached directly to the prism without any gap. This sensor has been applied in numerous studies for various kinds of detection such as dengue viruses [43–45], glucose [46,47], heavy metal ions [48–54], and phenol [55,56].

Hence, as part of this research, the NCC/PEDOT thin film was prepared via the spin-coating method, followed by its further characterization to analyze the structural and optical properties using atomic force microscopy (AFM), Fourier transform infrared (FTIR), and ultraviolet/visible (UV/vis) spectroscopy. Then, the composite material will be incorporated into a plasmonic sensor based on SPR to determine its potential as a sensing thin film. The potential sensing of the NCC/PEDOT thin film will be observed and proven based on the graph of reflectance against the incidence angle when in contact with various concentrations of the analyte.

2. Experimental Section

2.1. Materials and Reagents

Nanocrystalline cellulose (NCC) powder, mercury standard solution, and poly(3,4-ethylenedioxythiophene)-poly(styrenesulfonate) (PEDOT/PSS) were supplied by Sigma-Aldrich (St. Louis, MO, USA). All the chemicals used in this experiment were of analytical grade and were used without further purification. Then, a high refractive index prism of 1.77861 at 632.8 nm and glass coverslips (24 mm × 24 mm × 0.1 mm) were purchased from Menzel-Glaser (Braunschweig, Germany) and used as received.

2.2. Preparation of Thin Film and Analyte

To prepare the thin film, initially, 5 g of NCC was dissolved in 100 mL of deionized water and stirred for 24 h using a magnetic stirrer to ensure that the powered NCC was fully dissolved. Later, the composite solution of NCC/PEDOT was synthesized by mixing 1 mL of the prepared NCC solution with 1 mL of the commercially purchased PEDOT/PSS solution, and the resulting mixture was placed into a reagent bottle for storage purposes. Before depositing the composite solution onto the glass slip, the glass coverslips must be cleaned using acetone solution to eliminate any fingerprint marks or dirt. Then, the glass slip was coated with a 50 nm gold layer using the sputter coater model K575X with 20 mA of current and 2.2 kV of voltage. Eventually, another coating process was performed on the gold layer to accomplish the fabrication of the sensing thin film. Hence, the NCC/PEDOT solution was placed on the gold thin film and then spun for 30 s using the Spin Coating System, P-6708D, which operated at 3000 rpm, to ensure that a thin layer was coated and covered evenly on a gold film.

Continuing with the analyte, various concentrations of mercury ion were employed throughout the experiment. The deionized water and standard solution of mercury ion at a concentration of 1000 ppm were involved in the preparation of various analyte concentrations. To produce the concentrations of 0.01, 0.1, 1, and 10 ppm, the concentrated solution of mercury (Hg^{2+}) was diluted using the $M_1V_1 = M_2V_2$ formula.

2.3. Thin Film Characterization

The characterizations began with the structural properties by Fourier transform infrared (FTIR). The spectra measurements of the composite material were performed using the FTIR spectrophotometer (Perkin-Elmer Spectrum 100, Waltham, MA, USA) within a range from 400 to 4000 cm^{-1}. Then, the AFM images of the prepared thin films were examined by Bruker AFM (Multimode 8) in Scan Asyst mode, with scan sizes in a range of 1 μm × 1 μm. It will then proceed with the optical properties, where UV–vis spectroscopy (Shimadzu UV-3600) was utilized to measure the sample absorption spectra ranging from 220 to 700 nm. The absorption peak obtained from the UV–vis spectrophotometer can also be applied to evaluate the optical band gap energy. Meanwhile, for the potential plasmonic sensing properties, SPR spectroscopy was set up as depicted in Figure 1. Using an index matching liquid, the sensing layer was coated onto the gold layer, which was sandwiched between the cell and the prism on the optical stage. Next, the varied concentrations of Hg^{2+} solutions were instilled into the cell one at a time by a microsyringe during the analysis, and the O-ring was used for the sensing thin film to make contact with the targeted analyte. The cell and prism were attached on a rotational stage (Newport MM3000) operated by a stepper motor with a resolution of 0.001°. An He-Ne laser, as a light source, passed through the pin hole, filter, optical chopper, and polarizer, in order to allow p-polarized monochromatic light with a specific wavelength to strike the prism [57–59]. The reflected light is then recorded and captured by the sensitive photodiode connected by a lock-in amplifier [60–62].

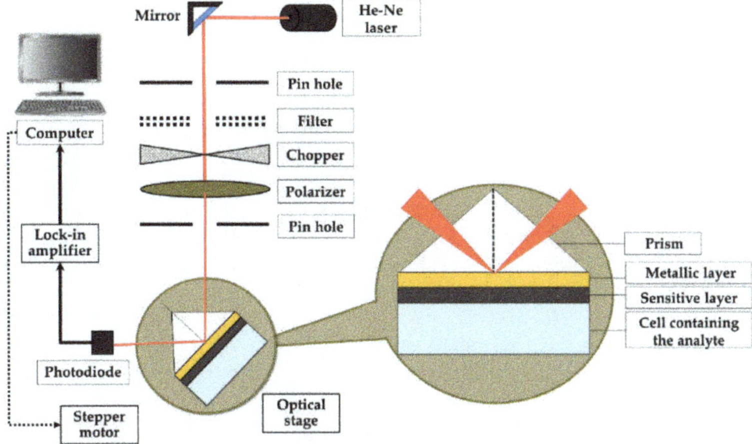

Figure 1. Experimental setup for exploring plasmonic sensing of the NCC/PEDOT thin film.

3. Results

3.1. Structural Properties

Figure 2 shows the comparison of FTIR spectrum for NCC, PEDOT, and NCC/PEDOT with NCC as a base material. The FTIR spectra are recorded within wavenumbers ranging from 500 to 4000 cm^{-1}. Based on the FTIR spectrum, the significant absorption peaks are assigned to identify the characteristics of the functional groups of each composite thin film.

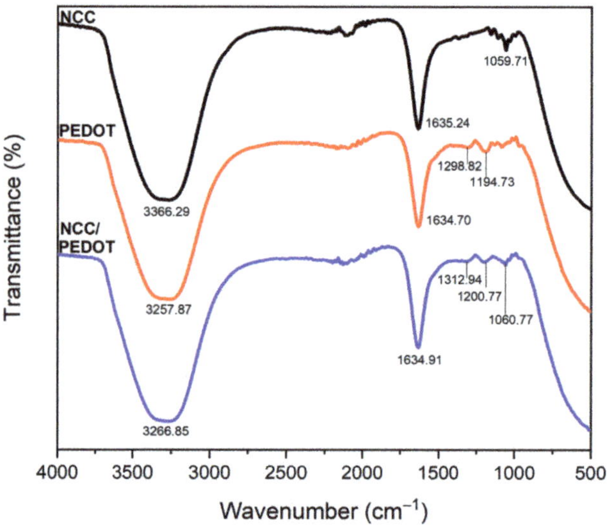

Figure 2. FTIR spectrum of NCC, PEDOT, and NCC/PEDOT thin film.

The FTIR spectrum of NCC thin film showed the characteristic peak of alcohol with O–H stretching bonds representing 3366.29 cm^{-1}. This characteristic peak is typically found in another previous study of NCC spectra [63]. The other peak at 1635.24 cm^{-1} represents the O–H bending of absorbed water, while the peak at 1059.71 cm^{-1} corresponds to C–O stretching bonds. Hence, this spectrum is in good agreement with previous studies of NCC [64,65].

Next, in the spectra of PEDOT thin film, the broad absorption peak that appeared at 3257.87 cm^{-1} can be indicated to the O–H stretching bond, which represents an alcohol group. Then, C=C stretching bond at 1634.70 cm^{-1} is attributed to the quinoidal structure of PEDOT [66]. Meanwhile, two peaks at 1298.82 cm^{-1} and 1194.73 cm^{-1} have explained the presence of ethylenedioxy and sulfonate groups in the PEDOT surface [67].

The spectrum of NCC/PEDOT and PEDOT thin film has displayed a few common properties, which the absorption bands at 3266.85 cm^{-1} referred to as O–H stretching vibration. Besides the absorption band displayed at 1634.91 cm^{-1} assigned to O–H bending of absorbed water, there are also characteristic absorbance bands centered at 1312.94 cm^{-1} and 1200.77 cm^{-1}, which correspond to sulfonate and ether groups, respectively. Lastly, the C–O stretching bond is located at 1060.07 cm^{-1}. The functional groups that exist match the spectra of the NCC/PEDOT thin film in a previous study by Ravit et al. (2019) [68].

The comparison of FTIR spectra of the NCC, PEDOT, and NCC/PEDOT thin film proved the combination band of NCC and PEDOT. In general, the peak around 3200 to 3400 cm^{-1} is assigned to the O–H group. Then, the O–H bending of absorbed water is located at around 1630 cm^{-1}. Meanwhile, ether and sulfonate groups only existed in PEDOT and NCC/PEDOT spectra.

3.2. Surface Morphology

The morphology features of the spin-coated NCC, PEDOT, and NCC/PEDOT thin films were identified by atomic force microscope (AFM), and it was carried out in tapping mode. In this study, a scan size was fixed (1 μm × 1 μm), for the imaging of the topographical difference in the thin films. From AFM images, the root mean square (RMS) roughness was obtained, which indicates the relative roughness and the standard deviation of the surface height. Figures 3a–c and 4a–c show the AFM images of the NCC, PEDOT, and NCC/PEDOT thin film, respectively. Then, the AFM results of NCC/PEDOT thin film after being in contact with the Hg^{2+} solution are depicted in Figures 3d and 4d.

Based on the two-dimensional (2D) and three-dimensional (3D) images of the NCC, PEDOT, and NCC/PEDOT thin film, it can be observed that the composite materials are well distributed on the whole scanned surface, the gold thin film. In Figure 3a, the images of NCC thin films show a rod-like structure with an RMS value of 2.90 nm. This morphology was related to another study done by Elazzouzi-Hafraoui et al. (2008) and the rod structure could be attributed to the crystallinity properties of NCC [69]. On the other hand, a sharp and relatively rough surface with RMS roughness of 2.79 nm for the PEDOT thin film was obtained. Then, from Figure 3c, it can be observed that, with the presence of NCC on the sensor surface, the rod-like shape reappeared in the NCC/PEDOT thin film, which leads to agglomeration. The incorporation between NCC and PEDOT results in the surface roughness increasing to 8.32 nm.

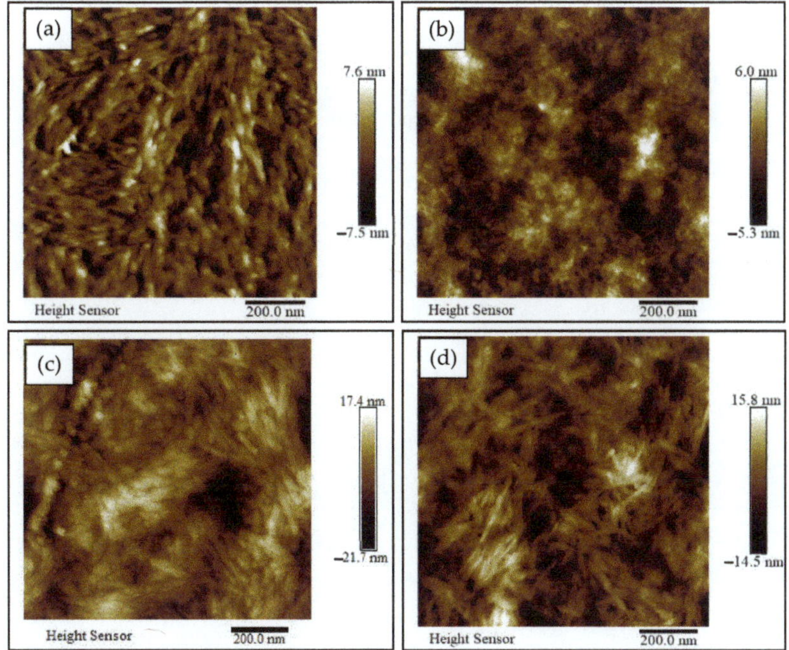

Figure 3. AFM topographic images in 2D of (**a**) NCC, (**b**) PEDOT, (**c**) NCC/PEDOT (before), and (**d**) NCC/PEDOT (after) thin films.

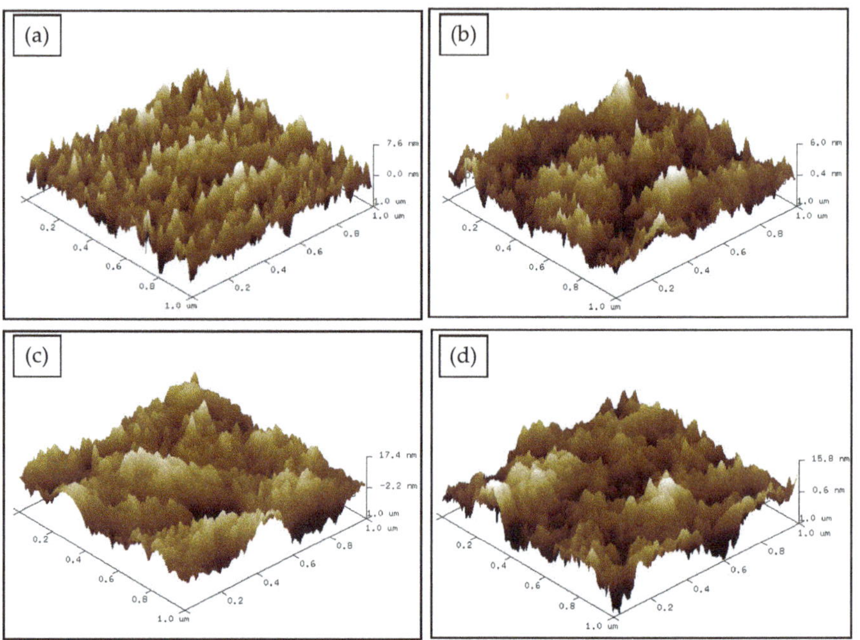

Figure 4. AFM topographic images in 3D of (**a**) NCC, (**b**) PEDOT, (**c**) NCC/PEDOT (before), and (**d**) NCC/PEDOT (after) thin films.

3.3. Optical Properties

To identify the optical properties of composite materials, the absorption spectra of the NCC, PEDOT, and NCC/PEDOT thin film were recorded at different wavelengths, ranging from 220 nm to 700 nm. The UV/vis absorption spectrum of three composite thin films is presented in Figure 5.

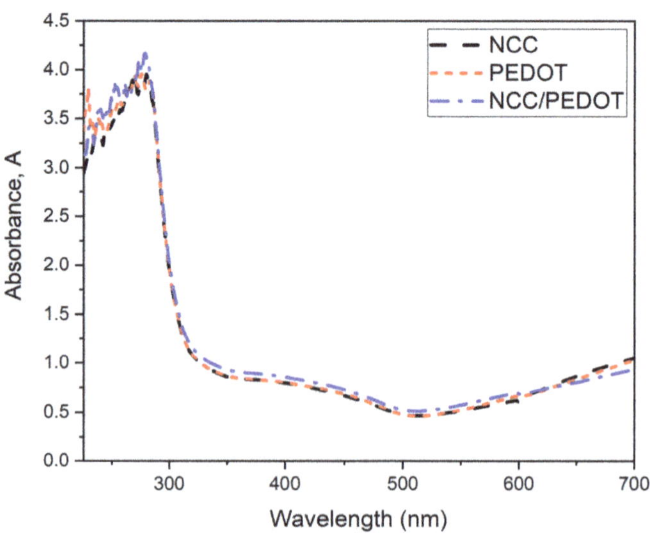

Figure 5. Absorbance spectrum of the NCC, PEDOT, and NCC/PEDOT thin film.

As shown in the figure, there is a distinction in the absorbance value for the absorption spectrum of each thin film. The spectra for NCC show the highest absorbance value of 3.90 at 268 nm and 280 nm, whereas the absorption band is around 260 nm to 280 nm. Then, PEDOT also shows the same value of highest absorbance as NCC at 269 nm, which can be assigned to the substituted phenyl groups of PEDOT [70]. Meanwhile, the NCC/PEDOT thin film exhibits the highest absorbance value among the others, of approximately 4.19 at the wavelength of 279 nm. Meanwhile, from the absorption spectra of UV/vis analysis, the relationship between the absorption coefficient and the energy of a photon can be studied via the Tauc equation, as expressed in the following equation [71]:

$$\alpha = \frac{k\left(h\upsilon - E_g\right)^{\frac{1}{2}}}{h\upsilon} \tag{1}$$

Further, it can be rearranged into

$$\left(\alpha h\upsilon\right)^2 = k\left(h\upsilon - E_g\right), \tag{2}$$

where α is the absorption coefficient, $h\upsilon$ is the energy of a photon, k is a proportionality constant, and E_g is the optical band gap. Then, an intersection from extrapolation of the plot of $(\alpha h\upsilon)^2$ against $h\upsilon$ will determine the optical energy band gap of the NCC, PEDOT, and NCC/PEDOT thin film, as illustrated in Figures 6–8, respectively [72]. Thus, the achieved values of the optical band gap, E_g, were obtained from the extrapolation of the straight plot on the x-axis from Tauc's relation.

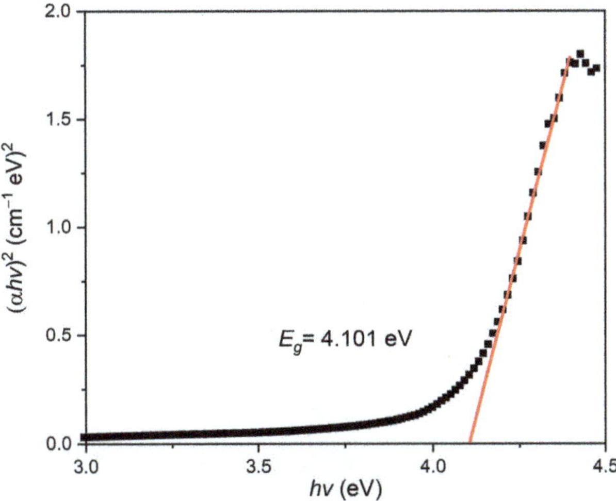

Figure 6. Energy band gap of the NCC thin film.

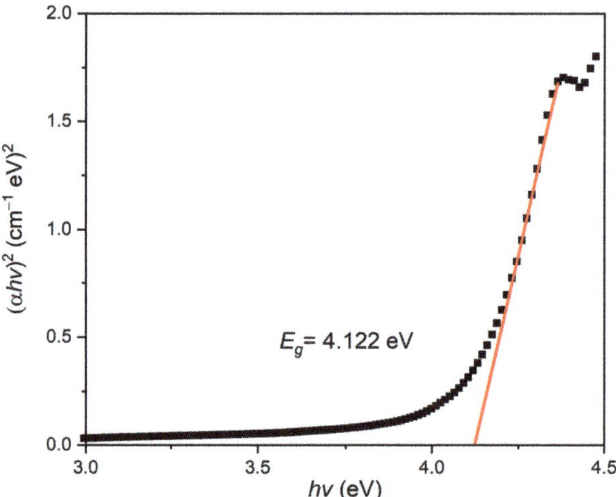

Figure 7. Energy band gap of the PEDOT thin film.

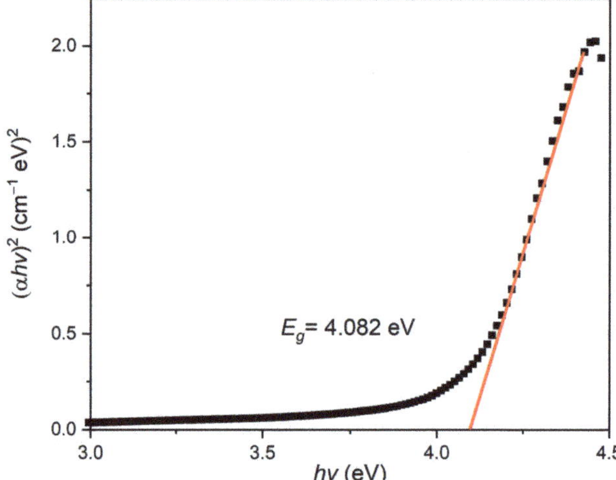

Figure 8. Energy band gap of the NCC/PEDOT thin film.

From Figures 6–8, the energy band gaps were experimentally obtained and plotted using linear fitting techniques. The plot of $(\alpha h v)^2$ versus $h v$ for NCC, PEDOT, and NCC/PEDOT displayed a slight difference in the value of the energy band gap, E_g, whereas PEDOT exhibit the highest optical band gap among all three thin films at 4.122 eV. Next, the energy band gap for the NCC thin film was found to be 4.101 eV, and the mixture of NCC and PEDOT produced a slight decrease in the band gap to 4.082 eV.

3.4. Potential Plasmonic Sensing Properties

The plasmonic sensing of the NCC/PEDOT thin film was investigated and tested with deionized water and various concentrations of analyte (0.001 ppm, 0.01 ppm, and 0.1 ppm of Hg^{2+}) via SPR. When the analyte was injected separately into the cell, the graph of reflectance against the incidence angle for each concentration was recorded and compared with deionized water (0 ppm) as the baseline. This SPR-based plasmonic sensor works by monitoring the shift in the resonance angle, which can be influenced by any changes in the refractive index of the sensing layer attached to the gold thin film [73–75]. Figure 9 reveals a clear trend of an increase in the resonance angle shift of the NCC/PEDOT thin film in contact with 0 ppm to 0.1 ppm of mercury ion. This sensor is very sensitive to any changes near the metal surface. The attachment of metal ions on the layer of the NCC/PEDOT thin film appears to have contributed to the changes in the refractive index as well as the resonance angle [76–79]. The SPR signals were shifted to a higher incidence angle from 54.0099°, 54.3782°, and 54.4086° for 0.001 ppm, 0.01 ppm, and 0.1 ppm respectively. Because there is an interaction between the NCC/PEDOT thin film and the analyte, the existence of more active sites on the surface of the thin film may lead more mercury ions to occupy and access these sites [80].

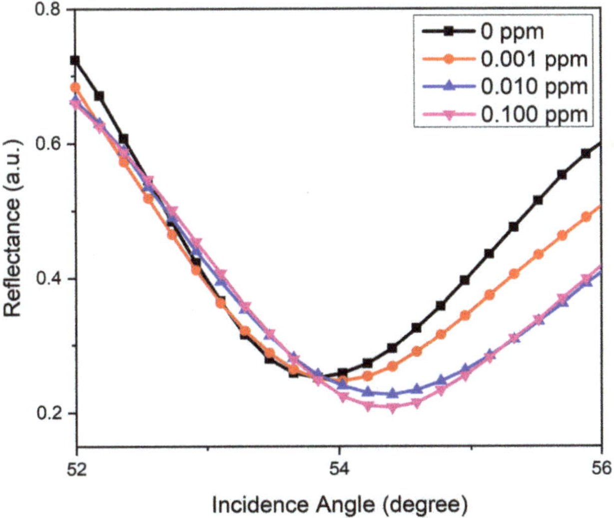

Figure 9. Reflectance as a function of the incidence angle for deionized water and various concentrations of mercury ion solution.

4. Discussions

In the characterization process, the structural and optical properties were observed based on three different thin films, NCC, PEDOT, and NCC/PEDOT. From the FTIR spectra, it is noticed that the functional groups of the NCC/PEDOT thin film have resulted in the incorporation between the biopolymer and conducting polymer solution. The presence of a C–O stretching band of alcohol in the NCC thin film reoccurred in the spectra of NCC/PEDOT. Moreover, the absorption peak of the S=O stretching band and C–O–C bond, marked at 1312.94 and 1200.77 cm^{-1}, clearly indicates the formation of two main functional groups of conducting polymers in the composite solution [81,82]. Thus, overall characteristic peaks exist in the NCC/PEDOT thin film, corresponding to the identical functional groups in both spectra of NCC and PEDOT.

Meanwhile, the NCC/PEDOT thin film was observed before and after contact with the analyte solution. The morphology surface of the NCC/PEDOT thin film prepared by the spin-coating technique showed a significant change after being in contact with mercury ions. This revealed a similar thin rod-like structure as the thin film before being exposed to the analyte solution. Despite that, there are changes in surface roughness where the value decreased from 8.32 nm to 6.88 nm. This indicated that the changes are probably affected by the formation of a pair of shared electrons between the thin film and the mercury solution. Anyway, the surface roughness of the thin film does not have any significant effect on the plasmonic performance, as the rough surfaces less than 20 nm do not influence the excitation of propagating surface plasmon along the metal surface [83].

Based on the absorption spectrum of three thin films, the results pointed out that the NCC/PEDOT thin film has the lowest optical band gap compared with the other two films. The results showed that a slight decrement in the energy band gap may be caused by the highest occupied molecular orbital (HOMO)–lowest unoccupied molecular orbital (LUMO) gap interaction. There is an interaction between two opposite surface charges carried by the NCC and PEDOT/PSS solution; therefore, this will influence the excitation of the electrons to be promoted from the valence band to the conduction band [84]. Hence, the incorporation process of the NCC and PEDOT solution reduces the band gap between the HOMO and LUMO band. The changes in the energy band gap of the thin films can also be attributed to the recrystallization of atoms in the crystal lattice when there is an addition of NCC into the composite material [85]. Then, the conducting properties of PEDOT/PSS are also believed effect to this outcome.

Importantly, the sensing potential of the NCC/PEDOT thin film toward mercury ions was observed by the changes in the resonance angle. As the concentrations increased, the reflectance curves were increased and shifted to the right. The deposited sensing layer on the gold thin film plays an important role in detecting the analyte. When the mercury ions are attached to the sensing layer, the analyte may interact with the NCC/PEDOT thin film owing to the formation of a pair of shared electrons between the positive charge from the mercury ion and the negative charge from the PEDOT. Thus, the NCC/PEDOT thin film showed a response in detecting mercury ions using surface plasmon resonance spectroscopy. Based on the response, the sensitivity of the sensors can also be evaluated by plotting a graph of resonance angle shift against concentration, and the slope of the linear regression plotting will define the sensitivity value [86–89]. In future work, it would be interesting to identify the sensitivity of other heavy metal ions as compared with mercury ions. The selectivity of the sensor towards mercury ion in the presence of various heavy metal ions could also be an interesting future study.

5. Conclusions

In a nutshell, the structural and optical properties of the nanocrystalline cellulose/poly (3,4-ethylenethiophene) (NCC/PEDOT) sensing thin film were synthesized and studied. Furthermore, the thin film was developed as an SPR sensing layer for the detection of mercury ions in an aqueous solution. Analysis from the characterization techniques of the prepared thin films by FTIR, AFM, and UV/vis showed the presence of functional groups in the composite materials, followed by the surface morphology and the rough mean standard of the NCC/PEDOT thin film after being in contact with mercury ions, which confirmed the interaction between the thin films and the analyte. The absorption peaks of UV/vis spectra reveal that the NCC/PEDOT thin film exhibits the lowest energy band gap and the PEDOT thin film has the highest value. As the main objective, the detection of mercury ions using the NCC/PEDOT thin film indicated a positive response according to the reflectance curves of the thin film in contact with various concentrations of mercury ions. Besides, this paper is the first report on the incorporation between two types of polymers, nanocrystalline cellulose and poly(3,4-ethylenethiophene), to detect heavy metal ions using an optical sensor. This study has extended the opportunity of developing biopolymers and conducting polymer-based materials as a sensing layer in sensor applications.

Author Contributions: Conceptualization, N.S.M.R.; methodology, N.A.S.O.; validation, Y.W.F.; formal analysis, N.A.A.A.; investigation, N.S.M.R.; resources, J.Y.C.L.; writing—original draft preparation, N.S.M.R.; writing—review and editing, Y.W.F.; visualization, W.M.E.M.M.D. and N.I.M.F.; supervision, Y.W.F. and J.Y.C.L.; funding acquisition, Y.W.F. All authors have read and agreed to the published version of the manuscript.

Funding: This research was financially supported by the Ministry of Higher Education, Malaysia, grant number FRGS/1/2019/STG02/UPM/02/1 and Putra Grant Universiti Putra Malaysia.

Institutional Review Board statement: Not applicable.

Informed Consent Statement: Not applicable.

Data Availability Statement: Data sharing is not applicable in this article.

Acknowledgments: The authors also acknowledged the laboratory facilities provided by the Department of Physics, Department of Chemistry and Functional Devices Laboratory, Universiti Putra Malaysia.

Conflicts of Interest: The authors declare no conflict of interest.

References

1. Tang, F.A.N.; Zhang, L.; Zhang, Z.; Cheng, Z.; Zhu, X. Cellulose filter paper with antibacterial activity from surface-Initiated ATRP. *J. Macromol. Sci.* **2009**, *46*, 989–996. [CrossRef]
2. Incani, V.; Danumah, C.; Boluk, Y. Nanocomposites of nanocrystalline cellulose for enzyme immobilization. *Cellulose* **2013**, *20*, 191–200. [CrossRef]
3. Edwards, J.V.; Prevost, N.; French, A.; Concha, M.; DeLucca, A.; Wu, Q. Nanocellulose-based biosensors: Design, preparation, and activity of peptide-linked cotton cellulose nanocrystals having fluorimetric and colorimetric elastase detection sensitivity. *Engineering* **2013**, *5*, 20–28. [CrossRef]
4. Rånby, B.G. The colloidal properties of cellulose micelles. *Discuss. Faraday Soc.* **1951**, *11*, 158–164. [CrossRef]
5. Habibi, Y.; Lucia, L.A.; Rojas, O.J. Cellulose nanocrystals: Chemistry, self-assembly, and applications. *Chem. Rev.* **2010**, *110*, 3479–3500. [CrossRef]
6. You, J.; Hu, H.; Zhou, J.; Zhang, L.; Zhang, Y.; Kondo, T. Novel cellulose polyampholyte-gold nanoparticle-based colorimetric competition assay for the detection of cysteine and mercury(II). *Langmuir* **2013**, *29*, 5085–5092. [CrossRef] [PubMed]
7. Manan, F.A.A.; Weng, W.; Abdullah, J.; Yusof, N.A.; Ahmad, I. Nanocrystalline cellulose decorated quantum dots based tyrosinase biosensor for phenol determination. *Mater. Sci. Eng. C* **2019**, *99*, 37–46. [CrossRef]
8. Chen, S.; Tao, H.; Wang, Y.; Ma, Z. Process optimization of soy protein isolate-based edible films containing nanocrystalline cellulose from sunflower seed hull and chitosan. *Trans. Chin. Soc. Agric. Eng.* **2016**, *32*, 306–314.
9. Tan, C.; Peng, J.; Lin, W.; Xing, Y.; Xu, K.; Wu, J.; Chen, M. Role of surface modification and mechanical orientation on property enhancement of cellulose nanocrystals/polymer nanocomposites. *Eur. Polym. J.* **2015**, *62*, 186–197. [CrossRef]
10. Golmohammadi, H.; Morales-Narváez, E.; Naghdi, T.; Merkoçi, A. Nanocellulose in sensing and biosensing. *Chem. Mater.* **2017**, *29*, 5426–5446. [CrossRef]
11. Listyanda, R.F.; Kusmono; Wildan, M.W.; Ilman, M.N. Extraction and characterization of nanocrystalline cellulose (NCC) from ramie fiber by hydrochloric acid hydrolysis. *AIP Conf. Proc.* **2020**, *2217*, 030069.
12. Jia, Y.; Guo, Y.; Wang, S.; Chen, W.; Zhang, J.; Zheng, W.; Jiang, X. Nanocrystalline cellulose mediated seed-growth for ultra-robust colorimetric detection of hydrogen sulfide. *Nanoscale* **2017**, *9*, 9811–9817. [CrossRef]
13. Heidari, H.; Karbalaee, M. Ultrasonic assisted synthesis of nanocrystalline cellulose as support and reducing agent for Ag nanoparticles: Green synthesis and novel effective nanocatalyst for degradation of organic dyes. *Appl. Organomet. Chem.* **2019**, *33*, e5070. [CrossRef]
14. Chen, Q.; Shi, Y.; Chen, G.; Cai, M. Enhanced mechanical and hydrophobic properties of composite cassava starch films with stearic acid modified MCC (microcrystalline cellulose)/NCC (nanocellulose) as strength agent. *Int. J. Biol. Macromol.* **2020**, *142*, 846–854. [CrossRef]
15. Daniyal, W.M.E.M.M.; Fen, Y.W.; Abdullah, J.; Saleviter, S.; Sheh Omar, N.A. Preparation and characterization of hexadecyltrimethylammonium bromide modified nanocrystalline cellulose/graphene oxide composite thin film and its potential in sensing copper ion using surface plasmon resonance technique. *Optik* **2018**, *173*, 71–77. [CrossRef]
16. Omar, N.A.S.; Fen, Y.W.; Ramli, I.; Azmi, U.Z.M.; Hashim, H.S.; Abdullah, J.; Mahdi, M.A. Cellulose and vanadium plasmonic sensor to measure Ni^{2+} ions. *Appl. Sci.* **2021**, *11*, 2963. [CrossRef]
17. Cui, X.; Martin, D.C. Electrochemical deposition and characterization of poly(3,4-ethylenedioxythiophene) on neural microelectrode arrays. *Sens. Actuators B* **2003**, *89*, 92–102. [CrossRef]
18. Yang, Y.; Jiang, Y.; Xu, J.; Yu, J. Conducting PEDOT-PSS composite films assembled by LB technique. *Colloids Surf. A Physicochem. Eng. Asp.* **2007**, *302*, 157–161. [CrossRef]
19. Harman, D.; Gorkin, R.; Stevens, L.; Thompson, B.; Wagner, K. Poly(3,4-ethylenedioxythiophene): Dextran sulfate (PEDOT: DS)—A highly processable conductive organic biopolymer. *Acta Biomater.* **2015**, *14*, 33–42. [CrossRef] [PubMed]
20. Cao, Z.; Chen, Z.; Escoubas, L. Optical, structural, and electrical properties of PEDOT:PSS thin films doped with silver nanoprisms. *Opt. Mater. Express* **2014**, *4*, 2525. [CrossRef]
21. Janmanee, R.; Chuekachang, S.; Sriwichai, S.; Baba, A.; Phanichphant, S. Functional conducting polymers in the application of SPR biosensors. *J. Nanotechnol.* **2012**, *2012*, 1–7. [CrossRef]

22. Zhao, Q.; Jamal, R.; Zhang, L.; Wang, M.; Abdiryim, T. The structure and properties of PEDOT synthesized by template-free solution method. *Nanoscale Res. Lett.* **2014**, *9*, 557. [CrossRef] [PubMed]

23. Pacios, R.; Marcilla, R.; Pozo-gonzalo, C.; Pomposo, J.A.; Grande, H.; Aizpurua, J.; Mecerreyes, D. Combined electrochromic and plasmonic optical responses in conducting polymer/metal nanoparticle films. *J. Nanosci. Nanotechnol.* **2007**, *7*, 2938–2941. [CrossRef]

24. Yemata, T.A.; Zheng, Y.; Kyaw, A.K.K.; Wang, X.; Song, J.; Chin, W.S.; Xu, J. Modulation of the doping level of PEDOT:PSS film by treatment with hydrazine to improve the Seebeck coefficient. *R. Soc. Chem.* **2020**, *10*, 1786–1792. [CrossRef]

25. McFarlane, S.L.; Deore, B.A.; Svenda, N.; Freund, M.S. A one-step, organic-solvent processable synthesis of PEDOT thin films via in situ metastable chemical polymerization. *Macromolecules* **2010**, *43*, 10241–10245. [CrossRef]

26. Horikawa, M.; Fujiki, T.; Shirosaki, T.; Ryu, N.; Sakurai, H.; Nagaoka, S.; Ihara, H. The development of a highly conductive PEDOT system by doping with partially crystalline sulfated cellulose and its electric conductivity. *J. Mater. Chem. C* **2015**, *3*, 8881–8887. [CrossRef]

27. Sui, L.; Zhang, B.; Wang, J.; Cai, A. Polymerization of PEDOT/PSS/Chitosan-coated electrodes for electrochemical bio-sensing. *Coatings* **2017**, *7*, 96. [CrossRef]

28. Gangopadhyay, R.; Das, B.; Molla, M.R. How does PEDOT combine with PSS? Insights from structural studies. *RSC Adv.* **2014**, *4*, 43912–43920. [CrossRef]

29. Najeeb, M.A.; Abdullah, S.M.; Aziz, F.; Ahmad, Z.; Rafique, S.; Wageh, S.; Al-Ghamdi, A.A.; Sulaiman, K.; Touati, F.; Shakoor, R.A.; et al. Structural, morphological and optical properties of PEDOT:PSS/QDs nano-composite films prepared by spin-casting. *Phys. E* **2016**, *83*, 64–68. [CrossRef]

30. Rattan, S.; Singhal, P.; Verma, A.L. Synthesis of PEDOT: PSS (Poly(3,4-ethylenedioxythiophene)/poly(4-styrene sulfonate))/NGPs (nanographitic platelets) nanocomposites as chemiresistive Sensors for detection of nitroaromatics. *Polym. Eng. Sci.* **2013**, *53*, 2045–2052. [CrossRef]

31. Wang, H.; Xu, C.; Yuan, B. Polymer-based electrochemical sensing platform for heavy metal ions detection—A critical review. *Int. J. Electrochem. Sci.* **2019**, *14*, 8760–8771. [CrossRef]

32. Ravit, R.; Abdullah, J.; Ahmad, I.; Sulaiman, Y. Electrochemical performance of poly(3,4-ethylenedioxythipohene)/nanocrystalline cellulose (PEDOT/NCC) film for supercapacitor. *Carbohydr. Polym.* **2019**, *203*, 128–138. [CrossRef] [PubMed]

33. Chauhan, K.; Singh, P.; Singhal, R.K. New Chitosan-Thiomer: An efficient colorimetric sensor and effective sorbent for mercury at ultralow concentration. *ACS Appl. Mater. Interfaces* **2015**, *7*, 26069–26078. [CrossRef]

34. Kumar, P.; Kim, K.; Bansal, V.; Lazarides, T.; Kumar, N. Progress in the sensing techniques for heavy metal ions using nanomaterials. *J. Ind. Eng. Chem.* **2017**, *54*, 30–43. [CrossRef]

35. Palermo, G.; Sreekanth, K.V.; Maccaferri, N.; Lio, G.E.; Nicoletta, G.; De Angelis, F.; Hinczewski, M.; Strangi, G. Hyperbolic dispersion metasurfaces for molecular biosensing. *Nanophotonics* **2020**, *10*, 295–314. [CrossRef]

36. Palermo, G.; Rippa, M.; Conti, Y.; Vestri, A.; Castagna, R.; Fusco, G.; Suffredini, E.; Zhou, J.; Zyss, J.; De Luca, A.; et al. Plasmonic metasurfaces based on pyramidal nanoholes for high-efficiency SERS biosensing. *ACS Appl. Mater. Interfaces* **2021**, *13*, 43715–43725. [CrossRef] [PubMed]

37. Lukosz, W.; Tiefenthaler, K. Embossing technique for fabricating integrated optical components in hard inorganic waveguiding materials. *Opt. Lett.* **1983**, *8*, 537–539. [CrossRef]

38. Cai, S.; Pan, H.; González-Vila, Á.; Guo, T.; Gillan, D.C.; Wattiez, R.; Caucheteur, C. Selective detection of cadmium ions using plasmonic optical fiber gratings functionalized with bacteria. *Opt. Express* **2020**, *28*, 19740–19749. [CrossRef]

39. Verma, R.; Gupta, B.D. Detection of heavy metal ions in contaminated water by surface plasmon resonance based optical fibre sensor using conducting polymer and chitosan. *Food Chem.* **2015**, *166*, 568–575. [CrossRef]

40. Homola, J.; Ctyroky, J.; Slavik, R.; Skalsky, M. Surface plasmon resonance sensors using optical waveguides. In Proceedings of the Third Conference on Photonic Systems for Ecological Monitoring, Prague, Czech Republic, 11 August 1997; pp. 100–106.

41. Eddin, F.B.K.; Fen, Y.W.; Omar, N.A.S.; Liew, J.Y.C.; Daniyal, W.M.E.M.M. Femtomolar detection of dopamine using surface plasmon resonance sensor based on chitosan/graphene quantum dots thin film. *Spectrochim. Acta-Part A Mol. Biomol. Spectrosc.* **2021**, *263*, 120202. [CrossRef]

42. Daniyal, W.M.E.M.M.; Fen, Y.W.; Abdullah, J.; Sadrolhosseini, A.R.; Mahdi, M.A. Design and optimization of surface plasmon resonance spectroscopy for optical constant characterization and potential sensing application: Theoretical and experimental approaches. *Photonics* **2021**, *8*, 361. [CrossRef]

43. Omar, N.A.S.; Fen, Y.W.; Abdullah, J.; Sadrolhosseini, A.R.; Kamil, Y.M.; Fauzi, N.I.M.; Hashim, H.S.; Mahdi, M.A. Quantitative and selective surface plasmon resonance response based on a reduced graphene oxide-polyamidoamine nanocomposite for detection of dengue virus E-proteins. *Nanomaterials* **2020**, *10*, 569. [CrossRef] [PubMed]

44. Omar, N.A.S.; Fen, Y.W.; Abdullah, J.; Mustapha Kamil, Y.; Daniyal, W.M.E.M.M.; Sadrolhosseini, A.R.; Mahdi, M.A. Sensitive detection of dengue virus type 2 E-Proteins signals using self-assembled monolayers/reduced graphene oxide-PAMAM dendrimer thin film-SPR optical sensor. *Sci. Rep.* **2020**, *10*, 2374. [CrossRef] [PubMed]

45. Omar, N.A.S.; Fen, Y.W.; Ramli, I.; Sadrolhosseini, A.R.; Abdullah, J.; Yusof, N.A.; Kamil, Y.M.; Mahdi, M.A. An optical sensor for dengue envelope proteins using polyamidoamine dendrimer biopolymer-based nanocomposite thin film: Enhanced sensitivity, selectivity, and recovery studies. *Polymers* **2021**, *13*, 762. [CrossRef] [PubMed]

46. Rosddi, N.N.M.; Fen, Y.W.; Anas, N.A.A.; Omar, N.A.S.; Ramdzan, N.S.M.; Daniyal, W.M.E.M.M. Cationically modified nanocrystalline cellulose/carboxyl-functionalized graphene quantum dots nanocomposite thin film: Characterization and potential sensing application. *Crystals* **2020**, *10*, 875. [CrossRef]

47. Rosddi, N.N.M.; Fen, Y.W.; Omar, N.A.S.; Anas, N.A.A.; Hashim, H.S.; Ramdzan, N.S.M.; Fauzi, N.I.M.; Anuar, M.F.; Daniyal, W.M.E.M.M. Glucose detection by gold modified carboxyl-functionalized graphene quantum dots-based surface plasmon resonance. *Optik* **2021**, *239*, 166779. [CrossRef]

48. Anas, N.A.A.; Fen, Y.W.; Yusof, N.A.; Omar, N.A.S.; Ramdzan, N.S.M.; Daniyal, W.M.E.M.M. Investigating the properties of cetyltrimethylammonium bromide/hydroxylated graphene quantum dots thin film for potential optical detection of heavy metal ions. *Materials* **2020**, *13*, 2591. [CrossRef]

49. Saleviter, S.; Fen, Y.W.; Omar, N.A.S.; Zainudin, A.A.; Daniyal, W.M.E.M.M. Optical and structural characterization of immobilized 4-(2-pyridylazo) resorcinol in chitosan-graphene oxide composite thin film and its potential for Co_2+ sensing using surface plasmon resonance technique. *Results Phys.* **2018**, *11*, 118–122. [CrossRef]

50. Anas, N.A.A.; Fen, Y.W.; Omar, N.A.S.; Daniyal, W.M.E.M.M.; Ramdzan, N.S.M.; Saleviter, S. Development of graphene quantum dots-based optical sensor for toxic metal ion detection. *Sensors* **2019**, *19*, 3850. [CrossRef]

51. Daniyal, W.M.E.M.M.; Fen, Y.W.; Saleviter, S.; Chanlek, N.; Nakajima, H.; Abdullah, J.; Yusof, N.A. X-ray photoelectron spectroscopy analysis of chitosan-graphene oxide-based composite thin films for potential optical sensing applications. *Polymers* **2021**, *13*, 478. [CrossRef]

52. Saleviter, S.; Fen, Y.W.; Daniyal, W.M.E.M.M.; Abdullah, J.; Sadrolhosseini, A.R.; Omar, N.A.S. Design and analysis of surface plasmon resonance optical sensor for determining cobalt ion based on chitosan-graphene oxide decorated quantum dots-modified gold active layer. *Opt. Express* **2019**, *27*, 32294–32307. [CrossRef]

53. Fauzi, N.I.M.; Fen, Y.W.; Omar, N.A.S.; Saleviter, S.; Daniyal, W.M.E.M.M.; Hashim, H.S.; Nasrullah, M. nanostructured chitosan/maghemite composites thin film for potential optical detection of mercury ion by surface plasmon resonance investigation. *Polymers* **2020**, *12*, 1497. [CrossRef] [PubMed]

54. Anas, N.A.A.; Fen, Y.W.; Yusof, N.A.; Omar, N.A.S.; Daniyal, W.M.E.M.M.; Ramdzan, N.S.M. Highly sensitive surface plasmon resonance optical detection of ferric ion using CTAB/hydroxylated graphene quantum dots thin film. *J. Appl. Phys.* **2020**, *128*, 083105. [CrossRef]

55. Hashim, H.S.; Fen, Y.W.; Omar, N.A.S.; Abdullah, J.; Daniyal, W.M.E.M.M.; Saleviter, S. Detection of phenol by incorporation of gold modified-enzyme based graphene oxide thin film with surface plasmon resonance technique. *Opt. Express* **2020**, *28*, 9738–9752. [CrossRef]

56. Daniyal, W.M.E.M.M.; Fen, Y.W.; Fauzi, N.I.M.; Hashim, H.S.; Ramdzan, N.S.M.; Omar, N.A.S. Recent advances in surface plasmon resonance optical sensors for potential application in environmental monitoring. *Sens. Mater.* **2020**, *32*, 4191–4200.

57. Saleviter, S.; Fen, Y.W.; Sheh Omar, N.A.; Daniyal, W.M.E.M.M.; Abdullah, J.; Mat Zaid, M.H. Structural and optical studies of cadmium sulfide quantum dot-graphene oxide-chitosan nanocomposite thin film as a novel SPR spectroscopy active layer. *J. Nanomater.* **2018**, *2018*, 4324072. [CrossRef]

58. Ramdzan, N.S.M.; Fen, Y.W.; Anas, N.A.A.; Omar, N.A.S.; Saleviter, S. Development of biopolymer and conducting polymer-based optical sensors for heavy metal ion detection. *Molecules* **2020**, *25*, 2548. [CrossRef]

59. Fen, Y.W.; Yunus, W.M.M.; Yusof, N.A. Surface plasmon resonance optical sensor for detection of Pb^{2+} based on immobilized p-tert-butylcalix [4] arene-tetrakis in chitosan thin film as an active layer. *Sens. Actuators B Chem.* **2012**, *171*, 287–293. [CrossRef]

60. Sadrolhosseini, A.R.; Naseri, M.; Kamari, H.M. Surface plasmon resonance sensor for detecting of arsenic in aqueous solution using polypyrrole-chitosan-cobalt ferrite nanoparticles composite layer. *Opt. Commun.* **2017**, *383*, 132–137. [CrossRef]

61. Fen, Y.W.; Yunus, W.M.M. Utilization of Chitosan-based sensor thin films for the detection of lead ion by surface plasmon resonance optical sensor. *IEEE Sens. J.* **2013**, *13*, 1413–1418. [CrossRef]

62. Fen, Y.W.; Yunus, W.M.M.; Moksin, M.M.; Talib, Z.A.; Yusof, N.A. Surface plasmon resonance optical sensor for mercury ion detection by crosslinked chitosan thin film. *J. Optoelectron. Adv. Mater.* **2011**, *13*, 279–285.

63. Huq, T.; Salmieri, S.; Khan, A.; Khan, R.A.; Le, C.; Riedl, B.; Fraschini, C.; Bouchard, J.; Uribe-calderon, J.; Kamal, M.R.; et al. Nanocrystalline cellulose (NCC) reinforced alginate based biodegradable nanocomposite film. *Carbohydr. Polym.* **2012**, *90*, 1757–1763. [CrossRef]

64. Tehrani, A.D.; Basiryan, A. Dendronization of cellulose nanowhisker with cationic hyperbranched dendritic polyamidoamine. *Carbohydr. Polym.* **2015**, *120*, 46–52. [CrossRef]

65. Azrina, Z.A.Z.; Beg, M.D.H.; Rosli, M.Y.; Ramli, R.; Alam, A.K.M.M. Modification of nanocrystalline cellulose (NCC) by hyperbranched polymer. *Indian J. Sci. Technol.* **2017**, *10*, 1–5. [CrossRef]

66. Selvaganesh, S.V.; Mathiyarasu, J.; Phani, K.L.N.; Yegnaraman, V. Chemical synthesis of PEDOT—Au nanocomposite. *Nanoscale Res. Lett.* **2007**, *2*, 546–549. [CrossRef]

67. Gaspar, D.; Fernandes, S.; de Oliveira, A.; Fernandes, J.; Grey, P.; Pontes, R.; Pereira, L.; Martins, R.; Godinho, M.; Fortunato, E. Nanocrystalline cellulose applied simultaneously as the gate dielectric and the substrate inflexible field effect transistors. *Nanotechnology* **2014**, *25*, 094008. [CrossRef]

68. Abidin, S.N.J.S.Z.; Azman, N.H.N.; Kulandaivalu, S.; Sulaiman, Y. Poly(3,4-ethylenedioxythiophene) Doped with Carbon Materials for High-Performance Supercapacitor: A Comparison Study. *J. Nanomater.* **2017**, *2017*, 13.

69. Elazzouzi-hafraoui, S.; Nishiyama, Y.; Putaux, J.; Heux, L.; Dubreuil, F.; Rochas, C. The shape and size distribution of crystalline nanoparticles prepared by acid hydrolysis of native cellulose. *Biomacromolecules* **2008**, *9*, 57–65. [CrossRef] [PubMed]

70. Sakunpongpitiporn, P.; Phasuksom, K.; Paradee, N.; Sirivat, A. Facile synthesis of highly conductive PEDOT: PSS via surfactant templates. *R. Soc. Chem.* **2019**, *9*, 6363–6378.

71. Roshidi, M.D.A.; Fen, Y.W.; Daniyal, W.M.E.M.M.; Omar, N.A.S.; Zulholinda, M. Structural and optical properties of chitosan–poly(amidoamine) dendrimer composite thin film for potential sensing Pb^{2+} using an optical spectroscopy. *Optik* **2019**, *185*, 351–358. [CrossRef]

72. Ramdzan, N.S.M.; Fen, Y.W.; Omar, N.A.S.; Anas, N.A.A.; Daniyal, W.M.E.M.M.; Saleviter, S.; Zainudin, A.A. Optical and surface plasmon resonance sensing properties for chitosan/carboxyl-functionalized graphene quantum dots thin film. *Optik* **2019**, *178*, 802–812. [CrossRef]

73. Fen, Y.W.; Yunus, W.M.M.; Talib, Z.A.; Yusof, N.A. Development of surface plasmon resonance sensor for determining zinc ion using novel active nanolayers as probe. *Spectrochim. Acta-Part A Mol. Biomol. Spectrosc.* **2015**, *134*, 48–52. [CrossRef] [PubMed]

74. Zainudin, A.A.; Fen, Y.W.; Yusof, N.A.; Al-Rekabi, S.H.; Mahdi, M.A.; Omar, N.A.S. Incorporation of surface plasmon resonance with novel valinomycin doped chitosan-graphene oxide thin film for sensing potassium ion. *Spectrochim. Acta-Part A Mol. Biomol. Spectrosc.* **2018**, *191*, 111–115. [CrossRef] [PubMed]

75. Omar, N.A.S.; Fen, Y.W.; Saleviter, S.; Daniyal, W.M.E.M.M.; Anas, N.A.A.; Ramdzan, N.S.M.; Roshidi, M.D.A. Development of a graphene-based surface plasmon resonance optical sensor chip for potential biomedical application. *Materials* **2019**, *12*, 1928. [CrossRef]

76. Fen, Y.W.; Yunus, W.M.M.; Yusof, N.A. Detection of mercury and copper ions using surface plasmon resonance optical sensor. *Sens. Mater.* **2011**, *23*, 325–334.

77. Fen, Y.W.; Yunus, W.M.M. Surface plasmon resonance spectroscopy as an alternative for sensing heavy metal ions: A review. *Sens. Rev.* **2013**, *33*, 305–314.

78. Daniyal, W.M.E.M.M.; Fen, Y.W.; Abdullah, J.; Sadrolhosseini, A.R.; Saleviter, S.; Omar, N.A.S. Exploration of surface plasmon resonance for sensing copper ion based on nanocrystalline cellulose-modified thin film. *Opt. Express* **2018**, *26*, 34880. [CrossRef]

79. Fen, Y.W.; Yunus, W.M.M.; Yusof, N.A.; Ishak, N.S.; Omar, N.A.S.; Zainudin, A.A. Preparation, characterization and optical properties of ionophore doped chitosan biopolymer thin film and its potential application for sensing metal ion. *Optik* **2015**, *126*, 4688–4692. [CrossRef]

80. Babakhani, B.; Ivey, D.G. Improved capacitive behavior of electrochemically synthesized Mn oxide/PEDOT electrodes utilized as electrochemical capacitors. *Electrochim. Acta* **2010**, *55*, 4014–4024. [CrossRef]

81. Chanthaanont, P.; Sirivat, A. Effect of transition metal ion-exchanged into the zeolite Y on electrical conductivity and response of PEDOT-PSS/MY composites toward SO_2. *Adv. Polym. Technol.* **2013**, *32*, 21367. [CrossRef]

82. Chen, W.C.; Liu, C.L.; Yen, C.T.; Tsai, F.C.; Tonzola, C.J.; Olson, N.; Jenekhe, S.A. Theoretical and experimental characterization of small band gap poly(3,4-ethylenedioxythiophene methine)s. *Macromolecules* **2004**, *37*, 5959–5964. [CrossRef]

83. Byun, K.M.; Yoon, S.J.; Kim, D.; Kim, S.J. Sensitivity analysis of a nanowire-based surface plasmon resonance biosensor in the presence of surface roughness. *J. Opt. Soc. Am. A* **2007**, *24*, 522–529. [CrossRef] [PubMed]

84. Ashery, A.; Said, G.; Arafa, W.A.; Gaballah, A.E.H.; Farag, A.A.M. Structural and optical characteristics of PEDOT/n-Si heterojunction diode. *Synth. Met.* **2016**, *214*, 92–99. [CrossRef]

85. Oluyamo, S.S.; Akinboyewa, L.O.; Fuwape, I.A.; Olusola, O.I.; Adekoya, M.A. Influence of nanocellulose concentration on the tunability of energy bandgap of cadmium telluride thin films. *Cellulose* **2020**, *27*, 8147–8153. [CrossRef]

86. Ramdzan, N.S.M.; Fen, Y.W.; Omar, N.A.S.; Anas, N.A.A.; Liew, J.Y.C.; Daniyal, W.M.E.M.M.; Hashim, H.S. Detection of mercury ion using surface plasmon resonance spectroscopy based on nanocrystalline cellulose/poly (3,4-ethylenedioxythiophene) thin film. *Measurement* **2021**, *182*, 109728. [CrossRef]

87. Shalabney, A.; Abdulhalim, I. Sensitivity-enhancement methods for surface plasmon sensors. *Laser Photonics Rev.* **2011**, *5*, 571–606. [CrossRef]

88. Anas, N.A.A.; Fen, Y.W.; Omar, N.A.S.; Ramdzan, N.S.M.; Daniyal, W.M.E.M.M.; Saleviter, S.; Zainudin, A.A. Optical properties of chitosan/hydroxyl-functionalized graphene quantum dots thin film for potential optical detection of ferric (III) ion. *Opt. Laser Technol.* **2019**, *120*, 105724. [CrossRef]

89. Daniyal, W.M.E.M.M.; Fen, Y.W.; Abdullah, J.; Sadrolhosseini, A.R.; Saleviter, S.; Omar, N.A.S. Label-free optical spectroscopy for characterizing binding properties of highly sensitive nanocrystalline cellulose-graphene oxide based nanocomposite towards nickel ion. *Spectrochim. Acta-Part A Mol. Biomol. Spectrosc.* **2019**, *212*, 25–31. [CrossRef]

Article

Performance of Surface Plasmon Resonance Sensors Using Copper/Copper Oxide Films: Influence of Thicknesses and Optical Properties

Dominique Barchiesi [1,*,†], Tasnim Gharbi [1,2,†], Deniz Cakir [3], Eric Anglaret [3], Nicole Fréty [4], Sameh Kessentini [5] and Ramzi Maâlej [2]

[1] Research Unit on Automatic Mesh Generation and Advanced Methods (GAMMA3), University of Technology of Troyes, 12 rue Marie Curie, CS 42060, CEDEX, F-10004 Troyes, France; tasnim.gharbi@utt.fr
[2] Georesources Materials Environment and Global Changes Laboratory (GEOGLOB), Faculty of Sciences of Sfax, University of Sfax, Sfax 3018, Tunisia; ramzi.maalej@fss.usf.tn
[3] Laboratoire Charles Coulomb, University Montpellier, Bâtiment 2—CC069 Place Eugène Bataillon, 34095 Montpellier, France; deniz.cakir@mersen.com (D.C.); eric.anglaret@umontpellier.fr (E.A.)
[4] Institut Charles Gerhardt Montpellier ICGM, University Montpellier, CNRS, Bât 17 CC1700, Place Eugène Bataillon, 34095 Montpellier, France; nicole.frety@umontpellier.fr
[5] Laboratory of Probability and Statistics, Faculty of Sciences of Sfax, University of Sfax, Sfax 3018, Tunisia; sameh.kessentini@fss.usf.tn
* Correspondence: dominique.barchiesi@utt.fr; Tel.: +33-325-715-826
† These authors contributed equally to this work.

Abstract: Surface plasmon resonance sensors (SPR) using copper for sensitive parts are a competitive alternative to gold and silver. Copper oxide is a semiconductor and has a non-toxic nature. The unavoidable presence of copper oxide may be of interest as it is non-toxic, but it modifies the condition of resonance and the performance of the sensor. Therefore, the characterization of the optical properties of copper and copper oxide thin films is of interest. We propose a method to recover both the thicknesses and optical properties of copper and copper oxide from absorbance curves over the $(0.9; 3.5)$ eV range, and we use these results to numerically investigate the surface plasmon resonance of copper/copper oxide thin films. Samples of initial copper thicknesses 10, 30 and 50 nm, after nine successive oxidations, are systematically studied to simulate the signal of a Surface Plasmon Resonance setup. The results obtained from the resolution of the inverse problem of absorbance are used to discuss the performance of a copper-oxide sensor and, therefore, to evaluate the optimal thicknesses.

Keywords: surface plasmon resonance; inverse problem; copper; copper oxide

Citation: Barchiesi, D.; Gharbi, T.; Cakir, D.; Anglaret, E.; Fréty, N.; Kessentini, S.; Maâlej, R. Performance of Surface Plasmon Resonance Sensors Using Copper/Copper Oxide Films: Influence of Thicknesses and Optical Properties. *Photonics* **2022**, *9*, 104. https://doi.org/10.3390/photonics 9020104

Received: 30 October 2021
Accepted: 2 February 2022
Published: 11 February 2022

Publisher's Note: MDPI stays neutral with regard to jurisdictional claims in published maps and institutional affiliations.

1. Introduction

Surface plasmon resonance sensors are widely used for their high-sensitivity in real-time detection. Many improvements of the basic SPR [1], using a single gold layer as the active part of the sensor, have been proposed to improve the sensor performance [2]. For example, adding a specific absentee layer of KCL or Si_3N_4 or an absentee porous silica film can improve the performance of the SPR sensor by 5%, 11% and more than 300% [3,4]. The performance of the SPR sensor can be evaluated by the depth of the resonance dip, the full width at half maximum (FWHM) of the dip, the sensitivity to changes of the refractive index of the medium of detection and the figure of merit (FOM). The figure of merit is the ratio of the sensitivity to the FWHM. Indeed, a smaller FWHM gives a higher signal-to-noise ratio and higher measurement accuracy [3].

Copper presents interesting properties for optics, nanotechnology [5–7] and, therefore, for SPR sensors [8–10]. A recent paper demonstrated that the performance of a Cu-SPR sensor is quite similar to Au-SPR ones [9]. Moreover, bovine serum albumin (BSA) and other proteins do not bind irreversibly onto thin Cu-films. Copper phthalocyanine Langmuir–Blodgett films have been characterized using surface plasmon resonance (SPR) [11]. Exposure to toluene resulted in a partially reversible shift in the resonance depth and position of the SPR curves [12]. Moreover, the high cost of silver and gold materials—widely used in nanotechnologies—makes copper an adequate alternative to them.

Copper layers are naturally oxidized and their oxidation can be controlled by annealing. Copper oxides have non-toxic nature. The performance of SPR sensors depends not only on the metal layer used but also on dielectric layers above. Therefore, the study of Cu-oxide-SPR sensor performance requires finding out the thickness and optical properties of both materials, optical properties of copper and copper oxide thin films being different from those of bulk materials [13]. Consequently, to investigate Cu-oxide-SPR sensors, we first determine simultaneously the thicknesses and optical properties of 30 bilayers, from the fitting of absorbance curves. We detail the characteristics of the samples (Appendices A and B), the methods used for fitting (Appendix C), and the results (Appendix C.4) in the Appendix. Then, we use the recovered data to analyze numerically and discuss the Cu-oxide-SPR sensor performance. We propose the same electromagnetic approach to simulate the SPR signal and the metric of SPR performance (Section 2.2). The calculated SPR signals and performances are given in Section 3 and discussed in Section 4. Two cases are considered: the dry case (the upper medium is air), in which the SPR is used to analyze molecules in air, and the wet case (the medium of detection is water), in which the SPR characterize analytes in solutions. The optimal thicknesses are deduced from the performance parameters.

2. Materials and Methods

First, we summarize the method used to simultaneously determine the thicknesses and optical properties from experimental absorbance curves. This is an opportunity to detail the electromagnetic model of the generalized Fresnel coefficients also used for modeling the SPR sensor. We also give insights into the performance of SPR sensors.

2.1. Determination of Thicknesses and Optical Properties

The investigated samples are thin copper layers of target deposition thicknesses 10, 30 and 50 nm. They are successively annealed to control their oxidation under specific temperatures and for a given time. Thirty samples are characterized through the measurement of absorbance. Appendix A provides details on the sample preparation and experimental characterization [14]. The methods we propose for absorbance fitting and results are detailed in Appendices C.1–C.4 and summarized in the following.

In Reference [15], we targeted the best fitting of UV-visible-NIR absorbance curves by a multilayer electromagnetic model using a function describing the optical properties over the range of investigated photon energies ($(0.9; 3.5)$ eV, i.e., $(350; 138)$ nm). Our methodology relied on using a single metaheuristic optimization method and the Drude–Lorentz model for the optical properties of copper. Consequently, the inverse problem was solved, revealing the best parameters of the electromagnetic and materials models. In this paper, we improve the method proposed in [15] to recover the thicknesses and optical properties of thin layers of copper/copper oxide from absorbance measurements as follows.

1. In the model of optical properties, we replace the Drude–Lorentz two-terms model of dispersion by a sum of N_{PF} partial fraction functions for fitting copper and copper oxide properties as a function of the photon energy $\hbar\omega$ ($N_{PF} = 4$ and $N_{PF} = 2$, respectively). N_{PF} is the order of the partial fractions model, also called the complex-conjugate pole-residue pair model [16]. N_{PF} is actually the number of poles in the function (see Equation (5)).
2. We replace the two-steps method using first the bulk properties by a single multi-objective method, involving the error of fit and the relative difference between the optical properties we found and the bulk material properties. The purpose is to decrease the computational time.

3. To achieve the fitting, we use three metaheuristic methods: Particle Swarm Method (PSO), Genetic Algorithm (GA), and Artificial Bee Colony method (ABC), with Gradient (GR) or Nelder–Mead simplex (NM) hybridization. Actually, metaheuristics are more or less efficient for achieving optimization, depending on the dimension of the problem (the number of input parameters of the model that have to be found) and on the topology of the objective function: the error of fit. Moreover, many realizations of the metaheuristics must be analyzed carefully to numerically characterize the stability of the solution and, hopefully, its uniqueness.

The calculation of the absorbance A is based on the electromagnetic model of the normal transmission of light across two plane material layers (with refractive index and thicknesses denoted n_{Cu} and h_{Cu} for copper, n_{ox}, h_{ox} for oxide). The copper layer is deposited on a dielectric glass substrate of refractive index n_g (medium 1). The absorbance A is deduced from the transmitted intensity across the successive material layers according to:

$$A = -\log\left(|t_{14}|^2 T_{AG}\right), \tag{1}$$

where:

$$T_{AG} = \left|\frac{2n_g}{n_g+1}\right|^2, \tag{2}$$

and

$$t_{14} = (8n_{Cu}n_{ox})\exp(ik_0(h_{Cu}(n_{Cu}-n_g)+h_{ox}(n_{ox}-n_g)))/D_0. \tag{3}$$

i is the pure imaginary number, and k_0 is the magnitude of the illumination wave vector. The denominator D_0 is:

$$
\begin{aligned}
D_0 \;=\; & (n_{Cu}+n_g)(n_{ox}+n_{Cu})(1+n_{ox}) \\
& +(n_{Cu}-n_g)(n_{ox}-n_{Cu})(1+n_{ox})\exp(2ik_0h_{Cu}n_{Cu}) \\
& +(n_{Cu}-n_g)(n_{ox}+n_{Cu})(1-n_{ox})\exp(2ik_0(h_{Cu}n_{Cu}+h_{ox}n_{ox})) \\
& +(n_{Cu}+n_g)(n_{ox}-n_{Cu})(1-n_{ox})\exp(2ik_0h_{ox}n_{ox}).
\end{aligned}
\tag{4}
$$

This electromagnetic model of generalized Fresnel coefficients includes the transmission of light across the Air-Quartz substrate T_{AG}. The optical properties n_g of the glass substrate are measured before its coating with copper. The optical properties of copper and copper oxide are modeled as follows.

We use the partial fraction model, also called the complex-conjugate pole-residue pair model [16], to describe the dispersion of materials [17,18]. In a recent paper, the partial fraction model of dispersion yielded more accurate results than a sum of Drude–Lorentz functions to fit the bulk copper [18]. The relative permittivity is a complex number that is the square of the refractive index n:

$$n^2 = n_\infty^2 + \sum_{m=1}^{N_{PF}} \frac{c_m}{i\omega - p_m} + \frac{c_m^*}{i\omega - p_m^*}, \tag{5}$$

where X^* denotes the complex conjugate of Xy, N_{PF} is the model order (number of poles), ω is angular frequency of the incoming light (the photon energy is $E = \hbar\omega$). n_∞^2 is the value of the limit of the permittivity when ω tends toward infinity and is actually the contribution of interband transitions. Therefore, it is spectrally located beyond the investigated energy range. p_i are poles and c_i are residues. The occurrence of poles in permittivity corresponds to light–matter damped resonances in the complex plane [19,20]. They can also be related to electronic transitions in materials. This partial fraction model is used for both copper and copper oxide. In this study, we use model order $N_{PF} = 2$ for copper oxide (semiconductor) and $N_{PF} = 4$ for copper (metal) to maintain sufficient accuracy on the optical properties.

Therefore, the number of unknowns of the problem is $D = 28$ by splitting complex numbers into two real numbers: two thicknesses and $1 + 4N_{PF}$ parameters for each PF dispersion model ($N_{PF} = 4$ for copper and $N_{PF} = 2$ for copper oxide). The model of absorbance (Equation (1)), including the models of optical properties (Equation (5)), is used within an evolutionary loop to find the best set of parameters for the model of absorbance. For the fitting of experimental UV-visible-NIR absorbance curves, a multi-objective function should be minimized to find the best input parameters of the fitting model.

The recovered parameters of the partial fraction model of dispersion (Equation (5)) are used to calculate the optical properties of copper and oxide at the wavelength of the laser incident light source (632.8 nm) used by the SPR sensor.

2.2. The SPR Model

The SPR setup configuration (Figure 1) was modeled by the electromagnetic interaction of light with a plane multilayer [21,22]. The same approach has been used to study the influence of the functionalization layer [21] and adhesion layer for nanostructured Au-EPR [23,24]. The light source is considered as a monochromatic plane wave of wavelength $\lambda_0 = 632.8$ nm (photon energy $E = \hbar\omega = 1.959$ eV). For simplicity, we choose the same refractive index for the SPR prism as that used for the glass supporting the copper/copper oxide sample.

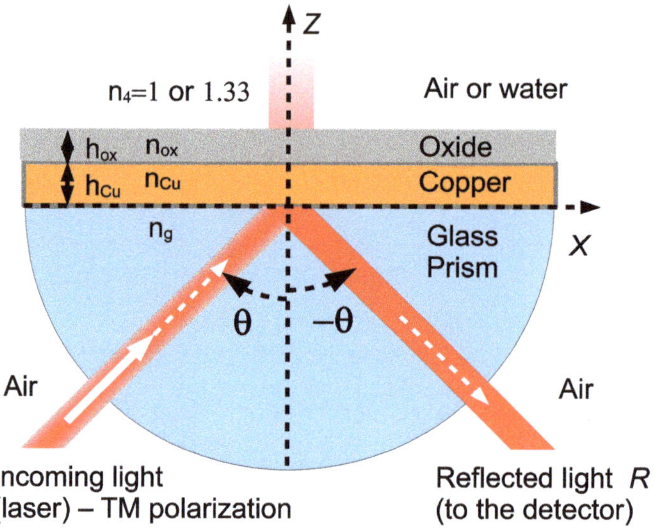

Figure 1. Schematic of the prism-based SPR sensor with the Kretschmann configuration. The plane material layers are deposited on the hemispherical lens.

The SPR detected signal R is a function of the incident angle of the plane wave from normal to the multilayer surface:

$$R(\theta) = |r_{14}|^2 T_{AG} T_{GA},\qquad(6)$$

where T_{AG} and T_{GA} are the transmitted intensities of the light incoming and outcoming from the SPR setup (see Equation (2)). The reflected amplitude is r_{14}:

$$
\begin{aligned}
r_{14} = \ & ((n_{Cu}^2 k_g^\perp - n_g^2 k_{Cu}^\perp)(n_{ox}^2 k_{Cu}^\perp + n_{Cu}^2 k_{ox}^\perp)(n_4^2 k_{ox}^\perp + n_{ox}^2 k_A^\perp) + \\
& (n_{ox}^2 k_{Cu}^\perp - n_{Cu}^2 k_{ox}^\perp)(n_{Cu}^2 k_g^\perp + n_g^2 k_{Cu}^\perp)(n_4^2 k_{ox}^\perp + n_{ox}^2 k_A^\perp) \\
& \exp(2ih_{Cu}k_{Cu}^\perp) + \\
& (n_4^2 k_{ox}^\perp - n_{ox}^2 k_A^\perp)(n_{Cu}^2 k_g^\perp + n_g^2 k_{Cu}^\perp)(n_{ox}^2 k_{Cu}^\perp + n_{Cu}^2 k_{ox}^\perp) \\
& \exp(2i(h_{ox}k_{ox}^\perp + h_{Cu}k_{Cu}^\perp)) + \\
& (n_{Cu}^2 k_g^\perp - n_g^2 k_{Cu}^\perp)(n_{ox}^2 k_{Cu}^\perp - n_{Cu}^2 k_{ox}^\perp)(n_4^2 k_{ox}^\perp - n_{ox}^2 k_A^\perp) \\
& \exp(2ih_{ox}k_{ox}^\perp))/D_\theta.
\end{aligned}
\tag{7}
$$

$k^\perp$ being the normal to the surface of multilayer component of the wave vector. In medium m, this component is $k_i^\perp = k_0(n_m^2 - n_g^2 \sin^2(\theta))^{1/2}$, with $k_0 = \frac{\omega}{c}$, c being the speed of light in vacuum. The refractive index of the detection medium of the SPR sensor is n_4. The denominator D_θ of the transmitted amplitude is:

$$
\begin{aligned}
D_\theta = \ & (n_{Cu}^2 k_g^\perp + n_g^2 k_{Cu}^\perp)(n_{ox}^2 k_{Cu}^\perp + n_{Cu}^2 k_{ox}^\perp)(n_4^2 k_{ox}^\perp + n_{ox}^2 k_A^\perp) + \\
& (n_{Cu}^2 k_g^\perp - n_g^2 k_{Cu}^\perp)(n_{ox}^2 k_{Cu}^\perp - n_{Cu}^2 k_{ox}^\perp)(n_4^2 k_{ox}^\perp + n_{ox}^2 k_A^\perp) \\
& \exp(2ih_{Cu}k_{Cu}^\perp) + \\
& (n_{Cu}^2 k_g^\perp - n_g^2 k_{Cu}^\perp)(n_4^2 k_{ox}^\perp - n_{ox}^2 k_A^\perp)(n_{ox}^2 k_{Cu}^\perp + n_{Cu}^2 k_{ox}^\perp) \\
& \exp(2i(h_{Cu}k_{Cu}^\perp + h_{ox}k_{ox}^\perp)) + \\
& (n_{ox}^2 k_{Cu}^\perp - n_{Cu}^2 k_{ox}^\perp)(n_4^2 k_{ox}^\perp - n_{ox}^2 k_A^\perp)(n_{Cu}^2 k_g^\perp + n_g^2 k_{Cu}^\perp) \\
& \exp(2ih_{ox}k_{ox}^\perp).
\end{aligned}
\tag{8}
$$

We obtain $D_\theta = D_0$ (Equation (4)) by considering normal incidence of the light ($\theta = 0$) and $n_4 = 1$. However, Equation (4) is used for the fitting of experimental UV-visible-NIR absorbance, as it requires less computational time.

The basics of SPR are given in Reference [25]. In that reference, we gave the conditions of SPR excitation on an interface between two mediums and the corresponding formula, also used in [9]. Considering a plane interface between two media (one of them is a metal), the SPR angle can be easily evaluated, n_i^2 and n_j^2 being the relative permittivities of materials on both sides of the interface, and n_g^2 being that of the hemispherical glass substrate, and $\Re$ the real part:

$$
\theta(\min(R)) = \Re\left\{\arcsin\left(\frac{n_i n_j}{(n_i^2 + n_j^2)^{1/2}}\frac{1}{n_g}\right)\right\}
\tag{9}
$$

We consider the angle interrogation mode of the SPR sensor [2,9,12]. At a given wavelength, in the Angular Interrogation Mode (AIM), with a changing refractive index of water (upper medium, analyte), the resonance angle shifts [4].

2.3. Performance of the SPR Sensor

The performance parameters of the SPR sensor are the sensitivity S_θ, the full width at half maximum ($FWHM$) and the Figure of Merit (FOM) [4]. The resonance angle found at the minimum of reflectance ($\theta(\min(R))$, and the depth of the resonance dip ($\min(R)$) are also of interest to characterize SPR sensors. The performance of SPR can help to determine the optimal thicknesses of the bilayer.

2.3.1. Sensitivity

The sensitivity of the SPR sensor is the angular shift that is found at the minimum of reflectance $R(\theta)$ by varying the refractive index of the medium of detection. The angular sensitivity is, therefore:

$$
S_\theta = \frac{\theta(\min(R(n_4'))) - \theta(\min(R(n_4)))}{RIU},
\tag{10}
$$

where RIU is the relative index of refraction unit: $2(n_4' - n_4)/(n_4' + n_4)$. The greater the sensitivity is, the better performance of SPR.

2.3.2. Full Width at Half Maximum

The full width at half maximum is:

$$FWHM = \max\left[\theta\left(\frac{\max(R(\theta)) + \min(R(\theta))}{2}\right)\right] - \min\left[\theta\left(\frac{\max(R(\theta)) + \min(R(\theta))}{2}\right)\right]. \tag{11}$$

The thinner the resonance dip is, the higher is the signal to noise ratio of the SPR sensor [3].

2.3.3. Figure of Merit

Therefore, the figure of merit of the SPR sensor is deduced from both the sensitivity and the full width at half maximum of the reflectance (FWHM) [4]:

$$\text{FOM} = \frac{S_\theta}{\text{FWHM}}. \tag{12}$$

The thinner the wells in the reflectance (small value of FWHM), the more enhanced the FOM. The greater the sensitivity deduced from the shift of the minimum in reflectance by a change of optical index of the upper medium, the greater the FOM. However, the FOM does not take into account the value of the minimum in reflectance $\min(R)$. This minimum should be as close as possible to zero to obtain high dynamic of detection.

2.4. Optimal Thicknesses

Interference conditions in the metal layer have been considered in Reference [26] to find the optimal thickness of a classical SPR sensor. This condition relies on the calculation of the copper thickness h_o that verifies constructive interference at the copper–oxide interface, after two reflections on the interfaces [27]:

$$2k_{Cu}^{\perp}h_o = \Re\left(m2\pi - \arg\left(\frac{1}{r_{g-Cu}r_{Cu-ox}}\right)\right), \tag{13}$$

with m an integer number and r_{g-Cu} and r_{Cu-ox} the Fresnel coefficients of reflection on each interface. The real part of h_o is actually an approximation of the optimal thickness, by neglecting the finite thickness of oxide. However, this criterion can be extended by calculating the argument of the sum of the illumination field and of the reflected one $\arg(1 + r_{14})$ (Equation (7)) that should be as close as possible to 0. In this case, we obtain a minimum of reflectance (destructive interference between the illumination and the reflected wave). This minimum corresponds to a pole (complex number) of the generalized Fresnel coefficient (Equation (7)) [19,28].

The maximum of the FOM (Equation (12)), the maximum of sensitivity (Equation (10)), the minimum of FWHM (Equation (11)), or even the minimum of reflectance $\min(R)$ can also reveal the best thicknesses of copper and copper-oxide. We discuss the results obtained for all these criterion in Section 4.

2.5. Investigated Samples

The manufacturing and characterization of the investigated samples are detailed in Appendices A and B. The fitting methods are described in Appendix C, and the full results are given in Appendix C.4 (thicknesses of the copper and oxide layers, optical properties @632.8 nm, 1.96 eV) for each of the thirty investigated samples. We summarize the main results shown in Tables A3–A5:

- Copper and copper oxide thicknesses are globally in agreement with those obtained from experimental measurements (see Appendix B).
- The real part of the relative permittivity of copper is smaller than the bulk one for almost all samples. The imaginary part of the relative permittivity of copper is about

twice the bulk one. The mean value of n^2_{Cu} over the samples with $h_{Cu} > 2$ nm is $-13.3 + 3.3i$ (standard deviation $1.4 + 1.5i$) compared to $n^2_{Cu}(bulk) = -11.6 + 1.6i$ [29].

- The real part and the imaginary parts of the relative permittivity of oxide are close to that of the bulk, except for full oxidized samples for which both parts decrease. This behavior is probably due to air inclusion in oxide (the grain size of oxide can reach 80 nm, see Appendix B). The mean value of n^2_{ox} over the samples with $h_{ox} > 2$ nm is $8.2 + 1.0i$ (standard deviation $0.4 + 0.3i$) compared to $n^2_{Cu_2O}(bulk) = 8.6 + 0.6i$ and $n^2_{CuO}(bulk) = 7.1 + 2.3i$ [29]. If we suppose a chemical mix of both oxides, we deduce that oxide may be made of 76% Cu_2O and 24% CuO. This result was confirmed by XPS measurements [14,30]. The decrease in the real part of the oxide permittivity may also be due to air inclusion in oxide.
- For samples with roughness varying from 2 to 14 nm [14], the electromagnetic model of generalized Fresnel coefficients could be accurate enough. The quality of absorbance fitting shown in Figures A1–A11 confirms the validity of the model, which can therefore be used to model the SPR.

3. Results

In this section, we use the thicknesses and optical properties that are simultaneously recovered from the fitting of absorbance curves (Tables A3–A5) to study the SPR sensor setup, of which the sensitive part is a copper and copper oxide bilayer. From these results, we calculate the signal of the SPR sensor working in angular interrogation mode for air and water as the upper medium: in the dry and wet cases, respectively. We also evaluate the performance of such set up. We use $RIU = 0.01$.

3.1. Dry Case

Figure 1 illustrates the SPR setup. The wavelength of laser illumination is $\lambda_0 = 632.8$ nm. The photon energy of the excitation light ($E = 1.96$ eV) is close to that of the transition from d states (valence band) to th es-p conduction band [18] (2.1 eV, see the vertical black line in Figures A1, A7 and A11). Therefore, we expect a good quality of the plasmon resonance for adequate thickness of copper: a sharp dip and a small minimum.

Figures 2–4 show the simulation of the SPR setup signal from the model of reflectance in Equation (6), considering air as the upper medium. The reflectance curves are plotted as functions of the incident angle of illumination (at $\lambda_0 = 632.8$ nm) for each investigated sample.

In Figure 2, for negligible thicknesses of copper (less than 2 nm), the reflectance is characteristic of a dielectric material. In the other cases, the wells in reflectance on the right correspond to the absorption of photon energy by the copper layer (resonance). Nevertheless, the quality of the surface plasmon resonance is low: the wells are wide and the depths of dip are greater than 0.26.

The surface plasmon dip for copper thicknesses close to 30 nm are thinner than in the previous case (Figure 3 vs. Figure 2). The smallest values of R (depth of dip) are close to each other for h_{Cu} nm near 30 nm. The SPR angles are close to 46°. This SPR angle is close to that of a copper–air interface. The influence of the thin layer of oxide on the SPR angle is negligible for oxide thicknesses below 3.5 nm. However, the shape of the curve seems to be clearly modified, even for small thicknesses of the oxide. Thus, we can anticipate that the thickness of oxide will have an influence on the FWMH. For thicker oxide layers, the dip is widened and the SPR resonance angle approaches 65°, which is not so far from that of the copper–oxide interface (72°).

The SPR dips are thinner for the six samples for which $h_{Cu} \approx 50$ nm and are smaller than that for the previous samples (Figure 4 compared to Figures 2 and 3). By decreasing e_{Cu}, the SPR dips are enlarged and shifted, as in the previous case. The performance parameters introduced above can be evaluated for the investigated samples.

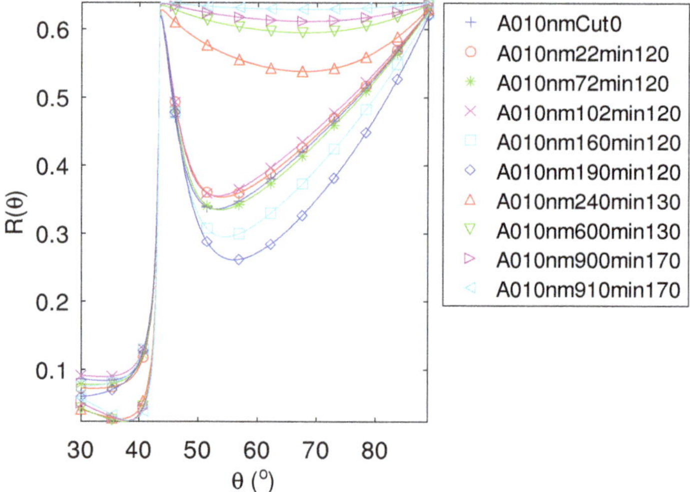

Figure 2. Simulated SPR setup signal for the samples of initial target thickness 10 nm. The calculation uses the parameters deduced from the fitting of UV-visible-NIR absorbance curves. The optical properties of copper and copper oxide are calculated at wavelength $\lambda_0 = 632.8$ nm. Upper medium is air.

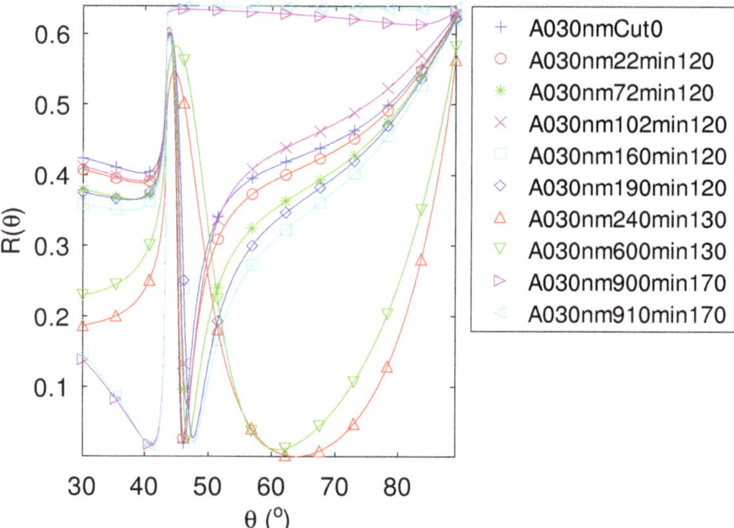

Figure 3. Simulated SPR setup signal for the samples of initial target thickness 30 nm. The calculation uses the parameters deduced from the fitting of UV-visible-NIR absorbance curves, at wavelength $\lambda_0 = 632.8$ nm. Upper medium is air.

The best performances of the SPR using the investigated bilayers as sensitive parts are shown in Table 1 [3,4]. We indicate both the retrieved thicknesses of copper and oxide (h_{Cu}, h_{ox}) from absorbance curves, and the corresponding optical properties (n_{Cu}^2, n_{ox}^2) 632.8 nm (extracted from Table A5). We give the resonance angle $\theta_{SPR} = \theta(\min(R))$, the depth of dip $\min(R)$, the full width at half maximum of the SPR dip (FWHM, Equation (11)), the sensitivity S_θ (Equation (10)), and the figure of merit (FOM, Equation (12)). The uncertainty on FOM is indicated in between brackets. This is the standard deviation of FOR for the recovered parameters of the multi-objective function $F \in [\min(F); 1.25 \times \min(F)]$

(see Appendix C). We also give the FOM calculated with the bulk optical properties for Cu and Cu_2O [29] and without the oxide layer for comparison.

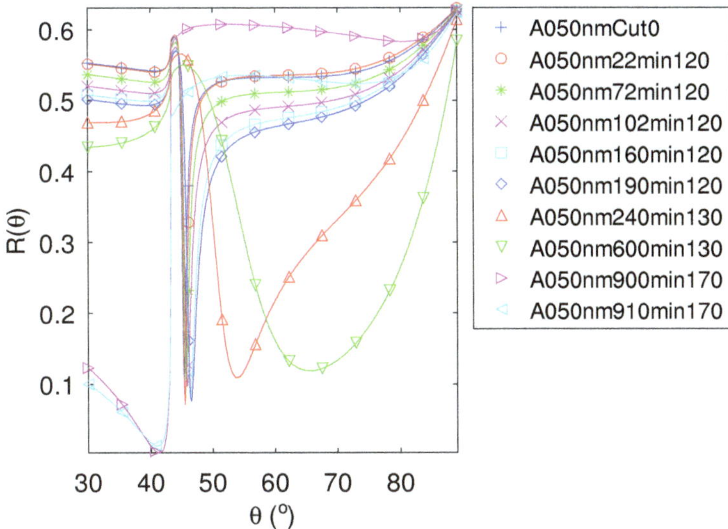

Figure 4. Simulated SPR setup signal for the samples of initial target thickness $h_{Cu}^0 = 50$ nm. The calculation uses the parameters deduced from the fitting of UV-visible-NIR absorbance curves, at wavelength $\lambda_0 = 632.8$ nm. Upper medium is air.

Table 1. Calculated performance of the SPR setup given the retrieved thicknesses and optical properties in the dry case: resonance angle, the depth of SPR dip, full width at half maximum, sensitivity and figure of merit (only FOM > 20 RIU^{-1} are indicated) and its uncertainty, FOM for optical properties of bulk materials, and for bare copper (without oxide).

h_{Cu} h_{ox} (nm)	n_{Cu}^2 n_{ox}^2	θ_{SPR} (°)	$\min(R)$	$FWHM$ °	S_θ (°/RIU)	FOM (RIU^{-1})	FOM Bulk (RIU^{-1})	FOM ox.Free (RIU^{-1})
49.1	$-12.2 + 2.3i$							
1.8	$8.3 + 0.7i$	46.8	0.08	3.2	67.5	20.9 (0.0)	27.8	25.8
49.2	$-12. + 2.3i$							
1.6	$8.4 + 0.8i$	46.7	0.08	2.8	67.0	23.5 (0.0)	28.6	28.5
49.8	$-13.2 + 2.3i$							
0.6	$8.8 + 0.5i$	46.2	0.09	2.1	65.0	31.2 (5.5)	32.9	33.5
49.8	$-14. + 2.2i$							
0.6	$8.3 + 0.9i$	46.0	0.10	1.5	63.9	41.4 (0.0)	33.1	44.2
49.8	$-14.6 + 1.8i$							
0.1	$8.2 + 0.8i$	45.8	0.07	1.1	63.4	57.3 (0.5)	34.9	57.7
50.0	$-15.4 + 2.0i$							
0.1	$7.9 + 1.0i$	45.6	0.11	1.1	62.7	58.5 (0.0)	35.1	59.3

The resonance angle θ_{SPR} is close to that of the interface between copper and the medium of detection (Equation (9), 45.3°). The minimum of reflectance is smaller than 0.1 and the sensitivity is about the same for all samples. The FWHM is smaller for copper thicknesses close to 50 nm; therefore, the value of FOM is greater. The uncertainty on FOM is small for almost all cases. The thickness of oxide being negligible, the FOM for copper without oxide is about the same. The values of S_θ are greater than 62.7. On the contrary, the FOM calculated from bulk optical properties of copper and copper oxide are different. This is due to the value of the real part of the copper relative permittivity, which is smaller than that of bulk copper ($n_{Cu}^2(bulk) = -11.6 + 1.6i$). This confirms the interest in simultaneously measuring the optical properties of copper and oxide for thin films.

In the general case, for a given thickness of copper and increasing thickness of oxide, the angle of resonance is shifted to the right, the FWHM increases as well as the depth of the SPR dip, and the FOM decreases.

The results for the investigated samples show that plasmon resonance can be launched for specific thicknesses. Therefore, the performance of the copper/copper oxide SPR setup deserves to be specifically studied, assuming that the medium of detection is water (wet case).

3.2. Wet Case

Table 2 restates the best thicknesses of copper and oxide, the retrieved relative permittivity (n_{Cu}^2 and n_{ox}^2) extracted from Tables A4 and A5 and the calculated performance of SPR for samples with FOM > 20.

Table 2. Calculated performance of the SPR setup given the retrieved thicknesses and optical properties in the wet case: resonance angle, the depth of SPR dip, full width at half maximum, sensitivity and figure of merit (only FOM > 20 RIU^{-1} are indicated) and its uncertainty, FOM for optical properties of bulk materials, and for bare copper (without oxide).

h_{Cu} h_{ox} (nm)	n_{Cu}^2 n_{ox}^2	θ_{SPR} (°)	$min(R)$	$FWHM$ °	S_θ (°/RIU)	FOM (RIU^{-1})	FOM Bulk (RIU^{-1})	FOM ox.Free (RIU^{-1})
27.9	$-13.5 + 2.9i$							
3.5	$8.4 + 0.6i$	81.4	0.04	11.1	258.0	23.3 (1.1)	9.5	19.1
28.2	$-12.8 + 3.1i$							
3.1	$8.3 + 0.5i$	81.4	0.03	11.3	253.0	22.4 (1.0)	10.1	18.7
29.4	$-13.0 + 2.0i$							
1.0	$8.1 + 1.1i$	78.6	0.11	10.0	231.4	23.1 (1.0)	5.8	22.1
29.5	$-12.5 + 2.9i$							
1.0	$8.2 + 1.0i$	78.9	0.04	10.9	229.6	21.1 (0.0)	5.9	20.0
29.9	$-13.3 + 2.8i$							
0.1	$8.6 + 0.7i$	77.3	0.06	10.3	214.5	20.8 (0.2)	10.1	20.8
30.0	$-14.1 + 2.7i$							
0.0	$8.3 + 1.2i$	76.8	0.06	9.9	210.7	21.4 (0.0)	5.9	21.3
49.1	$-12.2 + 2.3i$							
1.8	$8.3 + 0.7i$	83.3	0.26	8.7	200.2	22.9 (0.0)	11.9	34.4
49.2	$-12.5 + 2.3i$							
1.6	$8.4 + 0.8i$	82.8	0.23	8.6	252.8	29.2 (0.0)	16.4	35.5
49.8	$-13.2 + 2.3i$							
0.6	$8.8 + 0.5i$	80.3	0.15	8.3	305.7	36.8 (2.1)	38.7	36.5
49.8	$-13.2 + 2.3i$							
0.6	$8.8 + 0.5i$	80.3	0.15	8.3	305.7	36.8 (2.1)	38.7	36.5
49.8	$-14.2 + 2.2i$							
0.6	$8.3 + 0.9i$	78.9	0.13	7.3	292.6	40.0 (0.0)	39.4	40.2
49.8	$-14.6 + 1.8i$							
0.1	$8.2 + 0.8i$	77.8	0.08	5.9	277.7	47.2 (0.6)	43.7	47.3
50.0	$-15.4 + 2.0i$							
0.1	$7.9+1.0i$	77.0	0.11	5.7	262.2	45.7 (0.0)	43.7	45.8

The prism-SPR performance is evaluated as in the dry case. For water as upper medium, the SPR angle is greater than for air. The SPR angle falls between the SPR of the copper–oxide interface (72.4°) and that of the glass–copper interface (86.7°). On the contrary of the dry case, the samples with copper thicknesses around 28 and 30 nm give a reflectance smaller than 0.11 and FOM greater than 20. In this case, the values of both the sensitivity and the figure of merit are about the same whatever the oxide thickness is. The maximum of FOM is reached for thicker copper layers (near 50 nm), as in the dry case. The figure of merit is also calculated from the best values but by neglecting the oxide layer. In this case, we suppose that the bare copper layer is directly in contact with the above medium. Let us emphasize that the presence of oxide decreases the FOM (see FOM and FOM ox. free). This result is in agreement with that obtained for gold-SPR coated with porous silica film [4]. The performance of SPR working in dry and wet cases depends strongly on the thickness of both layers. The maximum sensitivity is obtained for the optimal thicknesses of 28 and 3 nm for copper and oxide, respectively, and for 49.8 and 0.6 nm.

4. Discussion

4.1. Cu-Oxide-SPR Performance

The performance of Cu-oxide-SPR in dry and wet cases are of the same order of magnitude as that mentioned in Reference [4] for silver/porous silica (wet case). The slight decrease in FOM by using a dielectric coating of metal can also be observed for our samples, for which the thickness of oxide is lower than 3.5 nm. The best values of FOM are obtained for oxide thicknesses lower than 3 nm. We observe that the FOM is highly sensitive to changes in the refractive index of copper. The FOM value is slightly greater than that obtained with Au, Au/Si$_3$N$_4$, Au/KCl materials [3]. The values of sensitivity and FWHM are greater in the wet case than in the dry case, leading to a smaller figure of merit.

In addition to its non-toxic nature, the oxide layer can be used to tune the position of the SPR resonance [9]. Indeed, the oxide layer could be used to adjust the resonance near the copper transition 2.1 eV that is characteristic of the inter-band transition from d states (valence band) to the 's-p' conduction band [18] (Figures A1, A7 and A11 where the vertical line shows the photon energy of illumination used in the SPR setup $E = 1.96$ eV). Let us note that cuprous copper oxide (Cu$_2$O) and cupric copper oxide (CuO) are p-type semiconductors with a bandgap of approximately 2.2–2.9 and 1.2–2.1 eV, respectively.

The optimal copper thickness (around 50 nm) is coherent with that found in Reference [9] (45 $\pm$ 5 nm), even if the substrate (prism) and the wavelength were not the same. However, these results show that more than one copper thickness may be used to design an efficient SPR sensor. This is the consequence of the high dependence of the optical properties on both the thickness and the sample elaboration mode (temperature and annealing time).

The SPR angle is around 46° in the dry case and 80° in the wet case. This last value is close to those obtained with a Al$_2$O$_3$ coating [10], and with BK7 substrate [9], both in the wet case. These SPR angles can also be compared to the SPR angles of single interfaces (Equation (9)). In the dry case, the shift of SPR angle is about 1–2° for oxide thicknesses below 2 nm. In the wet case, the shift is greater and falls between that of the glass/copper and copper/water interfaces. Therefore, the coupling of both SPR differs for two different mediums of detection.

Therefore, the performance parameters may reveal different properties of the SPR-sensor. In the next section, we discuss the optimal thickness obtained from the best performance parameters.

4.2. Optimal Thicknesses

The thicknesses of copper for the investigated samples are not sufficient to determine the optimal thickness. Nevertheless, the small dispersion of retrieved optical properties of copper and oxide leads us to use their mean value to determine the optimal thicknesses of copper and oxide layers. Actually, the mean value is $n_{Cu}^2 = -13.3 + 3.3i$ (standard deviation $1.4 + 1.5i$) and $n_{ox}^2 = 8.2 + 1.0i$ (standard deviation $0.4 + 0.3i$), see Section 2. They differ from the bulk values [29]. Therefore, we can use the mean values of the optical properties and a scan of copper thicknesses from 10 to 70 nm, and oxide thickness in (1;60) nm, in steps of 0.5 nm, to evaluate the optimal thicknesses from the best performance parameters and interference conditions, as explained in Section 2.4, in the dry and wet case. Tables 3 and 4 give the mean value and the standard deviation of the top 1% results for each performance parameter.

The argument of $1 + r_{14}$ should be close to 0 for these optimal values. This means that the conditions of destructive interference between incident and reflected waves are almost fulfilled and the minimum of reflectance is reached. The optimal thickness can be found for the maximum of FOM (Equation (12)) or sensitivity (Equation (10)), or for the minimum of FWHM and reflectance $\min(R)$.

In the dry case, the best thicknesses of copper are close together for $\min(R)$ and $\min(|\arg(1 + r_{14})|)$, on the one hand, and for $\max(FWHM)$, $\max(FOM)$ and $\max(S_\theta)$ on the other hand. If the conditions of constructive interference in the copper layer are fulfilled (Equation (13)), the optimal thicknesses of copper found are 29 nm for $m = 5$, 48 nm for $m = 8$. Therefore, we can deduce that all the copper thicknesses in Table 3 correspond to optimal ones, but for different optimal parameters. Moreover, these values are in agreement with that obtained from investigated samples in Section 3. The optimal thickness of copper for $\max(FOM)$ is between those for $\max(FWHM)$ and $\max(S_\theta)$. Therefore, the optimal thicknesses can be deduced from a selection of the solution with the minimum of $\min(R)$ among the top 1% solutions for $\max(FOM)$: $e_{Cu} = 44$ nm, $e_{ox} = 1$ nm, $\min(R) = 0.09$, $FOM = 12/, RIU^{-1}$, $FWHM = 5°$ $\theta_{SPR} = 46.7°$.

Table 3. Mean optimal thicknesses for the top 1% performance parameters of the SPR set up in the dry case according to the minimum of reflectance, sensitivity, full width at half maximum, figure of merit and argument of the total field on the glass-copper interface. In brackets, standard deviation of the top 1%.

h_{Cu} (nm)	h_{ox} (nm)	$\min(R)$	S_θ °/**RIU**$^{-1}$	$FWHM$ °	FOM **RIU**$^{-1}$	$\arg(1 + r_{14})$ °
			Top 1% $\min(R)$			
28.1 (5.6)	14.9 (9.5)	0.003 (0.003)	73.7 (26.2)	23.7 (6.3)	3.3 (1.5)	3.1 (1.8)
			Top 1% $\min(|D|)$			
31.9 (11.6)	51.0 (5.9)	1 (0)	0.0 (0.0)	90.0 (0.0)	0.0 (0.0)	0.0 (0.0)
			Top 1% $\min(|\arg(1 + r_{14})|)$			
31.9 (11.5)	50.3 (7.7)	1 (0.1)	1.4 (10.7)	88.9 (8.5)	0.1 (0.5)	0.0 (0.0)
			Top 1% $\max(FWHM)$			
46.0 (3.9)	1.0 (0.0)	0.1 (0.05)	65.4 (0.6)	5.5 (0.3)	12.0 (0.5)	14.5 (3.2)
			Top 1% $\max(FOM)$			
46.2 (1.7)	1.0 (0.0)	0.1 (0.02)	65.2 (0.6)	5.2 (0.1)	12.5 (0.0)	14.9 (1.4)
			Top 1% $\max(S_\theta)$			
54.1 (4.0)	16.9 (0.3)	0.3 (0.05)	125.0 (0.5)	30.1 (0.6)	4.2 (0.1)	30.9 (2.3)

Table 4. Optimal thicknesses for the top 1% performance parameters of the SPR set up in the wet case according to the minimum of reflectance, sensitivity, full width at half maximum, figure of merit and argument of the total field on the glass–copper interface. In brackets, standard deviation of the top 1%.

h_{Cu} (nm)	h_{ox} (nm)	$\min(R)$	S_θ °/**RIU**$^{-1}$	$FWHM$ °	FOM **RIU**$^{-1}$	$\arg(1 + r_{14})$ °
			Top 1% $\min(R)$			
24.8 (7.2)	6.5 (4.9)	0.004 (0.003)	199.5 (62.8)	10.7 (1.4)	18.2 (4.0)	3.6 (2.0)
			Top 1% $\min(|\arg(1 + r_{14})|)$			
31.9 (11.5)	50.4 (7.4)	1 (0.06)	0.9 (14.8)	89.7 (4.7)	0.1 (1.3)	0.0 (0.0)
			Top 1% $\min(|D|)$			
42.7 (17.0)	35.0 (15.3)	1 (0)	0.0 (0.0)	90.0 (0.0)	0.0 (0.0)	24.5 (18.2)
			Top 1% $\max(FWHM)$			
10.9 (0.9)	17.9 (1.3)	0.03 (0.03)	49.3 (30.8)	7.4 (0.3)	6.7 (4.2)	9.8 (6.5)
			Top 1% $\max(FOM)$			
43.5 (1.9)	1.0 (0.0)	0.1 (0.04)	265.5 (0.8)	11.0 (0.0)	24.1 (0.1)	27.3 (4.4)
			Top 1% $\max(S_\theta)$			
44.0 (2.2)	1.0 (0.0)	0.1 (0.05)	265.4 (0.8)	11.0 (0.0)	24.0 (0.1)	28.4 (5.0)

In the wet case, the best thicknesses of copper are close together for $\min(R)$ and $\min(|\arg(1 + r_{14})|)$, on the one hand, and for $\max(FOM)$ and $\max(S_\theta)$ on the other hand.

If the conditions of constructive interference in the copper layer are fulfilled (Equation (13)), the optimal thicknesses of copper found are 29 nm for $m = 5$, 42 nm for $m = 7$. Again, we can deduce that all the copper thicknesses in Table 3 correspond to optimal ones, but for different optimal parameters. Furthermore, these results are also in agreement with that obtained from investigated samples in Section 3. The optimal thickness of copper for $\max(FOM)$ is close to that for the $\max(S_\theta)$ performance parameter. As in the dry case, the optimal thicknesses can be deduced from selection of the solution with the minimum of $\min(R)$ among the top 1% solutions for $\max(FOM)$: $e_{Cu} = 41$ nm, $e_{ox} = 1$ nm, $\min(R) = 0.09$, $FOM = 24/, RIU^{-1}$, $FWHM = 11°$ $\theta_{SPR} = 86.9°$.

5. Conclusions

The performance of Cu-oxide-SPR has been studied, including the influence of copper oxide. To reach this goal, we have used the fitting of experimental UV-visible-NIR absorbance curves to simultaneously obtain the thicknesses and optical properties of copper/copper oxide samples. For this, more than one metaheuristic is of interest. We found that the optical properties of copper and copper oxide vary as a function of the thicknesses and differ from the bulk ones. The performance of Cu-oxide-SPR confirms that copper/copper oxide could be a valuable alternative to gold for SPR sensors. The performance parameters reveal different optimal thicknesses in the dry and wet cases. For a systematic study of the optimal thickness, the combination of the maximum of FOM and the minimum of reflectance could be an interesting alternative to find the optimal thickness. The relevance of the result depends on the accurate determination of thicknesses and optical properties of the bilayer. At $\lambda = 632.8$ nm, with $n_{Cu}^2 = -13.3 + 3.3i$, $n_{ox}^2 = -8.2 + 1.0i$, the optimal thicknesses of copper and oxide layers are about 44 and 1 nm in the dry case, and 41 and 1 nm in the wet case. They are determined to maximize the FOM and to minimize the dip magnitude. To optimize the SPR setup, we suggest the following process.

- Production series of samples with controlled annealing under low temperature;
- Characterization of samples (measurement of thickness and optical properties) from absorbance curves, for example;
- Characterization of samples in angular interaction mode;
- Fit of the SPR signal to verify the optical properties;
- Selection of the best sample and verification at regular time intervals that the thicknesses and optical properties remain the same.

Actually, the method of fit proposed in this paper could be applied to an SPR signal. In the future, we also intend to apply (and adapt if necessary) our method to characterize SPR with a multilayer sensitive part.

Author Contributions: Conceptualization: D.B. and T.G.; methodology, D.B. and T.G.; software: D.B.; validation, T.G.; formal analysis: D.B. and T.G.; investigation, D.B. and T.G.; experimental resources: D.C., E.A. and N.F., data curation: D.C., E.A. and N.F.; writing—original draft preparation: D.B.; writing—review and editing: D.B., T.G., S.K., E.A. and N.F.; supervision: D.B.; project administration: D.B.; funding acquisition: D.B., R.M., E.A. All authors have read and agreed to the published version of the manuscript.

Funding: This research was funded by the European Regional Development Fund grant number CUMIN CA0021200.

Institutional Review Board Statement: The study was conducted according to the guidelines of the European Charter for Researchers. https://euraxess.ec.europa.eu/jobs/charter/european-charter.

Informed Consent Statement: No Human Subjects of Research.

Data Availability Statement: Data available on request.

Abbreviations

The following abbreviations are used in this manuscript:

AIM	Angular Interrogation Mode
ABC	Artificial Bee Colony
EM	Evolutionary Method
FOM	Figure of Merit
FWHM	full width at half maximum
GR	method of gradient descent
NM	Nelder–Mead Simplex Method
PSO	Particle Swarm Optimization
SPR	Surface Plasmon Resonance

Appendix A. The Samples Preparation

Deniz Cakir prepared and characterized the copper/copper oxide [14] at Laboratoire Charles Coulomb, UMR CNRS 5221, Université de Montpellier, under the supervision of Eric Anglaret and Nicole Fréty [15]. The copper nanolayers were deposited by thermal evaporation on fused silica substrates (optical grade from Neyco). Before deposition, the substrates were cleaned in an acetone bath with ultrasounds for 5 min, and then plasma-treated in a 70% O2/30% N2 atmosphere for 6 min. The copper wire of purity of 99,999%, bought from Alfa Aesar, was placed 18 cm below the target substrate. The sublimation of the Cu wire was achieved under 120 A, at 10-5 mbar, using a tungsten crucible as the counter-electrode. The target thicknesses (nominal thickness before oxidation) of the copper layers are $h_{Cu} = 10, 30,$ and 50 nm. Copper thin films were annealed under air at atmospheric pressure. Low annealing temperatures were chosen to preferentially obtain Cu_2O [30–36]. Samples were annealed progressively for increasing times and temperatures following the thermal treatment, as detailed in Table A1. For each annealing condition and layer thickness, absorbance was measured with a UV-visible-NIR spectrometer (Cary 5000, Agilent, Les Ulys, France) over the spectral range 350–1380 nm, using a beam diameter of 1 mm. for each annealing condition. Therefore, 30 absorption spectra are available for fitting by the model described in this paper.

Table A1. Sample preparation. Three samples of copper initial target thicknesses $XY = 10, 30$ and 50 nm were successively annealed.

Annealing Time (Min)	Temperature (° C)	Reference Name of Data
0	20	A0XYnmCut0
22	120	A0XYnm22min120
72	120	A0XYnm72min120
102	120	A0XYnm102min120
160	120	A0XYnm160min120
10	120	A0XYnm190min120
240	130	A0XYnm240min130
600	130	A0XYnm600min130
900	170	A0XYnm900lin170
910	170	A0XYnm910min170

Appendix B. Measurement of Thicknesses

Cross-sections of the raw and fully oxidized samples were characterized using Scanning Electron Microscopy (SEM, Helios NanoLab 660 FEI and High resolution Hitachi S4800) and Atomic Force Microscopy (AFM, Dimension 3100 NanoScope IIIa, Bruker, Wissembourg, France). The AFM equipment was used in tapping mode using silicon nitride cantilevers with sharpened pyramidal tips. Multiple-Angle Incident (MAI) ellipsometry (Multiskop, Optrel GbR, Kleinmachnow, Germany), with green laser light $\lambda = 532$ nm and Spectroscopic Ellipsometry (SE, Nanofilm EP4, Accurion GmbH, Göttingen, Germany) were also used in addition to spectroscopic ellipsometry monochromatized light at 560,

660, 760, 860 and 960 nm. The results we obtain in this study are reported in Table A2 (UV-vis-NIR, NIR CARY 5000, Agilent, Les Ulys, France) extracted from Tables A3–A5 .

The transformation of a mole of metallic copper to oxide is leading to a volume increase that is evaluated to $r_v = 1.68$ for Cu_2O and $r_v = 1.77$ for CuO:

$$r_v = \frac{V_{ox}}{V_{Cu}} = \frac{1}{n_{Cu}} \frac{M_{ox}\rho_{Cu}}{\rho_{ox}M_{Cu}}, \tag{A1}$$

where n_{Cu} is the number of copper atoms in a molecule of oxide, $M_{ox} = 143.1$ g mol^{-1}, $M_{Cu} = 63.55$ g mol^{-1} are the molar mass, and $\rho_{ox} = 6$ g cm^{-3}, $\rho_{Cu} = 8.96$ g cm^{-3} are the densities. In Table A2, this metric is evaluated considering uniaxial growth with a fixed section:

$$r_v = \frac{e_{ox}(\text{A0XYnm900min170}) - e_{ox}(\text{A0XYnmCut0})}{e_{Cu}(\text{A0XYnmCut0}) - e_{Cu}(\text{A0XYnm900min170})} \tag{A2}$$

Cuprous oxide is expected to be highly dominant at low-temperatures of annealing [30].

Table A2. Measured thicknesses (nm) of raw and fully oxidized samples.

Sample	SEM $h_{Cu}+h_{ox}$	AFM $h_{Cu}+h_{ox}$	MAI h_{Cu} / h_{ox}	SE h_{Cu} / h_{ox}	UV-Vis-NIR h_{Cu} / h_{ox}
A010nmCut0	–	11 ± 3	7.2 ± 0.1	11.0 ± 0.2	9.8 ± 0.2
	–		0.2 ± 0.1	3.4 ± 0.2	1.8 ± 1.1
A030nmCut0	–	31 ± 5	28.2 ± 0.1	38.7 ± 1.3	30.0 ± 0.1
	–		2.9 ± 0.1	3.7 ± 0.2	0.0 ± 0.7
A050nmCut0	–	51 ± 8	–	–	50.0 ± 0.1
	–		–	5 ± 0.9	0.1 ± 0.2
A010nm900min170	34 ± 7	26 ± 3	5.6 ± 0.1	0.9 ± 0.4	1.2 ± 0.5
			19.6 ± 0.1	20.3 ± 3.1	19.7 ± 1.6
			$r_v = 12.2 \pm 1.1$	$r_v = 1.7 \pm 0.3$	$r_v = 2.1 \pm 0.3$
A030nm900min170	104 ± 29	98 ± 23	4.9 ± 0.1	0.3 ± 1.4	0.6 ± 4.2
			84.6 ± 0.1	63.1 ± 7.4	44.8 ± 6.6
			$r_v = 3.51 \pm 0.02$	$r_v = 1.6 \pm 0.2$	$r_v = 1.5 \pm 0.3$
A050nm900min170	136 ± 15	142 ± 30	–	0 ± 7.9	0.0 ± 0.6
			–	87.3 ± 3.7	89.6 ± 4.4
			–	–	$r_v = 1.79 \pm 0.09$

AFM and SEM measurements are the real thicknesses of samples, whereas MAI, SE spectroscopies and UV-vis-NIR give effective thicknesses. Copper films show a very fine microstructure, with a grain size of a few nm, and a very low roughness with a R_{ms} value of about 1 nm. The grain size of oxide and perhaps the slice process explain the values from AFM and SEM and the high magnitude of r_v. This microstructure remains thin after full oxidation with a grain size varying from a few to 80 nm according to the oxide thickness and a roughness varying from 2 to 14 nm. Our results are in agreement with AFM and SEM for sample A0XYnmCut0.

MAI fails to measure thicknesses of thick layers. MAI and SE use the optical properties of bulk materials to calculate effective thicknesses. MAI gives thicknesses smaller than those of SE. Moreover, the MAI copper thicknesses appear to be greater than those given by SE for fully oxidized samples. Our results are in agreement with SE except for A030nm900min170. The values of r_v for our results and by SE remain close to the theoretical value for all samples.

The agreement of our results and experimental measurement of thicknesses of copper and oxide for the sample before annealing (A0XYnmCut0) and after full oxidation (A0XYnm900min170) is satisfactory.

Appendix C. Method of Fitting and Results

The method of fitting is detailed in the next subsections: Particle Swarm Optimization (PSO), the Evolutionary Method (EM) and the Artificial Bee Colony (ABC), the domain of search and the multi-objective function.

Appendix C.1. Metaheuristics for UV-Visible-NIR Absorbance Fitting

The Particle Swarm Optimization (PSO), the Evolutionary Method (EM) and the Artificial Bee Colony (ABC) belong to metaheuristics methods of optimization. Metaheuristic optimization is based on the initial random sampling of the model parameters in the bounded domain of search. $N = 40$ parameter sets are generated within a bounded domain of search.

Then the objective function is evaluated and the evolutionary loop starts. Within this loop, the parameters are modified according to a transformation law, which can be defined as an evolution operator. If the parameter sets leave the domain of search, they are either randomly regenerated in the domain of search or stuck on the closest boundaries, according to a random number. The evolutionary loop stops either if the maximum number of iterations is reached ($T = 100$) or if the best value of the multi-objective function is lower than 0.001.

We shortly describe the evolution operators of each optimization method in the following.

- PSO: the parameters are moved along a vector of translation, which is the sum of three vectors [37–40]. More details are given in Reference [38]:
  - A uniform random contribution of the difference between the N parameter sets at the previous step of the evolutionary loop and the global best position of this swarm (weight $c_1 = 2.1$).
  - A uniform random contribution of the difference between the N parameter sets at the previous step of the evolutionary loop and their best positions obtained in all the previous steps (weight $c_2 = 2.1$).
  - A linearly decreasing contribution of the previous translation vector, with weight ranging from 0.99 at the beginning of the evolutionary loop to 0.43 at the maximum number of iterations.

  If the value of the multi-objective function is better than that obtained at the previous step of the evolutionary loop, then the best parameter set is updated for the next step. Classical PSO without hybridization was used to analyze aluminum oxidation from Turbadar experimental data [41] in the Kretschmann configuration. The strong dependence of Aluminum optical properties as a function of its nanometric thickness was demonstrated.

- GA: the input parameters are varied by using crossover, mutation and selection operators [38,40,42].
  - The crossover operator calculates the mean value of each parameter of two randomly selected sets. The crossover operates on $\mu = N/4$ parameter set.
  - The mutation operator modifies each parameter by adding a normally distributed random vector with zero expectation and self-adaption variance [43]. The mutation operates on a $\lambda = 3N/4$ parameter set.
  - The N parameter sets that result from the crossover, and mutation operators are evaluated.
  - The selection operator keeps the N best solutions among these parameter sets and the best set, obtained at the previous step of the evolutionary loop according to an elitist $(\lambda + \mu)$ strategy [43].

  GA was fully characterized in Reference [44] for inverse problem resolution. Details can be found in References [45,46]. The recovering of the unknown input parameters of a model from the fitting of experimental data is actually the resolution of an inverse problem [47].

- ABC: the parameter sets are divided into two families of size $N/2$. Specific operators are applied to these two families. The parameter set is updated by adding a translation vector [48]. More details on ABC can be found in [49].
  - First family: the translation vector is a uniform random contribution of the difference between the parameter set and another randomly picked one from the same family. The multi-objective function is evaluated and the family is

updated with a parameter set for improved values of the multi-objective function. A probability of attraction of these parameter sets is calculated to promote the best to guide the second family sets [48].

- Second family: the translation vector is the uniform random contribution of the difference between the parameter set and a random pickup of the most attractive parameter sets of the first family.

ABC presents few exogenous parameters, with settings discussed in many references, e.g., [50]. Here, we used the standard parametrization [48].

Metaheuristics subtly balance between the exploration of the whole domain of search and the exploitation of the best solutions. This balance must prevent rapid entrapment in local optima. The random regeneration of parameter sets inside the search domain, the mutation in GA, the third term of translation in PSO, and the first family in ABC, contributes to exploration. The two first terms of translation in PSO, the crossover in GA, and the second family in ABC, ensure the exploitation. In this study, we hybridize these metaheuristics with two unconstrained local minimum descent methods: the gradient and Nelder–Mead Simplex methods using the best solution as a starting point. This hybridization helps to determine if the best solution obtained from metaheuristics is close to a minimum. Moreover, it may improve the convergence speed. To test the reproducibility of the algorithms, especially as a function of the initial random generation of parameter sets, we run $nreal = 1000$ realizations of each algorithm. Typically, the maximum number evaluations of the multi-objective function is $nreal \times T \times N = 1000 \times 100 \times 30 = 3 \times 10^6$, which means about 7 min of computational time, only for the evaluation of the multi-objective function. The computational time of the whole algorithm of optimization is around one hour (Xeon E-2176M CPU @2.7 GHz, GNU octave x64) for each fit of experimental UV-visible-NIR absorbance data.

The best parameter sets of the model (thicknesses and parameters of the optical properties of copper and copper oxide) are the input parameters for the SPR model.

Appendix C.2. The Domain of Search

For the first samples (before annealing), the nominal thicknesses of copper deposition (or target) are $h^0_{Cu} = 10, 30$ and 50 nm. The starting domains of search for thicknesses are $h^0_{Cu} \pm 5$ nm for copper and $[0, 5]$ nm for oxide.

For the oxidized samples, the boundaries of the domain of search for thicknesses are set to search both increasing values of the copper oxide layer thickness and decreasing values of that of copper, when the samples are successively annealed. The domain of search of the partial fraction parameters is limited to 1% around those of the bulk materials [18,29]. The fitting basically requires the minimization of the error of fit. In our case, the goal is to minimize a multi-objective function to accelerate the convergence of the fitting method.

Appendix C.3. The Multi-Objective Function

Fitting of the experimental UV-visible-NIR absorbance curves with the above-mentioned model of absorbance consists of minimizing a multi-objective function. More specifically, we define a weighted sum, which scalarizes three objective functions by adding them pre-multiplied by a given weight [51]. The weights have been chosen after preliminary tests of convergence of the method for all investigated samples. The idea is to find solutions for optical properties that are close to the bulk values. The first term of the multi-objective function is the ratio of the absolute error of the calculated absorbance to the reference experimental UV-visible-NIR absorbance. The next two terms with weight $K_i = 1/25$ are the relative difference of the calculated relative permittivities to the bulk ones. We set the coefficients $K_{Cu} = K_{ox} = 1/25$ after preliminary trials to balance between the quality of fit and the closeness to bulk values and, therefore, to increase the speed of convergence of the algorithms used for fitting. The multi-objective function F is written as:

$$F = \sqrt{\frac{\sum\limits_{j=1}^{N_\omega}\left[\left|1 - \frac{A}{A_{exp}}\right|^2 + K_{Cu}\left|1 - \frac{n_{Cu}^2}{n_{Cu(Bulk)}^2}\right|^2 + K_{ox}\left|1 - \frac{n_{ox}^2}{n_{Cu_2O(Bulk)}^2}\right|^2\right]}{2N_\omega - (D+1)}}. \tag{A3}$$

The normalization of the objective function by $2N_\omega - (D+1)$ involves the number of photon energies considered in the UV-visible-NIR absorbance curves for fitting $N_\omega = 32$ and the dimension D of the problem. D is the number of input parameters of the model of absorbance. This term is the degree of freedom and helps to compare results by varying N_ω (the number of values of the experimental UV-visible-NIR absorbance used for fitting). Indeed, the computational time depends strongly on this exogenous parameter, more than one million of evaluation of the objective function being necessary to characterize each of the metaheuristic methods. Let us briefly outline the principles of the metaheuristics we use for finding the $D = 28$ inputs of the model that minimize the multi-objective function.

Appendix C.4. Detailed Results of Fitting

Tables A3–A5 give the recovered thicknesses of copper and copper oxide from fitting of the experimental UV-visible-NIR absorbance curves. We calculate the relative permittivity at photon energy used for SPR ($\hbar\omega = 1.96$ eV, for wavelength $\lambda_0 = 632.8$ nm), from the Partial Fraction model (which is a function of the photon energy) with the recovered parameters for both copper and oxide. Actually, the quality of fitting depends on the topological properties of the multi-objective function F and on the tuning of each metaheuristic. Figures A1, A7 and A11 show the absorbance spectra: in color, the absorbance calculated from the best parameters that are obtained from the fitting of the experimental ones (black). The vertical black line displays the photon energy that will be used for the SPR calculations. These Figures illustrate the goodness of fit, which is given in Tables A3–A5 (the value of the multi-objective function F).

In Tables A3–A5, we indicate the most efficient metaheuristic: ABC is the Artificial Bee Colony, PSO is the particle Swarm Optimization, and EM is the evolutionary method. If the best solution of these methods is improved with Gradient (GR) or Nelder–Mead Simplex methods (NM), the mention of this hybridization is indicated after the sign "+". We suppose that acceptable values of the multi-objective function are within the interval $F \in [\min(F); 1.25 \times \min(F)]$. Indeed, we select the 25% best values of the multi-objective function across the 1000 realizations of each algorithm, and we calculate the standard deviation of the corresponding family of model parameters. This helps to evaluate the sensitivity of the fitness function to the variations of each input parameter of the model. These values (between brackets) indicate the acceptable tolerance on these parameters (Tables A3–A5).

Figures A2, A3, A6, A8, A12 and A13 show the calculated permittivities from the best parameters of the partial fraction model for copper. The curves are the real and imaginary parts of the permittivity as a function of the photon energy. The corresponding plots for copper oxide are given in Figures A4, A5, A9, A10, A14 and A15. We also plot the bulk permittivity (solid black curve for copper and Cu_2O, dashed black curve for CuO). Consequently, from the results, we expect to deduce if CuO could be present.

Appendix C.4.1. Sample of Initial Copper Thickness $h_{Cu}^0 = 10$ nm

The EM+NM method seems to be more efficient than ABC+NM. The PSO fails to give acceptable values of fit in most cases. Actually, the tuning of PSO with $c_1 = c_2 = 2.1$ favors exploration of the domain of search, at the expense of exploration, but it does not verify the stochastic condition of convergence of the algorithm [52,53]. The thickness of the raw sample A010nmCut0 was measured with AFM (11 ± 3) nm [14]. We found (11.6 ± 1.1) nm. The thickness of copper decreases suddenly after the annealing for 600 min. The results obtained after 900 and 910 min are coherent: there is no more copper and the copper oxide is no more modified. The thickness of the fully oxidized sample A010nm900min170 was measured with AFM (26 ± 3) nm [14]. We found 19.7 ± 1.1 nm, which is at least 2.8 nm smaller.

The real part of the permittivity of copper is negative at $\lambda_0 = 632.8$ nm ($\hbar\omega = 1.96$ eV) and smaller than that of bulk. On the contrary, its imaginary part is about seven times greater than that of bulk. Figure A3 shows a global behavior of thin copper layers: for small photon energies, the imaginary part of the permittivity is smaller than that of bulk, contrary to the greater photon energies (toward 3.54 eV).

The real part of the permittivity of copper oxide is smaller than that of bulk Cu_2O for small oxide thicknesses and tends toward the bulk one (at $\lambda_0 = 632.8$ nm ($\hbar\omega = 1.96$ eV)). Figure A5 also shows a global behavior of thin copper oxide layers: the imaginary part of the permittivity is greater than that of bulk on the whole domain of photon energies. The real part is greater than that of bulk Cu_2O at high photon energies and smaller at low energies. The permittivity lays between those of CuO and Cu_2O. Therefore, we deduce that Cu_2O dominates at low annealing temperatures but that CuO is also present.

The smallest quality of fit (the greatest value of F) is obtained for the sample oxidized for 160 min under $120°$. This is revealed in Figure A1 as well in Table A3. The small value of the standard deviation of the best 25% solutions (6e-04) indicates that it does not result from the failure of metaheuristics.

Table A3. The recovered thickness of copper and copper oxide, and the relative permittivity of copper and copper oxide at $\lambda_0 = 632.8$ nm, by fitting experimental UV-visible-NIR absorbance curves. The value of the multi-objective function F (Equation (A3)) and the most efficient metaheuristics are also indicated. Values between brackets are standard deviations for acceptable solutions, i.e., obtained for F below a 25% threshold (for $F \in [\min(F); 1.25 \times \min(F)]$. The corresponding success of each metaheuristic is indicated in percents).

Sample	h_{Cu} (nm)	h_{ox} (nm)	n^2_{Cu} at $\hbar\omega = 1.96$ eV	n^2_{ox} at $\hbar\omega = 1.96$ eV	F	Method
A010nmCut0	9.8	1.8	$-12.5 + 3.6i$	$8.2 + 1.2i$	7.9×10^{-3}	EM + NM
	(0.2)	(1.1)	$(0.5 + 0.5i)$	$(0.4 + 0.5i)$	(5×10^{-4})	
			$n_{ABC} = 37\%, n_{EM} = 63\%, n_{PSO} = 0\%, n_{NM} = 100\%, n_{GR} = 0$			
A010nm22min120	9.7	2.8	$-11.9 + 3.3i$	$8.2 + 1.3i$	5.1×10^{-3}	ABC + NM
	(0.2)	(1.2)	$(0.5 + 0.5i)$	$(0.4 + 0.5i)$	(1×10^{-3})	
			$n_{ABC} = 100\%, n_{EM} = 0\%, n_{PSO} = 0\%, n_{NM} = 100\%, n_{GR} = 0$			
A010nm72min120	9.6	2.8	$-12.5 + 3.6i$	$7.8 + 1.1i$	7.1×10^{-3}	ABC + NM
	(0.3)	(1.4)	$(0.7 + 0.5i)$	$(0.5 + 0.4i)$	(8×10^{-4})	
			$n_{ABC} = 43\%, n_{EM} = 57\%, n_{PSO} = 0\%, n_{NM} = 100\%, n_{GR} = 0$			
A010nm102min120	9.5	3.2	$-14.1 + 3.2i$	$7.7 + 1.0i$	8.0×10^{-3}	ABC + NM
	(0.3)	(1.5)	$(0.8+0.5i)$	$(0.5 + 0.4i)$	(7×10^{-4})	
			$n_{ABC} = 100\%, n_{EM} = 0\%, n_{PSO} = 0\%, n_{NM} = 100\%, n_{GR} = 0$			
A010nm160min120	8.7	4.9	$-15.4 + 4.8i$	$8.0 + 1.0i$	1.1×10^{-2}	EM + NM
	(0.5)	(1.0)	$(1.1 + 1.0i)$	$(1.1 + 0.5i)$	(6×10^{-4})	
			$n_{ABC} = 36\%, n_{EM} = 45\%, n_{PSO} = 19\%, n_{NM} = 100\%, n_{GR} = 0$			
A010nm190min120	8.5	6.9	$-14.8 + 5.6i$	$7.5 + 1.2i$	9.3×10^{-3}	EM + NM
	(0.5)	(1.1)	$(1.2 + 1.0i)$	$(1.1 + 0.5i)$	(6×10^{-4})	
			$n_{ABC} = 60\%, n_{EM} = 40\%, n_{PSO} = 0\%, n_{NM} = 100\%, n_{GR} = 0$			
A010nm240min130	2.6	16.7	$-11.1 + 3.4i$	$9.6 + 0.6i$	6.2×10^{-3}	EM + NM
	(0.7)	(1.1)	$(0.4 + 0.6i)$	$(0.3 + 0.2i)$	(5×10^{-4})	
			$n_{ABC} = 24\%, n_{EM} = 29\%, n_{PSO} = 47\%, n_{NM} = 92\%, n_{GR} = 8$			
A010nm600min130	1.9	17.9	$-11.2 + 3.2i$	$8.6 + 0.1i$	7.6×10^{-3}	ABC + NM
	(0.5)	(0.9)	$(0.5 + 0.5i)$	$(0.4 + 0.0i)$	(5×10^{-4})	
			$n_{ABC} = 50\%, n_{EM} = 22\%, n_{PSO} = 28\%, n_{NM} = 100\%, n_{GR} = 0$			
A010nm900min170	1.2	18.5	$-14.2 + 4.2i$	$8.5 + 0.0i$	1.6×10^{-2}	PSO + NM
	(0.5)	(1.5)	$(1.1 + 1.5i)$	$(0.7 + 0.1i)$	(5×10^{-3})	
			$n_{ABC} = 25\%, n_{EM} = 44\%, n_{PSO} = 31\%, n_{NM} = 100\%, n_{GR} = 0$			
A010nm910min170	0.7	18.7	$-11.0+3.3i$	$8.3 + 0.0i$	6.4×10^{-3}	ABC + NM
	(0.5)	(1.0)	$(0.6 + 0.6i)$	$(0.4 + 0.0i)$	(8×10^{-4})	
			$n_{ABC} = 56\%, n_{EM} = 16\%, n_{PSO} = 28\%, n_{NM} = 100\%, n_{GR} = 0$			

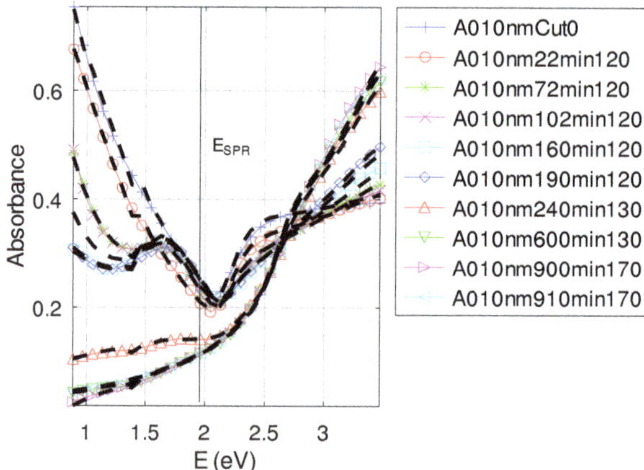

Figure A1. Absorbance as a function of the photon energy: experimental (black dashed) and best calculated solution (color) for each oxidized sample (initial thickness $h^0_{Cu} = 10$ nm).

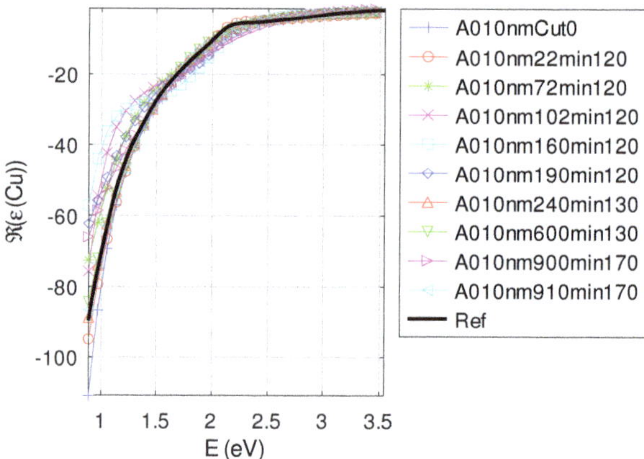

Figure A2. Real part of the relative permittivity of copper as a function of the photon energy: bulk (black) and best calculated (color) for each oxidized sample (initial thickness $h^0_{Cu} = 10$ nm).

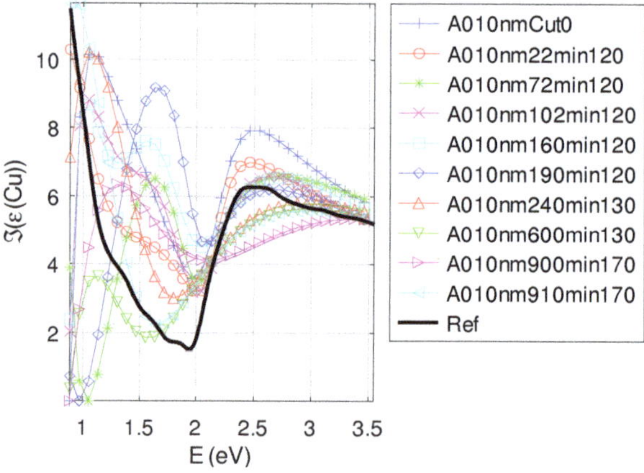

Figure A3. Imaginary part of the relative permittivity of copper as a function of the photon energy: bulk (black) and best calculated (color) for each oxidized sample (initial thickness $h^0_{Cu} = 10$ nm).

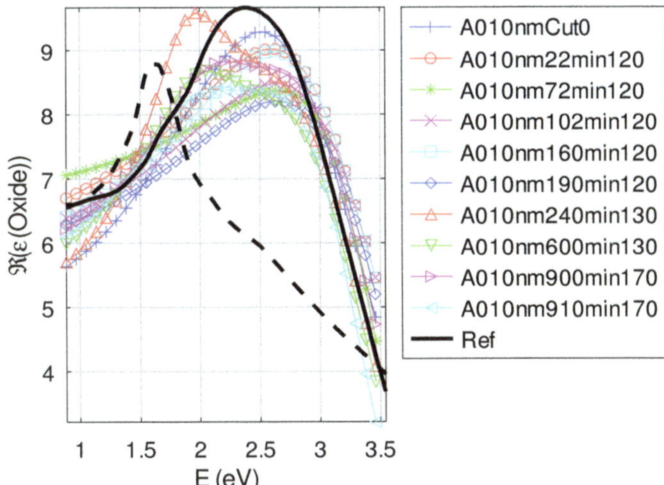

Figure A4. Real part of the relative permittivity of oxide as a function of the photon energy: bulk Cu$_2$O (black solid line), bulk CuO (black dashed line) and best calculated (color) for each oxidized sample (initial thickness $h^0_{Cu} = 10$ nm).

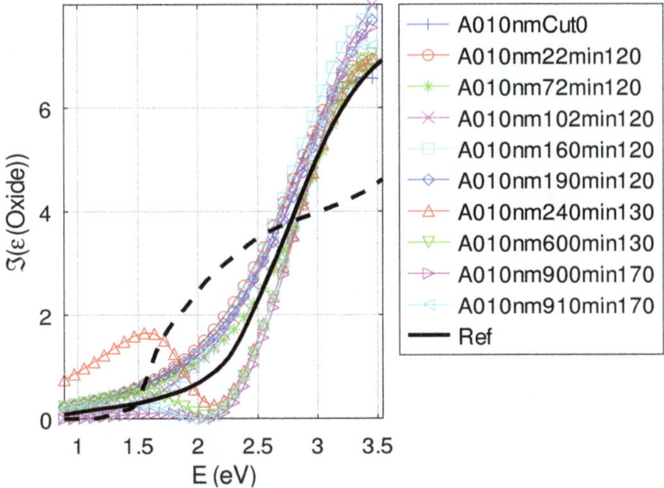

Figure A5. Imaginary part of the relative permittivity of oxide as a function of the photon energy: bulk Cu$_2$O (black solid line), bulk CuO (black dashed line) and best calculated (color) for each oxidized sample (initial thickness $h^0_{Cu} = 10$ nm).

Appendix C.4.2. Sample of Initial Copper Thickness $h^0_{Cu} = 30$ nm

Table A4 shows that the least efficient fittings are obtained for the sample with an annealing time greater than 240 min: it is about twice the best value considering all samples. Therefore, the visual inspection of Figure A7 reveals a good quality of fit. Again, the bad tuning of PSO could explain the superiority of ABC+NM and EM+NM. Let us note that neither of the Gradient (+GR) hybridized methods succeeded in reaching the best parameter set. The thickness of copper decreases gently up to the two last annealings. The results obtained after 900 and 910 min are in agreement: only copper oxide remains.

As for the previous sample (with an initial target thickness of 10 nm), the thickness of the raw sample A030nmCut0 was measured with AFM (31 ± 5) nm [14]. We found (30.0 ± 0.1) nm, which is in agreement. The thickness of the fully oxidized sample A030nm900min170 was also measured by AFM : 98 ± 23 nm. Our result is about half of this value. This is probably due to the grain material structure of oxide, which increases the true thickness compared to the effective thickness we calculate. The negative real part of the permittivity of copper is smaller than that of bulk at $\lambda_0 = 632.8$ nm ($\hbar\omega = 1.96$ eV). Its imaginary part is about seven times greater than that of bulk. Figure A8 shows a global behavior similar to that of thinner copper layers: for small photon energies, the imaginary part of the permittivity is smaller than that of bulk, contrary to the greater photon energies (toward 3.54 eV). Let us note that in the case of total oxidation (900, 910 min), the results for the optical properties of copper are not significant.

Figure A9 shows the same behavior as in Figure A4. The real part of the permittivity of copper oxide is smaller than that of bulk Cu$_2$O for small oxide thicknesses and tends toward the bulk one for thick layers (at $\lambda_0 = 632.8$ nm ($\hbar\omega = 1.96$ eV)). For thin copper oxide layers, the imaginary part of the permittivity is greater than that of bulk on the whole domain of photon energies. In the case of total oxidation (900–910 min), the real part is much smaller than that of bulk, and the shape of the curve is modified. This behavior is probably due to the inclusion of air in the thick oxide layers. Again, the permittivity remains far from that of CuO.

Table A4. The recovered thickness of copper and Cooper oxide, the relative permittivity of copper and copper oxide at $\lambda_0 = 632.8$ nm, by fitting experimental UV-visible-NIR absorbance curves. The recovered relative permittivities can be compared to those of bulk copper and copper oxide, resp. $-11.55 + 1.57i$ and $8.64 + 0.64i$ [29]. The value of the multi-objective function F (Equation (A3)) and the most efficient metaheuristics are also indicated. Values between brackets are standard deviation for acceptable solutions, i.e., obtained for F below the 25% threshold (for $F \in [\min(F); 1.25 \times \min(F)]$. The corresponding success of each metaheuristic is indicated in percentages).

Sample	h_{Cu} (nm)	h_{ox} (nm)	n_{Cu}^2 at $\hbar\omega = 1.96$ eV	n_{ox}^2 at $\hbar\omega = 1.96$ eV	F	Method
A030nmCut0	30.0	0.0	$-14.1 + 2.7i$	$8.3 + 1.2i$	8.7×10^{-3}	ABC + NM
	(0.1)	(0.7)	$(0.5 + 0.6i)$	$(0.2 + 0.3i)$	(4×10^{-4})	
	$n_{ABC} = 62\%, n_{EM} = 38\%, n_{PSO} = 0\%, n_{NM} = 100\%, n_{GR} = 0$					
A030nm22min120	29.9	0.1	$-13.3 + 2.8i$	$8.6 + 0.7i$	7.9×10^{-3}	ABC + NM
	(0.1)	(0.5)	$(0.5 + 0.5i)$	$(0.3 + 0.3i)$	(8×10^{-4})	
	$n_{ABC} = 42\%, n_{EM} = 50\%, n_{PSO} = 8\%, n_{NM} = 100\%, n_{GR} = 0$					
A030nm72min120	29.5	1.0	$-12.5 + 2.9i$	$8.2 + 1.0i$	6.2×10^{-3}	EM + NM
	(0.3)	(0.5)	$(0.4 + 0.4i)$	$(0.3 + 0.3i)$	(4×10^{-4})	
	$n_{ABC} = 44\%, n_{EM} = 56\%, n_{PSO} = 0\%, n_{NM} = 100\%, n_{GR} = 0$					
A030nm102min120	29.4	1.0	$-13.0 + 2.0i$	$8.1 + 1.1i$	5.2×10^{-3}	EM + NM
	(0.4)	(0.7)	$(0.5 + 0.3i)$	$(0.3 + 0.2i)$	(5×10^{-4})	
	$n_{ABC} = 14\%, n_{EM} = 86\%, n_{PSO} = 0\%, n_{NM} = 100\%, n_{GR} = 0$					
A030nm160min120	28.2	3.1	$-12.8 + 3.1i$	$8.3 + 0.5i$	4.9×10^{-3}	ABC + NM
	(0.6)	(1.0)	$(0.5 + 0.5i)$	$(0.3 + 0.3i)$	(5×10^{-4})	
	$n_{ABC} = 17\%, n_{EM} = 83\%, n_{PSO} = 0\%, n_{NM} = 100\%, n_{GR} = 0$					
A030nm190min120	27.9	3.5	$-13.5 + 2.9i$	$8.4 + 0.6i$	4.8×10^{-3}	ABC + NM
	(0.7)	(1.2)	$(0.5 + 0.5i)$	$(0.3 + 0.3i)$	(5×10^{-4})	
	$n_{ABC} = 50\%, n_{EM} = 50\%, n_{PSO} = 0\%, n_{NM} = 100\%, n_{GR} = 0$					
A030nm240min130	21.2	14.5	$-11.3 + 8.2i$	$7.7 + 1.3i$	1.1×10^{-2}	EM + NM
	(1.5)	(2.4)	$(1.1 + 1.5i)$	$(0.9 + 1.1i)$	(5×10^{-4})	
	$n_{ABC} = 9\%, n_{EM} = 70\%, n_{PSO} = 21\%, n_{NM} = 100\%, n_{GR} = 0$					
A030nm600min130	20.1	16.6	$-15.9 + 6.8i$	$7.7 + 0.6i$	1.2×10^{-2}	EM + NM
	(1.2)	(2.2)	$(0.9 + 0.8i)$	$(0.9 + 0.6i)$	(5×10^{-4})	
	$n_{ABC} = 18\%, n_{EM} = 57\%, n_{PSO} = 25\%, n_{NM} = 100\%, n_{GR} = 0$					
A030nm900min170	0.6	44.8	$-10.2 + 2.2i$	$7.0 + 0.0i$	1.1×10^{-2}	EM + NM
	(4.2)	(6.6)	$(0.9 + 1.0i)$	$(1.1 + 0.8i)$	(5×10^{-4})	
	$n_{ABC} = 4\%, n_{EM} = 96\%, n_{PSO} = 0\%, n_{NM} = 100\%, n_{GR} = 0$					
A030nm910min170	0.4	45.8	$-11.5 + 3.1i$	$7.1 + 0.0i$	1.1×10^{-2}	EM + NM
	(4.2)	(6.6)	$(0.9 + 1.0i)$	$(1.1 + 0.8i)$	(5×10^{-4})	
	$n_{ABC} = 0\%, n_{EM} = 40\%, n_{PSO} = 60\%, n_{NM} = 100\%, n_{GR} = 0$					

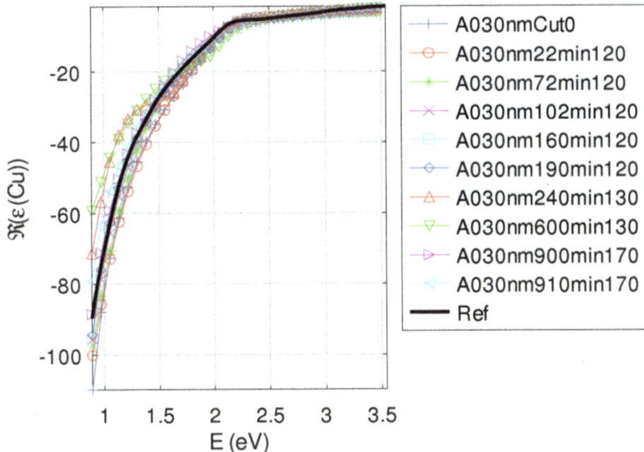

Figure A6. Real part of the relative permittivity of copper as a function of the photon energy: bulk (black) and best calculated (color) for each oxidized sample (initial thickness $h^0_{Cu} = 30$ nm).

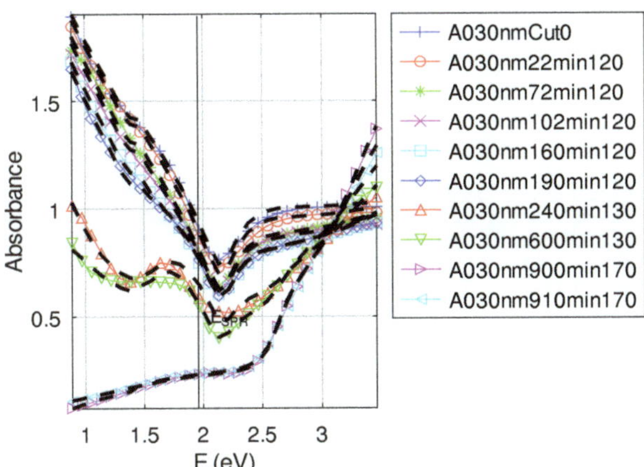

Figure A7. Absorbance as a function of the photon energy: experimental (black dashed) and best calculated solution (color) for each oxidized sample (initial thickness $h^0_{Cu} = 30$ nm).

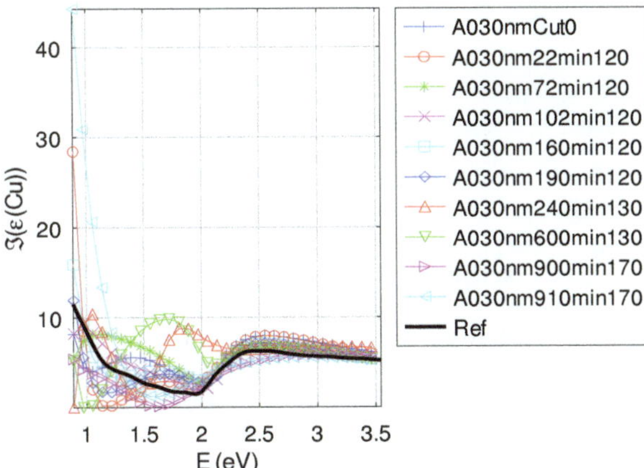

Figure A8. Imaginary part of the relative permittivity of copper as a function of the photon energy: bulk (black) and best calculated (color) for each oxidized sample (initial thickness $h^0_{Cu} = 30$ nm).

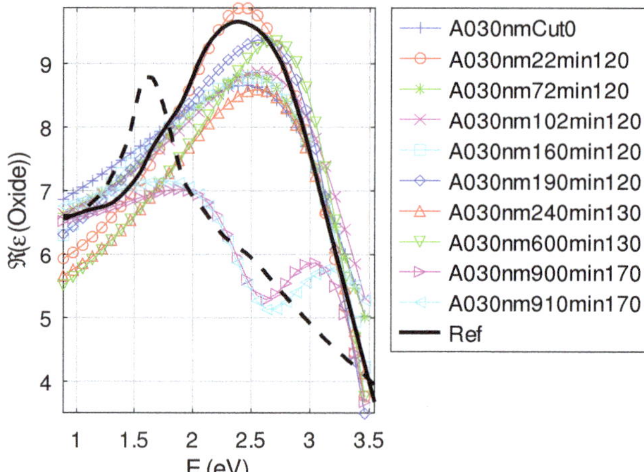

Figure A9. Real part of the relative permittivity of oxide as a function of the photon energy: bulk Cu$_2$O (black solid line), bulk CuO (black dashed line) and best calculated (color) for each oxidized sample (initial thickness $h^0_{Cu} = 30$ nm).

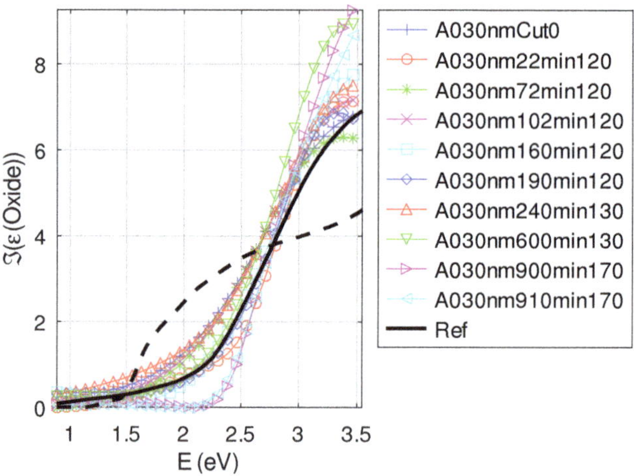

Figure A10. Imaginary part of the relative permittivity of oxide as a function of the photon energy: bulk Cu$_2$O (black solid line), bulk CuO (black dashed line) and best calculated (color) for each oxidized sample (initial thickness $h^0_{Cu} = 30$ nm).

Appendix C.4.3. Sample of Initial Copper Thickness $h^0_{Cu} = 50$ nm

Table A5 gives values of the multi-objective function F that are close to those obtained for the other samples. The stability of the methods also is of the same order of magnitude (smaller than 1×10^{-3}). The ABC + NM is the most efficient method even if $n_{EM} > n_{ABC}$. The PSO is more efficient for oxidized samples. The thickness of the raw sample A050nmCut0 was measured with AFM (51 ± 8) nm [14]. We found (50.1 ± 0.2) nm, which is in agreement. The thickness of the fully oxidized sample A050nm900min170 also was measured by AFM: 142 ± 30 nm. We found 82.3 ± 4.4 nm. As for the A030nm sample, we found a smaller value. The real part of the permittivity is about 50% smaller than that of the bulk, even for quasi non-oxidized samples. The oxidation appears to be slow to get going. This finding is coherent with that for the two previous samples. The relative permittivity of oxide at $\hbar\omega = 1.96$ eV appears to get closer to that of bulk, except for the two last samples. In these cases, the air inclusions in oxide may explain the strong decay seen in Figure A14. The increase in the imaginary part of the copper permittivity at small photon energies is clear in Figure A13, but for small thicknesses. In the same region of photon energies, the decrease in the real part is similar to that of the two previous samples.

Table A5. The recovered thickness of copper and copper oxide, and the relative permittivity of copper and copper oxide at $\lambda_0 = 632.8$ nm, by fitting experimental UV-visible-NIR absorbance curves. The recovered relative permittivities can be compared to those of bulk copper and copper oxide, resp. $-11.55 + 1.57$i and $8.64 + 0.64$i [29]. The value of the multi-objective function F (Equation (A3)) and the most efficient metaheuristics are also indicated. Values between brackets are standard deviation for acceptable solutions, i.e., obtained for F below 25% threshold (for $F \in [\min(F); 1.25 \times \min(F)]$. The corresponding success of each metaheuristic is indicated in percentages).

Sample	h_{Cu} (nm)	h_{ox} (nm)	n_{Cu}^2 at $\hbar\omega = 1.96$ eV	n_{ox}^2 at $\hbar\omega = 1.96$ eV	F	Method
A050nmCut0	50.0 (0.1)	0.1 (0.2)	$-15.4 + 2.0$i (0.5 + 0.4i)	$7.9 + 1.0$i (0.4 + 0.3i)	1.0×10^{-2} (5×10^{-4})	EM + NM
			$n_{ABC} = 38\%, n_{EM} = 62\%, n_{PSO} = 0\%, n_{NM} = 100\%, n_{GR} = 0$			
A050nm22min120	49.8 (0.1)	0.1 (0.2)	$-14.6 + 1.8$i (0.4 + 0.4i)	$8.2 + 0.8$i (0.3 + 0.3i)	9.3×10^{-3} (9×10^{-4})	EM + NM
			$n_{ABC} = 26\%, n_{EM} = 48\%, n_{PSO} = 26\%, n_{NM} = 100\%, n_{GR} = 0$			
A050nm72min120	49.8 (0.3)	0.6 (0.3)	$-14.2 + 2.2$i (0.4 + 0.4i)	$8.3 + 0.9$ i (0.3 + 0.3i)	7.0×10^{-3} (4×10^{-4})	EM + NM
			$n_{ABC} = 33\%, n_{EM} = 51\%, n_{PSO} = 15\%, n_{NM} = 100\%, n_{GR} = 0$			
A050nm102min120	49.8 (0.3)	0.6 (0.3)	$-13.2 + 2.3$i (0.3 + 0.5i)	$8.8 + 0.5$i (0.3 + 0.3i)	5.7×10^{-3} (7×10^{-4})	EM + NM
			$n_{ABC} = 33\%, n_{EM} = 56\%, n_{PSO} = 11\%, n_{NM} = 100\%, n_{GR} = 0$			
A050nm160min120	49.2 (0.5)	1.6 (0.6)	$-12.5 + 2.3$i (0.4 + 0.4i)	$8.4 + 0.8$i (0.3 + 0.2i)	5.6×10^{-2} (4×10^{-4})	EM + NM
			$n_{ABC} = 38\%, n_{EM} = 59\%, n_{PSO} = 3\%, n_{NM} = 100\%, n_{GR} = 0$			
A050nm190min120	49.1 (0.5)	1.8 (0.6)	$-12.2 + 2.3$i (0.4 + 0.4i)	$8.3 + 0.7$i (0.3 + 0.2i)	5.5×10^{-3} (3×10^{-4})	EM + NM
			$n_{ABC} = 43\%, n_{EM} = 57\%, n_{PSO} = 0\%, n_{NM} = 100\%, n_{GR} = 0$			
A050nm240min130	42.1 (3.0)	14.1 (5.5)	$-13.7 + 2.6$i (0.8 + 0.7i)	$7.4 + 1.3$i (0.4 + 0.4i)	7.1×10^{-3} (9×10^{-4})	PSO + NM
			$n_{ABC} = 0\%, n_{EM} = 38\%, n_{PSO} = 62\%, n_{NM} = 100\%, n_{GR} = 0$			
A050nm600min130	39.0 (3.8)	19.0 (6.7)	$-14.4 + 3.5$i (1.0 + 0.8i)	$8.1 + 1.4$i (0.3 + 0.7i)	8.4×10^{-3} (1×10^{-3})	PSO + NM
			$n_{ABC} = 29\%, n_{EM} = 31\%, n_{PSO} = 40\%, n_{NM} = 97\%, n_{GR} = 3$			
A050nm900min170	0.4 (0.6)	81.9 (4.6)	$-9.1 + 2.3$i (0.9 + 1.1i)	$6.1 + 0.1$i (0.5 + 0.1i)	1.4×10^{-2} (1×10^{-3})	EM + NM
			$n_{ABC} = 7\%, n_{EM} = 83\%, n_{PSO} = 11\%, n_{NM} = 100\%, n_{GR} = 0$			
A050nm910min170	0.0 (0.6)	89.6 (4.4)	$-11.5 + 2.8$i (0.8 + 1.1i)	$6.6 + 0.3$i (0.5 + 0.1i)	1.4×10^{-2} (8×10^{-4})	ABC + NM
			$n_{ABC} = 97\%, n_{EM} = 0\%, n_{PSO} = 3\%, n_{NM} = 100\%, n_{GR} = 0$			

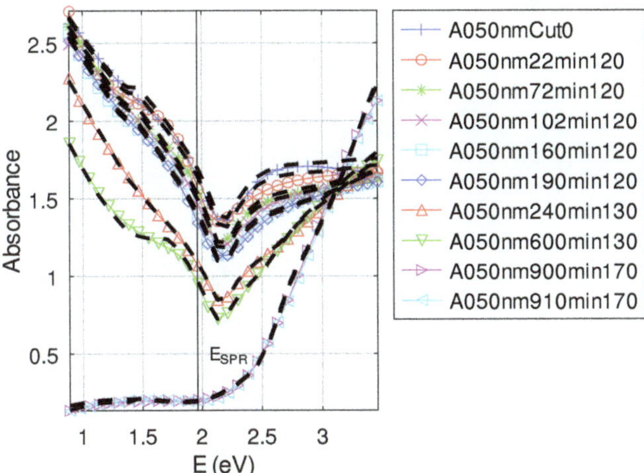

Figure A11. Absorbance as a function of the photon energy: experimental (black dashed) and best calculated solution (color) for each oxidized sample (initial thickness $h_{Cu}^0 = 50$ nm).

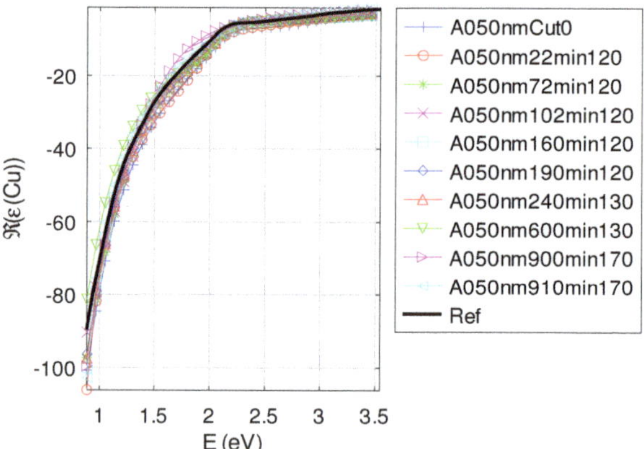

Figure A12. Real part of the relative permittivity of copper as a function of the photon energy: bulk (black) and best calculated (color) for each oxidized sample (initial thickness $h^0_{Cu} = 50$ nm).

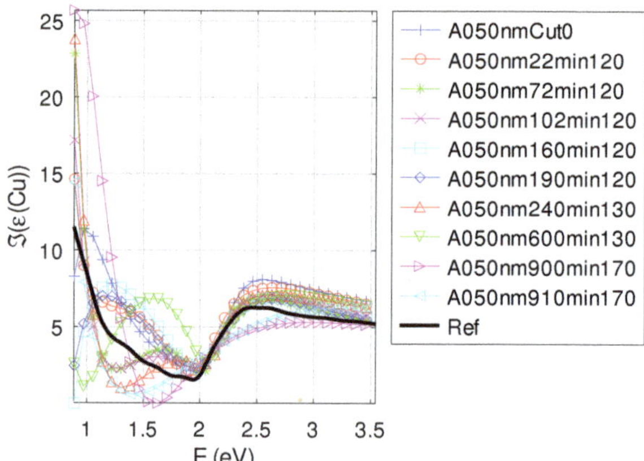

Figure A13. Imaginary part of the relative permittivity of copper as a function of the photon energy: bulk (black) and best calculated (color) for each oxidized sample (initial thickness $h^0_{Cu} = 50$ nm).

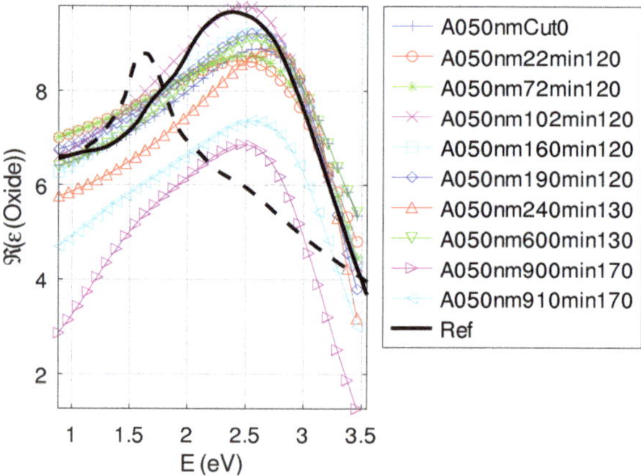

Figure A14. Real part of the relative permittivity of oxide as a function of the photon energy: bulk Cu_2O (black solid line), bulk CuO (black dashed line) and best calculated (color) for each oxidized sample (initial thickness $h^0_{Cu} = 50$ nm).

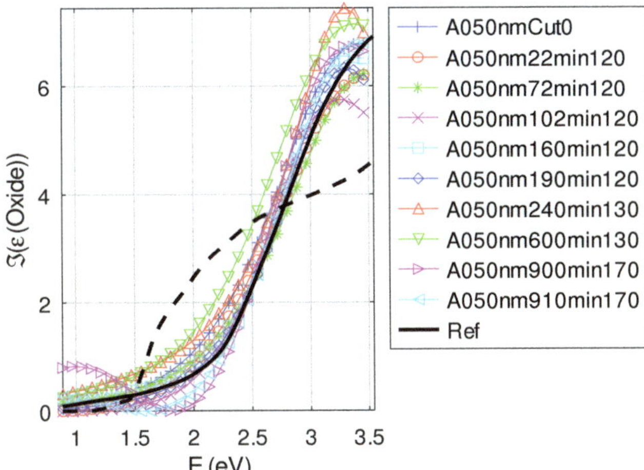

Figure A15. Imaginary part of the relative permittivity of oxide as a function of the photon energy: bulk Cu$_2$O (black solid line), bulk CuO (black dashed line) and best calculated (color) for each oxidized sample (initial thickness $h^0_{Cu} = 50$ nm).

Appendix C.5. Discussion on the Resolution of the Inverse Problem: the Fitting of UV-Visible-NIR Absorbance Curves

We used three metaheuristics to recover both thicknesses and optical properties of copper and copper oxide. ABC is globally the most efficient for thin layers, EM takes the second place, and PSO may succeed for oxidized samples. The hybridization with NM is the most efficient in all cases. Let us note that the success of each metaheuristic depends on the tuning of its exogenous parameters. In our case, the fitting requires $D = 28$ input parameters for the model; therefore, the fitting may be considered a hard problem. In this case, the "no free lunch theorem" can apply, the topology of the model being dependent on the balance between copper and copper oxide thicknesses [54]. Indeed, two metaheuristics are equivalent if their performances are comparable for all possible problems. Therefore, the resolution of a difficult inverse problem would require more than one metaheuristic to ensure the best result. The repeated realizations of the same algorithm and the use of a tolerance threshold for the values multi-objective function (here 25%) help assess the stability of the methods and the relevance of the outcome [55]. Using this careful approach leads to better results than those found in Reference [15]. The values obtained independently, for all samples, with the same tuning of optimization methods are physically sound.

The electromagnetic model of a plane multilayer combined with partial fraction models for copper (order 4) and oxide (order 2) allow fast calculation and seems to be sufficient to describe the oxidation of thin layers of copper. Actually, the roughness of the layers is small enough [14], and the possible air inclusion in thick oxide layers is reflected in the partial fraction model (Figures A9 and A14).

The model reproduces the dips in UV-visible-NIR absorbance curves. The main one in absorbance curves is close to 2.1 eV and characteristic of the inter-band transition from d states (valence band) to 's-p' conduction bands [18] (Figures A1, A7 and A11). The low boundary of photon energy is greater than the bandgap characteristic of the copper structure [56]. Figures A3, A8 and A13 show a global offset of the relative permittivity of copper relative to the bulk one, especially at low energies: the imaginary part is much smaller. On the contrary, copper is a more absorbing material at higher photon energies. The second dip near 1.25–1.4 eV in Figures A1 and A7 appears for thicknesses of copper and oxide of the same order of magnitude. In this case, the dip reveals the indirect gap of CuO. It disappears for complete oxidation, for which Cu$_2$O dominates. The observed bandgap near 2 and 2.2 eV are not those for CuO observed at a higher temperature of annealing (623 K) [57] but are closer to that of Cu_2O (near 2.1 eV) [56]). The slope change in absorbance of pure oxide samples appears near 2.2–2.4 eV and does not indicate a weakened CuO bandgap [57] but the Cu_2O bandgap near 2.5 eV [13,58]. This last reference separates CuO and Cu_2O bandgaps (respectively, 1.4–1.5 and 2.5 eV). Our results are intermediate between those in Reference [56] (2.1 eV) and [13,58] (2.5 eV) and are coherent with those in Reference [36]. This result is confirmed in Reference [59]: copper oxide phases are mainly Cu_2O at oxidation temperature below 400 °C for complete oxidation. CuO remains in the first oxidation steps. This inter-band transition peak (Cu_2O) appears and increases with both temperature and oxidation time. However, this latter is attenuated for increasing copper thicknesses. Indeed, for smaller thicknesses (<10 nm), the electric field is stronger. Therefore, the movement of Cu^+ ions through the oxide toward the reaction zone is impeded. For $h_{Cu} > 10$ nm, the effect of the electric field decreases, and the diffusion of the cations become limited. The optical property of copper oxide moves closer to the CuO bulk at low photon energies. For high photon energies, the imaginary part of the relative permittivity of oxide is greater than that of bulk Cu_2O. This may be due to the outbreak of nanoparticles (NPs) of Cu_2O, which red-shifts the dips.

References

1. Kretschmann, E.; Raether, H. Radiative Decay of Nonradiative Surface Plasmons Excited by Light. *Z. Naturforsch. A* **1968**, *23A*, 2135–2136. [CrossRef]
2. Ahn, H.; Song, H.; Choi, J.R.; Kim, K. A Localized Surface Plasmon Resonance Sensor Using Double-Metal-Complex Nanostructures and a Review of Recent Approaches. *Sensors* **2018**, *18*, 98. [CrossRef] [PubMed]
3. Chen, Z.; Zhao, X.; Lin, C.; Chen, S.; Yin, L.; Ding, Y. Figure of merit enhancement of surface plasmon resonance sensors using absentee layer. *Appl. Opt.* **2016**, *55*, 6832–6835. [CrossRef] [PubMed]
4. Meng, Q.Q.; Zhao, X.; Lin, C.Y.; Chen, S.J.; Ding, Y. C.; Chen, Z.Y. Figure of Merit Enhancement of a Surface Plasmon Resonance Sensor Using a Low-Refractive-Index Porous Silica Film. *Sensors* **2017**, *17*, 1846. [CrossRef]
5. Lee, S.; Kim, J.Y.; Lee, T.W.; Kim, W.K.; Kim, B.S.; Park, J.H.; Bae, J.S.; Cho, Y.C.; Kim, J.; Oh, M.W.; et al. Fabrication of high-quality single-crystal Cu thin films using radio-frequency sputtering. *Sci. Rep.* **2014**, *4*, 6230. [CrossRef]
6. Tripathi, A.; Dixit, T.; Agrawal, J.; Singh, V. Bandgap engineering in CuO nanostructures: Dual-band, broadband, and UV-C photodetectors. *Appl. Phys. Lett.* **2020**, *116*, 111102. [CrossRef]
7. Editorial Feature of AEO Nano. Copper (Cu) Nanoparticles-Properties, Applications. Available online: https://www.azonano.com/article.aspx?ArticleID=3271 (accessed on 9 May 2021).
8. Kesarwani, R.; Khare, A. Surface plasmon resonance and nonlinear optical behavior of pulsed laser-deposited semitransparent nanostructured copper thin films. *Appl. Phys. B* **2018**, *124*, 116. [CrossRef]
9. Rodrigues, E.P.; Oliveira, L.C.; Silva, M.L.F.; Moreira, C.S.; Lima, A.M.N. Surface Plasmon Resonance Sensing Characteristics of Thin Copper and Gold Films in Aqueous and Gaseous Interfaces. *IEEE Sens. J.* **2020**, *20*, 7701–7710. [CrossRef]
10. Stebunov, Y.V.; Yakubovsky, D.I.; Fedyanin, D.Y.; Arsenin, A.V.; Volkov, V.S. Superior Sensitivity of Copper-Based Plasmonic Biosensors. *Langmuir* **2018**, *34*, 4681–4687. [CrossRef]
11. Granito, C.; Wilde, J.; Petty, M.; Houghton, S.; Iredale, P. Toluene vapour sensing using copper and nickel phthalocyanine Langmuir-Blodgett films. *Thin Solid Films* **1996**, *284–285*, 98–101. [CrossRef]
12. Homola, J.; Yee, S.S.; Gauglitz, G. Surface Plasmon Resonance Sensors Review. *Sens. Actuators B* **1999**, *54*, 3–15. [CrossRef]
13. Murali, D.S.; Kumar, S.; Choudhary, R.J.; Wadikar, A.D.; Jain, M.K.; Subrahmanyam, A. Synthesis of Cu_2O from CuO thin films: Optical and electrical properties. *Int. Adv. Eng. Res. Sci. (IJAERS)* **2015**, *5*, 047143-1–047143-5. [CrossRef]
14. Cakir, D. Enhanced Raman Signatures on Copper Based-Materials. Ph.D. Thesis, Univesité de Montpellier, Montpellier, France, 2017.
15. Barchiesi, D.; Cakir, D.; Grosges, T.; Fréty, N.; Anglaret, E. Recovering effective thicknesses and optical properties of copper and copper oxide layers from absorbance measurements. *Opt. Mater.* **2019**, *91*, 138–146. [CrossRef]
16. Han, L.; Zhou, D.; Li, K.; Xun, l.; Huang, W.P. A Rational-Fraction Dispersion Model for Efficient Simulation of Dispersive Material in FDTD Method. *J. Light. Technol.* **2012**, *30*, 2216–2225. [CrossRef]
17. Michalski, K.A. On the Low-Order Partial-Fraction Fitting of Dielectric Functions at Optical Wavelengths. *IEEE Trans. Antennas Propag.* **2013**, *61*, 6128–6135. [CrossRef]
18. Gharbi, T.; Barchiesi, D.; Kessentini, S.; Maalej, R. Fitting optical properties of metals by Drude-Lorentz and partial-fraction models in the [0.5; /, 6] eV range. *Opt. Mater. Express* **2020**, *10*, 1129–1162. [CrossRef]
19. Barchiesi, D.; Otto, A. Excitations of surface plasmon polaritons by attenuated total reflection, revisited. *Riv. Nuovo C.* **2013**, *36*, 173–209.
20. Barchiesi, D.; Kremer, E.; Mai, V.P.; Grosges, T. A Poincaré's Approach for Plasmonics The Plasmon Localization. *J. Microsc.* **2008**, *229*, 525–532. [CrossRef]

21. Barchiesi, D.; Lidgi-Guigui, N.; Lamy de la Chapelle, M. Functionalization Layer Influence on the Sensitivity of Surface Plasmon Resonance (SPR) Biosensor. *Opt. Commun.* **2012**, *285*, 1619–1623. [CrossRef]

22. Salvi, J.; Barchiesi, D. Measurement of thicknesses and optical properties of thin films from Surface Plasmon Resonance (SPR). *Appl. Phys. A* **2014**, *115*, 245–255. [CrossRef]

23. Kessentini, S.; Barchiesi, D. Nanostructured Biosensors Influence of Adhesion Layer, Roughness and Size on the LSPR A Parametric Study. In *State of the Art in Biosensors-General Aspects*; Rinken, T., Ed.; INTECH Open Access: London, UK, 2013; Chapter 12, pp. 311–330, ISBN 978-953-51-1004-0.

24. Colas, F.; Barchiesi, D.; Kessentini, S.; Toury, T.; Lamy de la Chapelle, M. Comparison of adhesion layers of gold on silicate glasses for SERS detection. *J. Opt.* **2015**, *17*, 114010. [CrossRef]

25. Barchiesi, D. A classroom theory of the surface plasmon polariton. *Eur. J. Phys.* **2012**, *33*, 1345–1357. [CrossRef]

26. Lecaruyer, P.; Canva, M.; Rolland, J. Metallic Film Optimization in a Surface Plasmon Resonance Biosensor by the Extended Rouard Method. *Appl. Opt.* **2007**, *46*, 2361–2369. [CrossRef] [PubMed]

27. Barchiesi, D. Improved method based on S matrix for the optimization of SPR biosensors. *Opt. Commun.* **2013**, *286*, 23–29. [CrossRef]

28. Barchiesi, D. Numerical retrieval of thin aluminium layer properties from SPR experimental data. *Opt. Express* **2012**, *20*, 9064–9078. [CrossRef]

29. Software Spectra, I. Optical Data from Sopra SA. Available online: http://www.sspectra.com/sopra.html (accessed on 9 May 2021).

30. Ramirez, M.; Henneken, L.; Virtanen, S. Oxidation kinetics of thin copper films and wetting behaviour of copper and Organic Solderability Preservatives (OSP) with lead-free solder. *Appl. Surf. Sci.* **2011**, *257*, 6481–6488. [CrossRef]

31. Shanidand, M.; Abdul Khadar, M.; Sathe, V.G. Fröhlich interaction and associated resonance enhancement in nanostructured copper oxide films. *J. Raman Spectrosc.* **2011**, *42*, 1769–1773. [CrossRef]

32. Figueiredo, V.; Elangovan, E.; Goncalves, G.; Barquinha, P.; Pereira, L.; Franco, N.; Alves, E.; Martins, R.; Fortunato, E. Effect of post-annealing on the properties of copper oxide thin films obtained from the oxidation of evaporated metallic copper. *Appl. Surf. Sci.* **2008**, *254*, 3949–3954. [CrossRef]

33. Gao, W.; Hong, H.; He, J.; Thomas, A.; Chan, L.; Li, S. Oxidation behaviour of Cu thin films on Si wafer at 175–400 ° C. *Mater. Lett.* **2001**, *51*, 78–84. [CrossRef]

34. Iijima, J.; Lim, J.W.; Hong, H.S.; Suzuki, S.; Mimura, K.; Isshiki, M. Native oxidation of ultra high purity Cu bulk and thin films. *Appl. Surf. Sci.* **2006**, *253*, 2825–2829. [CrossRef]

35. Cabrera, N.; Mott, N.F. Theory of the oxidation of metals. *Rep. Prog. Phys.* **1949**, *12*, 163. [CrossRef]

36. Mugwang'a, F.; Karimi, P.; Njoroge, W.; Omayio, O.; Waita, S. Optical characterization of Copper Oxide thin films prepared by reactive dc magnetron sputtering for solar cell applications. *Int. J. Thin Film Sci. Technol.* **2012**, *2*, 15–24.

37. Shi, Y.; Eberhart, R.C. A modified particle swarm optimizer. In Proceedings of the IEEE Congress on Evolutionary Computation (CEC'98), Anchorage, Alaska, 4–9 May 1998; pp. 69–73.

38. Rahmat-Samii, Y. Genetic algorithm (GA) and particle swarm optimization (PSO) in engineering electromagnetics. In Proceedings of the 17th International Conference on Applied Electromagnetics and Communications, Dubrovnik, Republika Hrvatska, 1–3 October 2003; pp. 1–5. [CrossRef]

39. Robinson, J.; Rahmat-Samii, Y. Particle swarm optimization in electromagnetics. *IEEE Trans. Antennas Propag.* **2004**, *52*, 397–407. [CrossRef]

40. Barchiesi, D. Numerical optimization of plasmonic biosensors. In *New Perspectives in Biosensors Technology and Applications*; Serra, P.A., Ed.; INTECH Open Access: London, UK, 2011; Chapter 5, pp. 105–126, ISBN 978-953-307-448-1.

41. Turbadar, T. Complete Absorption of Light by Thin Metal Films. *Proc. Phys. Soc.* **1959**, *73*, 40–44. [CrossRef]

42. Holland, J.H. *Adaptation in Natural and Artificial Systems: An Introductory Analysis with Applications to Biology, Control, and Artificial Intelligence*; Michigan University: Ann Arbor, MI, USA, 1975.

43. Bäck, T.; Hammel, U.; Schwefel, H.P. Evolutionary Computation Comments on the History and Current State. *IEEE Trans. Evol. Comput.* **1997**, *1*, 3–17. [CrossRef]

44. Macías, D.; Vial, A.; Barchiesi, D. Evolution Strategies Approach for the Solution of an Inverse Problem in Near-Field Optics. In *Lecture Notes in Computer Science (6e European Workshop on Evolutionary Computation in Image Analysis and Signal Processing)*; Raidl, G., Cagnoni, S., Branke, J., Corne, D.W., Drechsler, R., Jin, Y., Johnson, C., Machado, P., Marchiori, E., Rothlauf, F., et al., Eds.; Springer: Berlin/Heidelberg, Germany, 2004; Volume 3005/2004, pp. 329–338.

45. Johnson, J.; Rahmat-Samii, V. Genetic algorithms in engineering electromagnetics. *IEEE Antennas Propag. Mag.* **1997**, *39*, 7–21. [CrossRef]

46. Rahmat-Samii, Y.; Michielssen, E. *Electromagnetic Optimization by Genetic Algorithms*, 1st ed.; John Wiley & Sons, Inc.: Hoboken, NJ, USA, 1999.

47. Macías, D.; Vial, A.; Barchiesi, D. Application of Evolution Strategies for the Solution of an Inverse Problem in Near-Field Optics. *J. Opt. Soc. Am. A* **2004**, *21*, 1465–1471. [CrossRef]

48. Karaboga, D. *An Idea Based on Honey Bee Swarm for Numerical Optimization*; Technical Report tr06; Erciyes University, Engineering Faculty, Computer Engineering Department: Kayseri, Turkey, 2005.

49. Zaman, M.; Gaffar, M.; Alam, M.M.; Mamun, S.; Matin, M. Synthesis of Antenna Arrays Using Artificial Bee Colony Optimization Algorithm. *Int. J. Microw. Opt. Technol.* **2011**, *6*, 234–241.

50. Kessentini, S.; Naãs, I. Absolute versus stochastic stability of the artificial bee colony in synchronous and sequential modes. *Natural Comput.* **2021**, *20*, 443–470. [CrossRef]
51. Chiandussi, G.; Codegone, M.; Ferrero, S.; Varesio, F. Comparison of multi-objective optimization methodologies for engineering applications. *Comput. Math. Appl.* **2012**, *63*, 912–942. [CrossRef]
52. Jiang, M.; Luo, Y.P.; Yang, S.Y. Stochastic convergence analysis and parameter selection of the standard particle swarm optimization algorithm. *Inf. Process. Lett.* **2007**, *102*, 8–16. [CrossRef]
53. Trelea, I.C. The particle swarm optimization algorithm convergence analysis and parameter selection. *Inf. Process. Lett.* **2003**, *85*, 317–325. [CrossRef]
54. Wolpert, D.H.; Macready, W. No Free Lunch Theorems for Optimization. *IEEE Trans. Evol. Comput.* **1997**, *1*, 67–82. [CrossRef]
55. Barchiesi, D.; Grosges, T. Propagation of uncertainties and applications in numerical modeling: tutorial. *J. Opt. Soc. Am. A* **2017**, *34*, 1602–1619. [CrossRef] [PubMed]
56. Korzhavyi, P.A.; Johansson, B. *Literature Review on the Properties of Cuprous Oxide Cu_2O and the Process of Copper Oxidation*; Technical Report SKB TR-11-08, Svensk Kärnbränslehantering AB; Swedish Nuclear Fueland Waste Management Co.: Stockholm, Sweden, 2011.
57. Alhassan, W.A.A.; Wadi, I.A. Determination of Optical Energy Gap fo rCopper oxide at Different Temperatures. *Int. Adv. Eng. Res. Sci. (IJAERS)* **2018**, *5*, 255–258. [CrossRef]
58. Wang, Y.; Lany, S.; Ghanbaja, J.; Fagot-Revurat, Y.; Chen, Y.P.; Soldera, F.; Horwat, D.; Mücklich, F.; Pierson, J.F. Electronic structures of Cu_2O, Cu_4O_3, and CuO: A joint experimental and theoretical study. *Phys. Rev. B* **2016**, *94*, 245418. [CrossRef]
59. Hu, Y.Z.; Sharangpani, R.; Tay, S.P. Kinetic investigation of copper film oxidation by spectroscopic ellipsometry and reflectometry. *J. Vac. Sci. Technol. A* **2000**, *18*, 2527–2532. [CrossRef]

Article

Two-Colour Sum-Frequency Generation Spectroscopy Coupled to Plasmonics with the CLIO Free Electron Laser

Christophe Humbert [1,*], Olivier Pluchery [2], Emmanuelle Lacaze [2], Bertrand Busson [1] and Abderrahmane Tadjeddine [1]

[1] Université Paris-Saclay, CNRS, Institut de Chimie Physique, UMR8000, 91405 Orsay, France; bertrand.busson@universite-paris-saclay.fr (B.B.); abderrahmane.tadjeddine@universite-paris-saclay.fr (A.T.)
[2] Sorbonne Université, CNRS, Institut des NanoSciences de Paris (INSP), 4 Place Jussieu, 75005 Paris, France; olivier.pluchery@insp.jussieu.fr (O.P.); emmanuelle.lacaze@insp.jussieu.fr (E.L.)
* Correspondence: christophe.humbert@universite-paris-saclay.fr

Abstract: Nonlinear plasmonics requires the use of high-intensity laser sources in the visible and near/mid-infrared spectral ranges to characterise the potential enhancement of the vibrational fingerprint of chemically functionalised nanostructured interfaces aimed at improving the molecular detection threshold in nanosensors. We used Two-Colour Sum-Frequency Generation (2C-SFG) nonlinear optical spectroscopy coupled to the European CLIO Free Electron Laser in order to highlight an energy transfer in organic and inorganic interfaces built on a silicon substrate. We evidence that a molecular pollutant, such as thiophenol molecules adsorbed on small gold metal nanospheres grafted on silicon, was detected at the monolayer scale in the 10 μm infrared spectral range, with increasing SFG intensity of three specific phenyl ring vibration modes reaching two magnitude orders from blue to green–yellow excitation wavelengths. This observation is related to a strong plasmonic coupling to the thiophenol molecules vibrations. The high level of gold nanospheres aggregation on the substrate allows us to dramatically increase the presence of hotspots, revealing collective plasmon modes based on strong local electric fields between the gold nanoparticles packed in close contact on the substrate. This configuration favors detection of Raman active vibration modes, for which 2C-SFG spectroscopy is particularly efficient in this unusual infrared spectral range.

Keywords: gold; nanoparticles; thiophenol; silicon; nonlinear optics; sum-frequency generation; plasmonics; UV-vis spectroscopy; atomic force microscopy; CLIO free electron laser

Citation: Humbert, C.; Pluchery, O.; Lacaze, E.; Busson, B.; Tadjeddine, A. Two-Colour Sum-Frequency Generation Spectroscopy Coupled to Plasmonics with the CLIO Free Electron Laser. *Photonics* **2022**, *9*, 55. https://doi.org/10.3390/photonics9020055

Received: 22 December 2021
Accepted: 17 January 2022
Published: 20 January 2022

Publisher's Note: MDPI stays neutral with regard to jurisdictional claims in published maps and institutional affiliations.

1. Introduction

Nonlinear optical Two-colour Sum-Frequency Generation (2C-SFG) spectroscopy of metal [1–3] and semiconducting [4–6] nanoparticles constitutes a reliable probe of the surface molecular chemistry of nanostructured samples in catalysis and (bio)chemical sensing [7]. For instance, in such systems, the evolution of various physico-chemical properties impacted by surface plasmons and excitons can be analysed [8,9]. In a general way, whatever the probed scale, 2C-SFG spectroscopy gives access to intramolecular vibronic couplings [10–12], molecule/substrate interactions [13–16] and molecule/nanostructure interactions [17–26]. Similar to most of the spectroscopy techniques from the nonlinear optics family [27], 2C-SFG spectroscopy works only with high-intensity laser sources. Here we need two high-intensity laser sources in the infrared (IR) and in the visible spectral ranges. While it is now possible from almost 20 years [28,29] to take profit from a visible wavelength tunable from violet to red in such spectroscopy, highlighting the effect of electronic properties of any kind of interface, the major drawback was the access of the infrared wavelength in optical parametric oscillators/amplifiers (OPO/OPA). Indeed, it is quite difficult to extract enough power from nonlinear crystals (e.g., $AgGaS_2$) beyond 8 μm without quick damage. As a consequence, the great majority of the SFG spectroscopy literature lies in the 2.5–8 μm spectral range. Vibration modes from OH, NH, CH, CN,

CO chemical bonds lie in this spectral range and were extensively scrutinised with SFG. Nevertheless, probing C-C bonds around 10 μm, such as in phenyl rings (as in thiophenol molecules, for instance, in the present work), is highly challenging, although they are fundamental markers as evidenced in Raman spectroscopy [30] and for modelling by DFT calculations [31]. This will be our topic here.

In order to address these issues, technical improvements have been carried out for years in SFG spectroscopy by switching the pulse duration from the ps temporal regime to the fs one, in order to broaden—as much as possible in one high-intensity laser pulse—the infrared spectral range. While the fs scale allows us to reach the 10 μm range, the delivered power remains limited for applications and, generally, to the detriment of the visible counterpart, limited to a fixed visible wavelength in the ps scale. While this promising approach is the most relevant for the next future of 2-Dimensional (2D) nonlinear spectroscopy for any kind of application [32], giving access to dynamical processes of surface chemistry, a successful alternative technical route for 2C-SFG spectroscopy was investigated from the beginning: the use of a high-intensity tunable infrared beam provided by Free Electron Lasers (FEL) in catalysis and electrochemistry [33–35], leading more recently to an experimental breakthrough by the SFG imaging of surface phonons [36].

In the present work, thanks to the use of the CLIO FEL [37], we successfully performed 2C-SFG spectroscopy in the 10 μm IR spectral range on small gold AuNPs functionalised by thiophenol molecules, acting as a molecular surface probe, for different visible wavelengths from blue to red. In this way, we highlighted and quantified an amplification of the SFG signal of three specific thiophenol vibration modes due to an optical coupling to surface plasmon resonance. This coupling effect is related to the presence of numerous hotspots due to the strong aggregation of AuNPs. This close-packed AuNPs configuration on the substrate induces the existence of strong local electric fields between them, leading to the presence of collective plasmon modes in the yellow–orange spectral range, overwhelming the LSPR (Localised Surface Plasmon Resonance) contribution of isolated AuNPs in the green. This framework leads to the amplification of the thiophenol nonlinear vibrational response by two orders of magnitude by switching from blue (480 nm) to yellow–orange (560–600 nm). To the best of the authors' knowledge, this is the first time that such strong optical plasmonic coupling has been observed and calculated in this vibrational fingerprint spectral range (~1000 cm^{-1}) on nanostructured interface with 2C-SFG spectroscopy.

2. Materials and Methods

2.1. Sample Preparation

The sample preparation is based on the same reliable and easily reproducible protocol described in our previous publications [38–40] as summarised hereafter and sketched in Figure 1, the purpose being for this specific study to lead to a large number of close-packed AuNPs. All chemicals were purchased from Sigma-Aldrich (France). AuNPs were synthesised from 1.7 mg of gold salts (HAuCl$_4$.3H$_2$O, 99.999%) dissolved in 20 mL Millipore water following Turkevich protocol. After the solution was heated up until boiling point and stirred vigorously, 0.8 mL of 8.5×10^{-4} M trisodium citrate (Na$_3$C$_6$H$_5$O$_7$) was added all at once while heating and stirring for 30 min. The solution then turned successively from light yellow to deep gray, looking finally dark red after ~10 min. The corresponding UV-visible absorbance spectrum of the resulting water solution of AuNPs was therefore obtained to monitor the position of the absorption peak. In order to graft AuNPs on the silicon Si(100) substrate, wafers of ultrasonically cleaned n-doped silicon (1 $\times$ 1 cm^2, Siltronix) were silanised in an absolute methanol solution through 3-aminopropyl-triethoxysilane (APTES, H$_2$N(CH$_2$)$_3$Si(OC$_2$H$_5$)$_3$, 10% vol.) over 90 min. Afterwards, they were dipped in the AuNPs colloidal solution over 24 h in order to ensure a large surface density of AuNPs Ns~8×10^{10}/cm^2, corresponding to a filling factor f ~15% as deduced by AFM measurements [40]. After grafting, AuNPs were functionalised during 18 h with a 10^{-3} M thiophenol (C$_6$H$_5$SH) solution dissolved in dichloromethane (CH$_2$Cl$_2$). We finally obtained a monolayer of AuNPs grafted to the APTES-Si surface and covered with thiophenol.

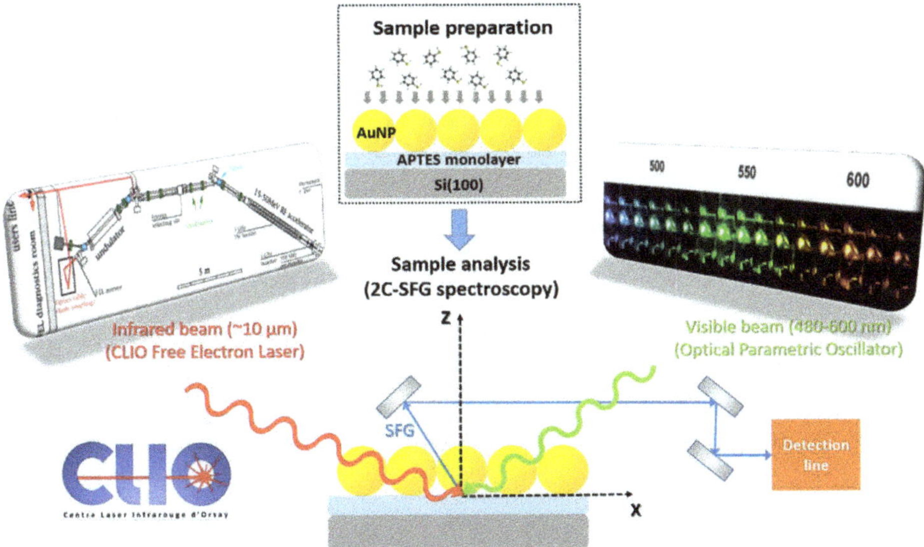

Figure 1. Sketch of the sample preparation and of the 2C-SFG experimental configuration used in counterpropagating geometry. See text for details.

2.2. AFM Microscopy

In order to analyse the surface density Ns of AuNPs grafted on the sample, an AFM imaging (Digital Instrument, DI3100, USA) was used in tapping mode. The working frequency of the silicon tip (curvature radius at the apex ~10 nm) was 130 KHz. This allowed us to count the AuNPs and measure their diameter (17 ± 2 nm) equal to their height with respect to the Si(100) surface, even if it did not have a good lateral resolution to image isolated AuNPs.

2.3. UV-Vis Spectroscopy

Absorption (solution) and reflection (sample) spectroscopy measurements were performed through a Cary 5 spectrophotometer (Varian). On the silicon substrate, the spectra were recorded with an incidence angle of $10°$. In order to detangle the AuNPs signature from the silicon contribution, the bare Si(100) reflectivity R_0 was first measured, prior to the sample reflectivity R in the same conditions. Within this differential measurement procedure, it was therefore possible to access the sample reflectance $-\log(R/R_0)$ with a reproducibility ensured within 1×10^{-3} A.U. (Absorbance Units).

2.4. 2C-SFG Spectroscopy with the CLIO Free Electron Laser Facility

A Sum-Frequency Generation (SFG) tabletop setup based on a Nd:YLF laser source at 1047 nm wavelength (repetition rate for macropulses: 25 Hz; for micropulses: 62.5 MHz) was coupled to the CLIO Free Electron Laser (FEL) Facility with the same temporal features [37]. The CLIO FEL was used to take profit of its intense and pulsed IR laser beam (~1 ps pulse duration, ~10 µJ energy per pulse from which ~2 µJ were used in order to avoid sample damage, ~5 cm^{-1} spectral resolution) to probe the vibration fingerprint of thiophenol molecules (ring deformations) in the ~10 µm spectral range (975–1100 cm^{-1}). The incident visible laser beam (~5 ps, 4 µJ per pulse, ~6 cm^{-1} spectral resolution) was tuned from 480 nm to 600 nm by ~20 nm steps, resulting in 8 SFG spectra corresponding to 8 different incidence visible wavelengths. The tunable visible beam was obtained from a visible Optical Parametric Oscillator (OPO) based on the pumping of a rotating BBO (Barium Borate) nonlinear crystal inside an optical cavity by the 3rd harmonic (UV wavelength = 349 nm) of the Nd:YLF laser source obtained after: first, the frequency-doubling (2nd harmonic visible wavelength = 523.5 nm) through a BBO (Barium Borate) nonlinear crystal: second, the mixing of the remaining IR part with the resulting 2nd harmonic in a LBO (Lithium Borate) nonlinear crystal. The IR and visible incident laser beams were mixed at the same point of the probed interface (~2 mm laser beam diameter) during the experiment. It is worth noting that the SFG experiments were performed in a counter-propagative geometry in order to well discriminate, spatially, the SFG beam from the reflected visible beam (by the silicon surface sample) going through a monochromator for spectral filtering. This original and unusual SFG spectroscopy configuration, aimed at avoiding light saturation of the photomultiplier located at the monochromator output, was derived from previous publications [29,33,41–44], allowing us to fix the detection direction for collecting SFG photons, whatever the IR or visible wavelengths. We improved the detection efficiency thanks to the use of rotating mirrors placed in the SFG path between the sample and the monochromator. Indeed, by considering the energy ($h\omega_{SFG} = h\omega_{Vis} + h\omega_{IR}$) and momentum ($k_{SFG} = k_{Vis} + k_{IR}$: valid for the parallel component only, i.e., in x-direction, see Figure 1) conservation rules in SFG spectroscopy at ~10 μm, whatever the chosen visible beam color, it was possible to well separate spatially the SFG beam from the latter. To meet these conditions, the IR and visible incidence angles of the laser beams were 65° and −55° with respect to the sample surface normal (z-direction) as sketched in Figure 1. All beams were p-polarised. SFG data were normalised by the IR and visible powers in order to compensate for eventual laser fluctuations. All data are also corrected by the optical responses of photomultiplier and monochromator gratings which evolved from blue to orange–red.

3. Results

3.1. AFM Characterisation

Figure 2 displays the typical AFM image of our Si(100)/AuNPs/thiophenol surface. In order to deduce the AuNPs surface density $Ns{\sim}8 \times 10^{10}/cm^2$, several scans were performed for different scales: 2×2 μm^2 and 500×500 nm^2, allowing us to identify and count each individual nanoparticle (17 ± 2 nm diameter) by probing and averaging on at least five areas of the sample surface. The results obtained for the sample structural characterisation were consistent with the specific preparation (dipping of the grafted covered Si for 24 h in the AuNPs solution [38–40]) chosen to lead to AuNPs that were aggregated in close contact. This drastically modified the UV-Vis optical fingerprint of the sample with respect to the optical response in solution as shown thereafter.

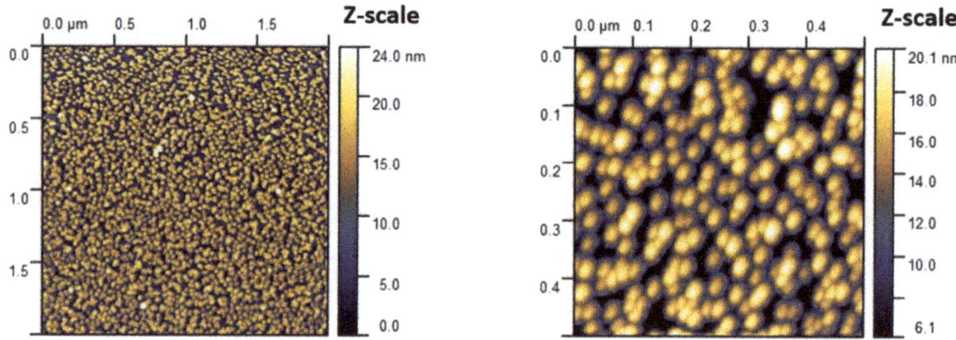

Figure 2. AFM image of the Silicon/APTES/AuNPs/Thiophenol interface. Left: 2×2 μm^2 scale. Right: 500×500 nm^2 scale. Z-scale intensity: height.

3.2. UV-Visible Measurements

Figure 3a displays the UV-Vis absorption spectrum of the AuNPs solution while Figure 3b displays the UV-Vis reflectance spectrum of the Si(100)/AuNPs/thiophenol sample. The AuNPs absorbance spectrum shows a peak at 520 nm, corresponding to their Localised Surface Plasmon Resonance (LSPR). When AuNPs were grafted on the Si(100) surface, the LSPR is dip-shaped and slightly red-shifted to 525 nm. A second strong dip-shaped band appeared at 650 nm in the optical response. The latter one is directly related to AuNPs aggregation occurring during the deposition process on the surface (see Figure 2). This is a well-known phenomenon, demonstrated both theoretically and experimentally for metal nanoparticles in colloidal solutions or deposited with high density on substrates. In fact, this band results from coupling processes of surface plasmons between nanoparticles in close contact (interparticles coupling) [45]. To observe this kind of collective plasmon modes, the AuNPs were physically connected through aggregation, leading to a dipolar coupling between neighboring plasmonic oscillators, which was the case when the spacing between them was smaller than ca. 30 nm. [46]. For instance, this kind of plasmonic collective coupling is used in biosensing to obtain a fine monitoring of the plasmon resonance position by playing at will for instance with gold nanorods size (aspect ratio) and spacing [47,48]. Besides, the two negative spectral shapes observed in Figure 3b on the silicon surface are in relation with the reflectivity measurement mode, because the reflectivity of the silicon was strongly modulated by the presence of a AuNPs layer. Qualitatively, the more gold on the surface, the higher the reflectivity R, compared to the reference R_0. Therefore, the presence of AuNPs induced negative values for the reflectance as observe in Figure 3b. This optical phenomenon is explained in previous works [40,49] based on the calculation of the dielectric constant of the AuNPs layer in the framework of the effective medium theory (Maxwell-Garnett and Bruggeman formalisms of low and high density composite layer, respectively).

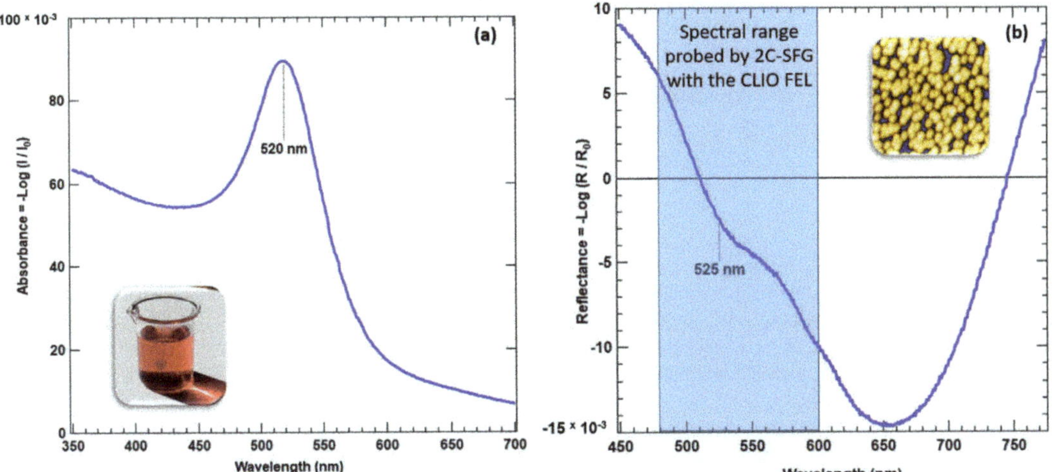

Figure 3. UV-Vis absorption (**a**) and reflectance (**b**) spectra of AuNPs colloidal solution (insert) and Si(100)/AuNPs/thiophenol sample (insert), respectively. The peak or valley shape at 520 and 525 nm wavelengths correspond to the LSPR of the AuNPs.

3.3. 2C-SFG Measurements

2C-SFG measurements of our sample made of 17 nm diameter AuNPs grafted on silicon and functionalised by thiophenol are recorded in counter-propagating geometry and depicted in Figure 4 for 8 different visible laser wavelengths from blue to orange–red as pointed out on the right of the panel. For each visible wavelength, three vibration modes of thiophenol molecules were observed at 1000, 1023 and 1071 cm^{-1}, respectively.

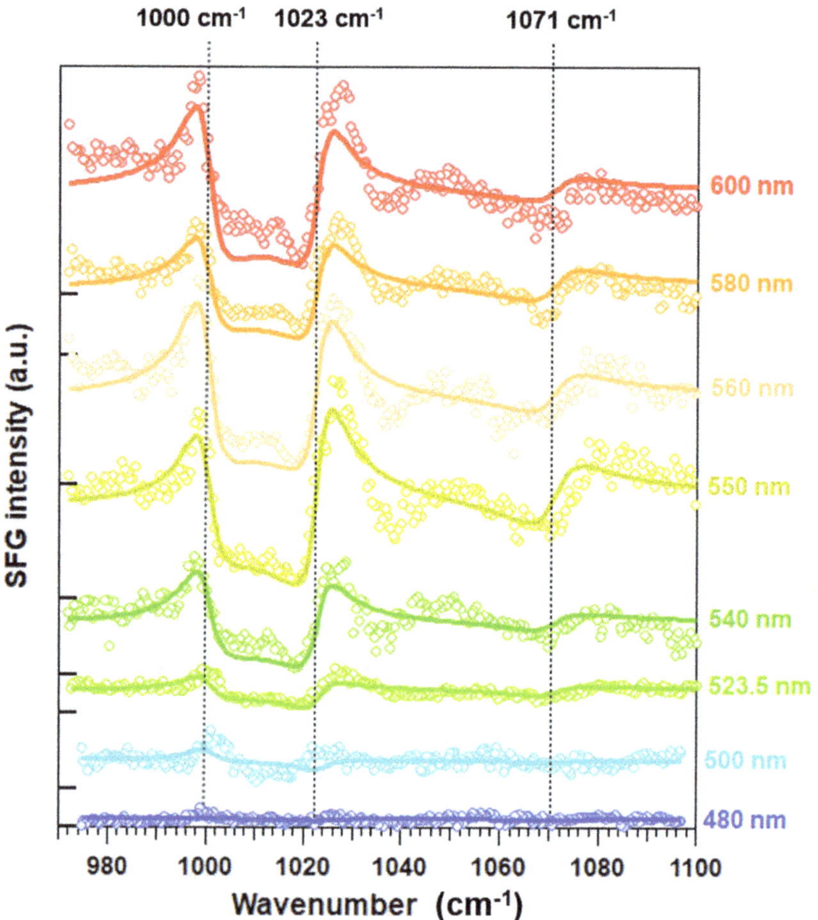

Figure 4. Experimental SFG spectra (open circles) obtained with the CLIO FEL for the Si(100)/AuNPs/thiophenol sample at 8 different visible beam wavelengths of the incident laser beam, ranging from blue to orange–red as noted on the right of the panel. On the left axis, the 0 of each curve is sketched by a short black line. While the SFG spectra are all shifted to ease view, they are all drawn within the same intensity scale to have a direct qualitative comparison. The three main vibration modes tagged at 1000 cm^{-1}, 1023 cm^{-1} and 1071 cm^{-1} correspond to specific phenyl ring deformations of the thiophenol molecules as established in the literature [31,39]. Colored lines are fitted to the SFG experimental data, as detailed and explained in the text. The SFG spectrum at 480 nm visible wavelength will act as the reference to calculate the plasmonic coupling effect and the amplification factor of the thiophenol vibrational fingerprint.

In 2C-SFG spectroscopy, a second order nonlinear optical process, it is counter intuitive to observe a signal from functionalised spherical objects because they are centrosymmetric media. In other words, there is a symmetry inversion center of the electronic properties (charge distribution) for an observer placed at the center of the gold spheres. A quick look on the sample structure given in Figure 1 shows us the reason for this apparent contradiction. The grafting of the AuNPs induced a symmetry breaking of the electronic properties in the z-direction where the ppp polarisation scheme of the three beams (SFG, Visible, IR) was the most efficient for 2C-SFG spectroscopy. Moreover, the AuNPs were mostly functionalised by thiophenol molecules on their upper surface, while they were not on their lower surface, grafted on silicon through APTES molecules. While AuNPs are not, in principle, the most efficient objects for second order nonlinear optics, the current work shows that it is again possible to detect such molecular species on small metal nanospheres and, more interestingly, to evidence the positive role of the plasmonic coupling effect in improving the detection efficiency of 2C-SFG spectroscopy. Besides, it has been shown that the SFG vibration mode amplitude of thiophenol molecules increased linearly with Ns (AuNPs surface density): the more the AuNPs were aggregated, the higher the SFG process efficiency [40]. In fact, the first striking feature observed in Figure 4 is the effect of the selected visible laser beam wavelengths on the SFG signal of the thiophenol vibrational fingerprint. It is clear that the three vibration modes were difficult to observe in the blue spectral range for a common intensity scale but they became more and more apparent with the increasing visible wavelength to the orange–red. The second striking feature is the Fano-like shape of these three vibration modes, i.e., different from valleys or peaks as observed in IR and Raman spectroscopy. This is specific to 2C-SFG spectroscopy as a sensitive tool of the interface constituting the sample. As a result, the two predominant contributions originate from thiophenol molecules and the AuNPs surface where they were adsorbed. In a general way, the SFG intensity writes: $I(\omega_{SFG}) = \frac{8\pi^3 \omega_{SFG}^2 \sec^2 \theta_{SFG}}{c^3 n_1(\omega_{SFG}) n_1(\omega_{IR}) n_1(\omega_{vis})} \left| \chi_{eff}^{(2)} \right|^2 I_{IR} I_{vis}$, with I_{IR} and I_{Vis} the intensities of the incident IR and visible laser beams, respectively. ω_{IR}, ω_{vis} and ω_{SFG} are the IR, visible and sum frequencies (energies) of the three beams involved in the nonlinear second order process. The incoming angles of incidence of the IR and visible beams are θ_{IR} and θ_{vis} while θ_{SFG} is the outgoing angle of incidence of the SFG beam travelling to the monochromator as depicted in Figure 1. From the sample point of view, $n_1(\omega)$ is the refractive index of the upper medium (ambient air) at frequency ω and $\chi_{eff}^{(2)}$ the effective nonlinear second order susceptibility of the probed interface. The latter includes Fresnel factors (sample reflectivity) and microscopic local-field corrections to the (thiophenol) molecular nonlinear 2nd order susceptibilities $\chi_{ijk}^{(2)}$. In these conditions, we model $\chi_{eff}^{(2)} = \parallel C_{NR} \parallel e^{i\Phi_{NR}} + \sum_n \frac{\parallel a_n \parallel e^{i\varphi_n}}{(\omega_{IR} - \omega_n + i\Gamma_n)}$ leading to the colored lines in good agreement with the experimental data points depicted in Figure 4. The first complex term (amplitude C_{NR} and phase shift Φ_{NR}) of the above equality corresponds to the substrate (AuNPs) contribution, which is non-resonant (NR) with IR energy. The second complex term is the sum of the three Lorentzian oscillators of the thiophenol molecules (n = 3) observed in the 975–1100 cm^{-1} spectral range, with a_n, φ_n, ω_n and Γ_n the amplitude, phase, frequency and damping constant of each vibration mode, respectively. It is worth noting that the complex value a_n is a combination of the IR transition dipole moments and Raman polarisability tensors of the molecule. The physical meaning and role of those fundamental parameters are exhaustively detailed in recent publications covering both a classical point of view [27] and advanced quantum formalism [50]. For the fitting of the set of SFG data, we used, for all spectra fitting procedure, the known fixed fit parameters for the SFG spectrum at 523.5 nm. Indeed, for similar samples, it has been established previously [39] that the thiophenol frequencies are ω_n (cm^{-1}) = 1000, 1023, 1071, while the damping constants are set to Γ_n (cm^{-1}) = 3.5, 3.5, 5 and φ_n (°) = 0 for n = 1,2,3, respectively. The results of the fitting procedure are summarised in Table 1. The observed variation of SFG data with visible wavelength is thus mainly related to a significant increase of the amplitude absolute value a_n of the three thiophenol vibration modes when moving from blue to orange–red. A similar behavior was observed for AuNPs amplitude C_{NR} but on a smaller relative scale. We define the intensity ratio $I_n^{ratio} = \parallel a_n / C_{NR} \parallel^2$ as calculated in Table 1 for the three thiophenol vibration modes and for each visible wavelength. I_n^{ratio} increased by a factor of ~100 from blue (480 nm) to green–yellow (540–560 nm), and remained almost stable in the orange–red spectral range. In the current work, we have chosen to consider that the reference SFG measurement was the one at 480 nm (blue) in order to calculate the plasmonic coupling effect on the amplification factor of the SFG vibrational molecular fingerprint. Indeed, no plasmonic effect and no significant s-d interband electronic transition competed with the SFG process efficiency when looking at the UV-Visible reflectance curve in the 480 nm spectral range depicted in Figure 3b. Comparative reference SFG measurements between an identical sample with respect to a flat gold surface covered by thiophenol performed in a previous work [38] demonstrated that the amplification factor by the AuNPs was around ~21 (for thiophenol SFG intensity) at a fixed 532 nm visible wavelength. A comparison with thiophenol adsorbed on silicon or glass is not relevant from the chemical and optical point of view because: (i) thiophenol does not adsorb on silicon or glass; (ii) only a comparison between a flat vs. a curved gold surface takes account of the plasmonic coupling effect for similar chemical conditions of molecular adsorption. In other words, we observed, in the best conditions of our SFG measurements of Figure 4, an amplification factor of two orders of magnitude for the SFG intensity of the thiophenol vibration modes as a function of the laser color impinging the AuNPs with respect to the reference SFG spectrum with 480 nm visible wavelength. Besides, a ~65° phase shift Φ_{NR} parameter evolving from blue to orange–red in the fitting procedure allowed us to adjust the interference pattern evolution of the thiophenol vibration modes modulated by gold electronic properties, as explained hereafter.

Table 1. Fit parameters for the 2C-SFG spectra of thiophenol adsorbed on AuNPs as obtained from Figure 4. For φ_n, ω_n and Γ_n, we refer to fixed parameters for the three thiophenol vibration modes as established in [39]. See text for details.

		Visible Incidence Wavelength (nm)							
		480	500	523.5	540	550	560	580	600
AuNPs	C_{NR}	1.4310	2.6538	2.3970	3.5530	4.8961	4.63128	4.0848	4.9943
	Φ_{NR} (°)	69.8753	59.4616	28.0546	12.6793	11.6879	5.5153	7.5741	4.6585
	a_1 (a.u.)	−0.2376	−1.8922	−3.6383	−9.5138	−10.931	−14.104	−9.3308	−12.517
	φ_1 (°)					0			
Vibr 1	ω_1 (cm^{-1})					1000			
	Γ_1 (cm^{-1})					3.5			
	I_1^{ratio}	0.02757	0.5084	2.3039	7.17	4.9845	9.2743	5.2179	6.2813
	a_2 (a.u.)	0.5194	1.6344	5.2915	10.8236	16.4952	14.3315	10.3851	11.763
	φ_2 (°)					0			
Vibr 2	ω_2 (cm^{-1})					1023			
	Γ_2 (cm^{-1})					3.5			
	I_2^{ratio}	0.1317	0.3793	4.8733	9.2801	11.3505	9.5759	6.4637	5.5474
	a_3 (a.u.)	0.2721	0.7215	2.6407	3	7.8051	5.2497	5	3
	φ_3 (°)					0			
Vibr 3 (Mode 4 in [31,39])	ω_3 (cm^{-1})					1071			
	Γ_3 (cm^{-1})					5			
	I_3^{ratio}	0.0361	0.0739	1.2137	0.7129	2.5413	1.2849	1.4983	0.3608

4. Discussion

4.1. Phase Shift Parameter: Gold s-d Interband Electronic Transitions

The evolving spectral shapes take their origin in the interference phenomenon between the SFG signals of the molecular adsorbate (thiophenol) and the substrate (AuNPs). This is well explained in the literature on gold surfaces and related to its s-d interband electronic transitions [51]. On a flat gold surface covered by a monolayer of dodecanethiol (DDT), a similar decreasing phase shift of ~65° Φ_{NR} was observed between 480 and 600 nm, inducing a shape reversal of the vibration modes when moving from a blue visible incidence beam to a red one. This is related to the different electronic contributions of the gold surface and bulk properties (free and bound electrons of the electronic density of states of gold, see

Figure 5, right) as exhaustively developed elsewhere [52,53], giving access to the dispersion of the absolute phase and the relative amplitude of its $\chi^{(2)}_{\text{eff}}$ in the visible spectral range. It allows us to model quantitatively SFG spectra of molecules adsorbed on metal surfaces. Our SFG measurements are performed and analysed in the same theoretical framework, leading to consistent interpretations and conclusions.

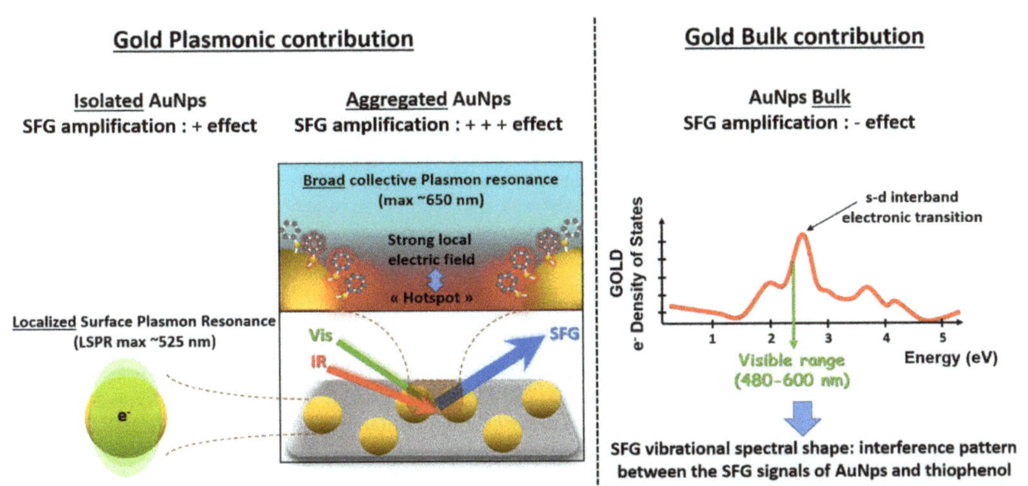

Figure 5. Sketch of the physical processes competing positively (+) or negatively (−) in the amplification of the thiophenol vibrational fingerprint for 2C-SFG spectroscopy coupled to plasmonics based on small 17 nm diameter AuNPs grafted on a silicon substrate. The LSPR and collective plasmon modes of AuNPs contributed favourably (left), especially as the number of hotspots increased on the sample. The surface and bound electrons of the gold bulk of AuNPs contributed unfavourably (right) due to the high electronic density of states in the probed visible (480–600 nm) spectral range. The latter physical process also explains the interference pattern between the SFG signals of the thiophenol molecules and the AuNPs constituting the interface. See text and references [51–53] for details on the influence of gold bulk electronic properties.

4.2. Co- vs. Counter-Propagating Geometry: SFG Spectra Shape Reversal

A notable difference is observed for the SFG spectrum at 523.5 nm wavelength with respect to reference [39], where a similar measurement was recorded, but in co-propagating classical geometrical configuration for the three beams involved in the SFG process: the shape of the three vibration modes displayed a different interference pattern. In practice, Φ_{NR} = ~140° in [39] while Φ_{NR} = ~30° in our current work in counter-propagating geometry. The only difference between these two works lies in the setup geometry; we can explain the phase shift difference $\geq \pi/2$ by an inversion symmetry of the visible beam incidence angle as depicted in Figure 1. The sign of the x-components of the probed $\chi^{(2)}_{\text{eff}}$ is inverted with respect to this previous work in such a way that the SFG vibrational spectral shapes are, therefore, reversed.

4.3. Amplification of Thiophenol Vibration Modes: Collective Plasmon Modes Coupling

In order to interpret the striking amplification factor, increasing as a function of laser color, we have to consider the optical properties of our 17 nm diameter AuNPs where thiophenol molecules are grafted, that are characterised by plasmon resonances (LSPR for isolated particles and collective modes for aggregated particles) and s-d interband electronic transitions (see Figure 5). In our case, due to the strong aggregation, the collective plasmon modes already coupled very efficiently to the SFG process (thiophenol vibration modes intensity) showing a continuous significant enhancement of the SFG intensity when the color was tuned from blue to orange–red. Nevertheless, its efficiency became slightly damped/stabilised beyond 560 nm, as observed in Figure 4 and Table 1, because of the more prominent role played by the surface and bound electrons of bulk gold in this spectral range. These two competing physical processes are summarised in Figure 5. The plasmonic coupling effect is similar to those observed and modeled in previous works on alkane thiol chains [1,17,23,24] but with a greater efficiency in the present case because we were able to probe selectively better conditions with the tunable visible beam, which was not the case in the aforementioned works. Indeed, we observed here a vibrational SFG intensity amplification of two orders of magnitude for thiophenol molecules, whereas it was only of around one order of magnitude for alkanethiol chains (DDT). This result can be explained by the combination of the following factors. Firstly, by considering our AFM images in Figure 2 and UV-Visible spectrum in Figure 3b, we had the substrate surface completely filled with AuNPs in tight contact, forming numerous aggregates. This constituted the best conditions to obtain numerous hotspots between AuNPs (Figure 5), with strong local electric fields where thiophenol molecules were bathing. These numerous hotspots added to the LSPR enhancement process alone [1] due to the resulting existence of collective plasmon modes as explained in Section 3.2. In our case, it was highlighted by the UV-Vis intensity continuous increase observed in Figure 3b, beyond the local LSPR maximum (525 nm) from blue to red colors in our SFG framework. Secondly, thiophenol molecules are of shorter length than alkanethiol molecules. The three vibration modes observed in Figure 4 underwent a stronger influence from the local electric field than the methyl end-groups (CH_3) of DDT. Thirdly, the thiophenol phenyl rings involved in the SFG process have a rich π-electronic structure with respect to DDT. All these factors come into play to explain our observations, as well as the fact that phenyl rings present well-known intense Raman active modes in the ~1000 cm^{-1} spectral range [30,31]. Since vibro-electronic 2C-SFG spectroscopy depends on IR transition dipole moments and Raman polarisability of molecules, further investigation will necessitate different polarisation combinations of the three beams involved in the nonlinear process in order to give access to the different components of the molecular nonlinear second order susceptibilities $\chi_{ijk}^{(2)}$.

To go further, considering the AuNPs' size effect on these two physical values would reveal the following. For the aggregation effects (650 nm), it would only shift the position and broaden the maximum absorption in the optical range of Figure 3b. From the optical point of view, it corresponds to tune the visible wavelength in the SFG process to find the best amplification factor for the plasmonic coupling effect. It could be very interesting for dedicated applications requiring a fine monitoring of the plasmon resonance position in the visible range as encountered for gold nanorods where it is possible to play with their aspect ratio and their spacing, such as in biosensing [47,48], as already mentioned in Section 3.2. For the s–d interband electronic transition, it will increase the non-resonant contribution from blue to red (C_{NR} parameter in Table 1) of the gold SFG signal because it is a bulk effect, with no change in the position of the s-d interband electronic transition as illustrated in Figure 5. It is worth noting that an increasing value of C_{NR} (with an increasing AuNPs size) will compete negatively with the plasmonic amplification factor.

4.4. Technical Prospects for 2C-SFG Spectroscopy

In order to analyse more precisely the role of AuNPs LSPR and hotspots, it will be necessary to gain a wider visible spectral range in direction of red colors. Besides, it should be noted that recording data beyond 10 μm in 2C-SFG spectroscopy will require better detection (spatial and spectral filtering) conditions, because the SFG beam is emitted in a direction almost colinear to the reflected visible beam with separation angles <1° in co-propagating geometry. It is consequently difficult to extract one SFG photon from the surrounding visible noise 10^6 times more intense that the nonlinear processes. The counter-propagating geometry proposed here gives an original alternative to sweep this problem in the mid/far IR range but remains demanding from the experimental point of view. Besides, it is worth noting that just increasing the visible laser beam power to compensate for weak

IR power in the 10 μm spectral range is not adapted for such nanostructured samples, explaining the lack of results published in the scientific literature. Indeed, increasing the visible beam energy will quickly damage (heating and burning) the sample, the aromatic rings of thiophenol molecules first, then the AuNPs and the silicon substrate next. This is related to the fact that the required mean visible energy distributed for large laser beam scale (2 mm diameter for IR and Visible beams in our case) on the sample to generate a nonlinear process becomes very important. There is a power balance to find with the required IR energy threshold which enhances specifically the vibration modes of thiophenol molecules (and is Non-Resonant for gold), knowing that the two IR and visible beams have the same spot size at the same point of the probed interface. In these conditions, the local electric field between hotspots being amplified by a factor of 10^3–10^6 is commonly observed in Raman spectroscopy at the molecular scale, so it is by far more interesting to use plasmonic amplification coupled to enough IR power to favour and enhance the SFG process to have a good spectral resolution. In other words, for nonlinear optics coupled to plasmonics, in order to have enough visible energy is a necessary condition but not sufficient for vibrational spectroscopy. A continuous effort in the development of table-top IR fs laser sources remains mandatory and challenging.

5. Conclusions

The CLIO Free Electron Laser has been revealed as a powerful and pioneering tool when coupled to SFG spectroscopy, giving a unique optical setup that led to numerous experimental firsts during 25 years of operation, thanks to its unique delivered power in the infrared spectral range, opening the door to nonlinear optical spectroscopy in the vibrational fingerprint as illustrated in the present work. We highlighted a plasmon-molecular coupling by 2C-SFG spectroscopy with an intensity amplification factor around two orders of magnitude from blue to green–yellow due to the presence of numerous hotspots between aggregates of small gold nanospheres functionalised by molecular thiophenol probes. Thanks to our work, if we combine the amplification factor obtained by comparing the SFG signal of thiophenol adsorbed on a flat gold surface with respect to thiophenol adsorbed on AuNPs (~20), with the one calculated here (~100), we tend to reach a global amplifcation factor of the SFG intensity of 10^3. This must be put in perspective with the commonly and easily reached better Raman amplification factors (10^3–10^6), knowing that: (i) our sample is made of small spherical objects, far from being the best conditions for 2C-SFG spectroscopy (centrosymmetry to be broken); (ii) conversely, the Raman spectroscopy efficiency become limited with decreasing size of nano-objects. Indeed, for Raman, smaller means greater light diffusion cross section, especially below 100 nm, damping drastically the molecular sensitivity in an intrinsic Raman signal background from gold nanospheres. Our prominent and promising results show that nonlinear plasmonics constitutes an efficient probe of constantly increasing threshold sensitivity. The golden age of surface nonlinear optical spectroscopy is only at its beginning.

Author Contributions: Conceptualisation of the 2C-SFG setup coupled to the CLIO FEL, A.T., B.B. and C.H.; sample preparation, O.P. and C.H.; UV-Vis and AFM characterisation, O.P. and E.L.; 2C-SFG measurements, C.H. and B.B.; 2C-SFG data treatment and analysis, C.H.; writing—original draft preparation, C.H.; writing—review and editing, C.H., O.P., E.L., B.B. and A.T.; funding acquisition of the 2C-SFG setup, A.T. All authors have read and agreed to the published version of the manuscript.

Funding: This research was financially supported by the EU Horizon 2020 Programme (CALIPSOPlus under grant number 730872).

Data Availability Statement: The data presented in this study are available on reasonable request from the corresponding author.

Acknowledgments: The authors acknowledge Research Engineer C. Six, Assistant Engineer A. Gayral for their continuous technical assistance in the proper functioning of the 2C-SFG spectroscopy setup, Research Engineer J.-P. Berthet, Research Director J.-M. Ortega for their technical and scientific support for FEL operation. C. Humbert also acknowledges all the past and present scientific (F. Glotin,

R. Prazeres) and technical (J. Vieira, G. Périlhous, F. Gobert, V. Fèries, B. Rieul and N. Jestin, Technical Head of the CLIO FEL) staff met at the CLIO Facility since year 1997 during his Master Internship until this day.

Conflicts of Interest: The authors declare no conflict of interest.

References

1. Dalstein, L.; Humbert, C.; Ben Haddada, M.; Boujday, S.; Barbillon, G.; Busson, B. The Prevailing Role of Hotspots in Plasmon-Enhanced Sum-Frequency Generation Spectroscopy. *J. Phys. Chem. Lett.* **2019**, *10*, 7706–7711. [CrossRef]
2. Lis, D.; Caudano, Y.; Henry, M.; Demoustier-Champagne, S.; Ferain, E.; Cecchet, F. Selective Plasmonic Platforms Based on Nanopillars to Enhance Vibrational Sum-Frequency Generation Spectroscopy. *Adv. Opt. Mater.* **2013**, *1*, 244–255. [CrossRef]
3. Lis, D.; Cecchet, F. Localized surface plasmon resonances in nanostructures to enhance nonlinear vibrational spectroscopies: Towards an astonishing molecular sensitivity. *Beilstein J. Nanotechnol.* **2014**, *5*, 2275–2292. [CrossRef] [PubMed]
4. Noblet, T.; Dreesen, L.; Boujday, S.; Méthivier, C.; Busson, B.; Tadjeddine, A.; Humbert, C. Semiconductor quantum dots reveal dipolar coupling from exciton to ligand vibration. *Commun. Chem.* **2018**, *1*, 76. [CrossRef]
5. Noblet, T.; Boujday, S.; Méthivier, C.; Erard, M.; Hottechamps, J.; Busson, B.; Humbert, C. Two-Dimensional Layers of Colloidal CdTe Quantum Dots: Assembly, Optical Properties, and Vibroelectronic Coupling. *J. Phys. Chem. C* **2020**, *124*, 25873–25883. [CrossRef]
6. Sengupta, S.; Bromley, L.; Velarde, L. Aggregated States of Chalcogenorhodamine Dyes on Nanocrystalline Titania Revealed by Doubly Resonant Sum Frequency Spectroscopy. *J. Phys. Chem. C* **2017**, *121*, 3424–3436. [CrossRef]
7. Dreesen, L.; Sartenaer, Y.; Humbert, C.; Mani, A.A.; Lemaire, J.-J.; Méthivier, C.; Pradier, C.-M.; Thiry, P.A.; Peremans, A. Sum-frequency generation spectroscopy applied to model biosensors systems. *Thin Solid Films* **2004**, *464–465*, 373–378. [CrossRef]
8. Humbert, C.; Noblet, T.; Dalstein, L.; Busson, B.; Barbillon, G. Sum-Frequency Generation Spectroscopy of Plasmonic Nanomaterials: A Review. *Materials* **2019**, *12*, 836. [CrossRef] [PubMed]
9. Noblet, T.; Dreesen, L.; Tadjeddine, A.; Humbert, C. Spatial dependence of the dipolar interaction between quantum dots and organic molecules probed by two-color sum-frequency generation spectroscopy. *Symmetry* **2021**, *13*, 1636. [CrossRef]
10. Dreesen, L.; Humbert, C.; Sartenaer, Y.; Caudano, Y.; Volcke, C.; Mani, A.A.; Peremans, A.; Thiry, P.A.; Hanique, S.; Frère, J.-M. Electronic and molecular properties of an adsorbed protein monolayer probed by two-color sum-frequency generation spectroscopy. *Langmuir* **2004**, *20*, 7201–7207. [CrossRef]
11. Raab, M.; Becca, J.C.; Heo, J.; Lim, C.K.; Baev, A.; Jensen, L.; Prasad, P.N.; Velarde, L. Doubly resonant sum frequency spectroscopy of mixed photochromic isomers on surfaces reveals conformation-specific vibronic effects. *J. Chem. Phys.* **2019**, *150*, 114704. [CrossRef] [PubMed]
12. Busson, B.; Farhat, M.; Nini Teunda, P.J.; Roy, S.; Jarisz, T.; Hore, D.K. All-experimental analysis of doubly resonant sum-frequency generation spectra: Application to aggregated rhodamine films. *J. Chem. Phys.* **2021**, *154*, 224704. [CrossRef] [PubMed]
13. Peremans, A.; Caudano, Y.; Thiry, P.A.; Dumas, P.; Zhang, W.Q.; Le Rille, A.; Tadjeddine, A. Electronic Tuning of Dynamical Charge Transfer at an Interface: K Doping of C60/Ag(111). *Phys. Rev. Lett.* **1997**, *78*, 2999–3002. [CrossRef]
14. Caudano, Y.; Silien, C.; Humbert, C.; Dreesen, L.; Mani, A.A.; Peremans, A.; Thiry, P.A. Electron-phonon couplings at C60 interfaces: A case study by two-color, infrared-visible sum-frequency generation spectroscopy. *J. Electron. Spectros. Relat. Phenom.* **2003**, *129*, 139–147. [CrossRef]
15. Elsenbeck, D.; Das, S.K.; Velarde, L. Substrate influence on the interlayer electron-phonon couplings in fullerene films probed with doubly-resonant SFG spectroscopy. *Phys. Chem. Chem. Phys.* **2017**, *19*, 18519–18528. [CrossRef]
16. Chou, K.C.; Westerberg, S.; Shen, Y.R.; Ross, P.N.; Somorjai, G.A. Probing the charge-transfer state of CO on Pt(111) by two-dimensional infrared-visible sum frequency generation spectroscopy. *Phys. Rev. B Condens. Matter Mater. Phys.* **2004**, *69*, 1–4. [CrossRef]
17. Busson, B.; Dalstein, L. Sum-Frequency Spectroscopy Amplified by Plasmonics: The Small Particle Case. *J. Phys. Chem. C* **2019**, *123*, 26597–26607. [CrossRef]
18. Dalstein, L.; Ben Haddada, M.; Barbillon, G.; Humbert, C.; Tadjeddine, A.; Boujday, S.; Busson, B. Revealing the Interplay between Adsorbed Molecular Layers and Gold Nanoparticles by Linear and Nonlinear Optical Properties. *J. Phys. Chem. C* **2015**, *119*, 17146–17155. [CrossRef]
19. Linke, M.; Hille, M.; Lackner, M.; Schumacher, L.; Schlücker, S.; Hasselbrink, E. Plasmonic Effects of Au Nanoparticles on the Vibrational Sum Frequency Spectrum of 4-Nitrothiophenol. *J. Phys. Chem. C* **2019**, *123*, 24234–24242. [CrossRef]
20. Kawai, T.; Neivandt, D.J.; Davies, P.B. Sum frequency generation on surfactant-coated gold nanoparticles. *J. Am. Chem. Soc.* **2000**, *122*, 12031–12032. [CrossRef]
21. Tourillon, G.; Dreesen, L.; Volcke, C.; Sartenaer, Y.; Thiry, P.A.; Peremans, A. Total internal reflection sum-frequency generation spectroscopy and dense gold nanoparticles monolayer: A route for probing adsorbed molecules. *Nanotechnology* **2007**, *18*, 415301. [CrossRef]
22. Tourillon, G.; Dreesen, L.; Volcke, C.; Sartenaer, Y.; Thiry, P.A.; Peremans, A. Close-packed array of gold nanoparticles and sum frequency generation spectroscopy in total internal reflection: A platform for studying biomolecules and biosensors. *J. Mater. Sci.* **2009**, *44*, 6805–6810. [CrossRef]

23. Weeraman, C.; Yatawara, A.K.; Bordenyuk, A.N.; Benderskii, A.V. Effect of nanoscale geometry on molecular conformation: Vibrational sum-frequency generation of alkanethiols on gold nanoparticles. *J. Am. Chem. Soc.* **2006**, *128*, 14244–14245. [CrossRef]

24. Bordenyuk, A.N.; Weeraman, C.; Yatawara, A.; Jayathilake, H.D.; Stiopkin, I.; Liu, Y.; Benderskii, A.V. Vibrational sum frequency generation spectroscopy of dodecanethiol on metal nanoparticles. *J. Phys. Chem. C* **2007**, *111*, 8925–8933. [CrossRef]

25. Alyabyeva, N.; Ouvrard, A.; Zakaria, A.M.; Bourguignon, B. Probing Nanoparticle Geometry down to Subnanometer Size: The Benefits of Vibrational Spectroscopy. *J. Phys. Chem. Lett.* **2019**, *10*, 624–629. [CrossRef]

26. Molinaro, C.; Cecchet, F. Label-free, quantitative and sensitive detection of nanoparticle/membrane interactions through the optical response of water. *Sens. Actuators B Chem.* **2019**, *289*, 169–174. [CrossRef]

27. Humbert, C.; Noblet, T. A unified mathematical formalism for first to third order dielectric response of matter: Application to surface-specific two-colour vibrational optical spectroscopy. *Symmetry* **2021**, *13*, 153. [CrossRef]

28. Raschke, M.B.; Hayashi, M.; Lin, S.H.; Shen, Y.R. Doubly-resonant sum-frequency generation spectroscopy for surface studies. *Chem. Phys. Lett.* **2002**, *359*, 367–372. [CrossRef]

29. Dreesen, L.; Humbert, C.; Celebi, M.; Lemaire, J.J.; Mani, A.A.; Thiry, P.A.; Peremans, A. Influence of the metal electronic properties on the sum-frequency generation spectra of dodecanethiol self-assembled monolayers on Pt (111), Ag (111) and Au (111) single crystals. *Appl. Phys. B Lasers Opt.* **2002**, *74*, 621–625. [CrossRef]

30. Carron, K.T.; Hurley, L.G. Axial and azimuthal angle determination with surface-enhanced Raman spectroscopy: Thiophenol on copper, silver, and gold metal surfaces. *J. Phys. Chem.* **1991**, *95*, 9979–9984. [CrossRef]

31. Feugmo, C.G.T.; Liégeois, V. Analyzing the vibrational signatures of thiophenol adsorbed on small gold clusters by DFT calculations. *ChemPhysChem* **2013**, *14*, 1633–1645. [CrossRef]

32. Hosseinpour, S.; Roeters, S.J.; Bonn, M.; Peukert, W.; Woutersen, S.; Weidner, T. Structure and Dynamics of Interfacial Peptides and Proteins from Vibrational Sum-Frequency Generation Spectroscopy. *Chem. Rev.* **2020**, *120*, 3420–3465. [CrossRef] [PubMed]

33. Braun, R.; Casson, B.D.; Bain, C.D.; Van Der Ham, E.W.M.; Vrehen, Q.H.F.; Eliel, E.R.; Briggs, A.M.; Davies, P.B. Sum-frequency generation from thiophenol on silver in the mid and far-IR. *J. Chem. Phys.* **1999**, *110*, 4634–4640. [CrossRef]

34. Bozzini, B.; D'Urzo, L.; Mele, C.; Busson, B.; Humbert, C.; Tadjeddine, A. Doubly resonant sum frequency generation spectroscopy of adsorbates at an electrochemical interface. *J. Phys. Chem. C* **2008**, *112*, 11791–11795. [CrossRef]

35. Bozzini, B.; Abyaneh, M.K.; Busson, B.; Pietro De Gaudenzi, G.; Gregoratti, L.; Humbert, C.; Amati, M.; Mele, C.; Tadjeddine, A. supported Pt-Part III: Monitoring of electrodeposited-Pt catalyst ageing by in situ Fourier transform infrared spectroscopy, in situ sum frequency generation spectroscopy and ex situ photoelectron spectromicroscopy. *J. Power Sources* **2016**, *231*, 6–17. [CrossRef]

36. Kiessling, R.; Tong, Y.; Giles, A.J.; Gewinner, S.; Schöllkopf, W.; Caldwell, J.D.; Wolf, M.; Paarmann, A. Surface Phonon Polariton Resonance Imaging Using Long-Wave Infrared-Visible Sum-Frequency Generation Microscopy. *ACS Photonics* **2019**, *6*, 3017–3023. [CrossRef]

37. Ortega, J.M.; Glotin, F.; Prazeres, R. Extension in far-infrared of the CLIO free-electron laser. *Infrared Phys. Technol.* **2006**, *49*, 133–138. [CrossRef]

38. Pluchery, O.; Humbert, C.; Valamanesh, M.; Lacaze, E.; Busson, B. Enhanced detection of thiophenol adsorbed on gold nanoparticles by SFG and DFG nonlinear optical spectroscopy. *Phys. Chem. Chem. Phys.* **2009**, *11*, 7729–7737. [CrossRef]

39. Humbert, C.; Pluchery, O.; Lacaze, E.; Tadjeddine, A.; Busson, B. A multiscale description of molecular adsorption on gold nanoparticles by nonlinear optical spectroscopy. *Phys. Chem. Chem. Phys.* **2012**, *14*, 280–289. [CrossRef]

40. Humbert, C.; Pluchery, O.; Lacaze, E.; Tadjeddine, A.; Busson, B. Optical spectroscopy of functionalized gold nanoparticles assemblies as a function of the surface coverage. *Gold Bull.* **2013**, *46*, 299–309. [CrossRef]

41. Bain, C.D. Sum-frequency Vibrational Spectroscopy of the Solid/Liquid Interface. *J. Chem. Soc. Faraday Trans.* **1995**, *91*, 1281–1296. [CrossRef]

42. Eliel, E.R.; van der Ham, E.W.M.; Vrehen, Q.H.F. Enhancing the yield in surface sum-frequency generation by the use of surface polaritons. *Appl. Phys. B Lasers Opt.* **1999**, *68*, 349–353. [CrossRef]

43. van der Ham, E.W.M.; Vrehen, Q.H.F.; Eliel, E.R. Self-dispersive sum-frequency generation at interfaces. *Opt. Lett.* **1996**, *21*, 1448–1450. [CrossRef] [PubMed]

44. Mani, A.A.; Dreesen, L.; Humbert, C.; Hollander, P.; Caudano, Y.; Thiry, P.A.; Peremans, A. Development of a two-color picosecond optical parametric oscillator, pumped by a Nd:YAG laser mode locked using a nonlinear mirror, for doubly-resonant sum frequency generation spectroscopy. *Surf. Sci.* **2002**, *502–503*, 261–267. [CrossRef]

45. Grabar, K.C.; Hommer, M.B.; Natan, M.J.; Freeman, R.G. Preparation and Characterization of Au Colloid Monolayers. *Anal. Chem.* **1995**, *67*, 735–743. [CrossRef]

46. Su, K.H.; Wei, Q.H.; Zhang, X.; Mock, J.J.; Smith, D.R.; Schultz, S. Interparticle coupling effects on plasmon resonances of nanogold particles. *Nano Lett.* **2003**, *3*, 1087–1090. [CrossRef]

47. Funston, A.M.; Novo, C.; Davis, T.J.; Mulvaney, P. Plasmon coupling of gold nanorods at short distances and in different geometries. *Nano Lett.* **2009**, *9*, 1651–1658. [CrossRef] [PubMed]

48. Pellas, V.; Hu, D.; Mazouzi, Y.; Mimoun, Y.; Blanchard, J.; Guibert, C.; Salmain, M.; Boujday, S. Gold Nanorods for LSPR Biosensing: Synthesis, Coating by Silica, and Bioanalytical Applications. *Biosensors* **2020**, *10*, 146. [CrossRef] [PubMed]

49. Bossard-Giannesini, L.; Cruguel, H.; Lacaze, E.; Pluchery, O. Plasmonic properties of gold nanoparticles on silicon substrates: Understanding Fano-like spectra observed in reflection. *Appl. Phys. Lett.* **2016**, *109*, 111901. [CrossRef]

50. Noblet, T.; Busson, B.; Humbert, C. Diagrammatic theory of linear and nonlinear optics for composite systems. *Phys. Rev. A* **2021**, *104*, 063504. [CrossRef]
51. Dalstein, L.; Revel, A.; Humbert, C.; Busson, B. Nonlinear optical response of a gold surface in the visible range: A study by two-color sum-frequency generation spectroscopy. I. Experimental determination. *J. Chem. Phys.* **2018**, 148. [CrossRef] [PubMed]
52. Busson, B.; Dalstein, L. Nonlinear optical response of a gold surface in the visible range: A study by two-color sum-frequency generation spectroscopy. II. Model for metal nonlinear susceptibility. *J. Chem. Phys.* **2018**, *149*, 034701. [CrossRef] [PubMed]
53. Busson, B.; Dalstein, L. Nonlinear optical response of a gold surface in the visible range: A study by two-color sum-frequency generation spectroscopy. III. Simulations of the experimental SFG intensities. *J. Chem. Phys.* **2018**, *149*, 154701. [CrossRef] [PubMed]

Review

Surface Enhanced Raman Spectroscopy: Applications in Agriculture and Food Safety

Yuqing Yang, Niamh Creedon, Alan O'Riordan and Pierre Lovera *

Nanotechnology Group, Tyndall National Institute, T12 R5CP Cork, Ireland; yuqing.yang@tyndall.ie (Y.Y.); niamh.creedon@umail.ucc.ie (N.C.); alan.oriordan@tyndall.ie (A.O.)

* Correspondence: pierre.lovera@tyndall.ie

Abstract: Recent global warming has resulted in shifting of weather patterns and led to intensification of natural disasters and upsurges in pests and diseases. As a result, global food systems are under pressure and need adjustments to meet the change—often by pesticides. Unfortunately, such agrochemicals are harmful for humans and the environment, and consequently need to be monitored. Traditional detection methods currently used are time consuming in terms of sample preparation, are high cost, and devices are typically not portable. Recently, Surface Enhanced Raman Scattering (SERS) has emerged as an attractive candidate for rapid, high sensitivity and high selectivity detection of contaminants relevant to the food industry and environmental monitoring. In this review, the principles of SERS as well as recent SERS substrate fabrication methods are first discussed. Following this, their development and applications for agrifood safety is reviewed, with focus on detection of dye molecules, melamine in food products, and the detection of different classes of pesticides such as organophosphate and neonicotinoids.

Keywords: Surface Enhanced Raman Scattering (SERS); fabrication; application; agriculture; food safety

Citation: Yang, Y.; Creedon, N.; O'Riordan, A.; Lovera, P. Surface Enhanced Raman Spectroscopy: Applications in Agriculture and Food Safety. *Photonics* **2021**, *8*, 568. https:// doi.org/10.3390/photonics8120568

Received: 27 October 2021
Accepted: 26 November 2021
Published: 10 December 2021

Publisher's Note: MDPI stays neutral with regard to jurisdictional claims in published maps and institutional affiliations.

1. Introduction

Climate change is manifesting itself with increased temperatures more favorable to spread of pests and diseases. This has resulted in more challenging food production as well as higher numbers of foodborne disease outbreaks. To try to protect yield and ensure food safety, farmers have little choice other than treating their crop with a range of pesticides [1]. Unfortunately, there is more and more evidence showing that these phytosanitary products would do harm to humans, and also lead to the loss of biodiversity, leading to accumulation in soil and deleterious effects on indigenous microbiome [2]. As a result, it is necessary to monitor biological and chemical contamination in food systems throughout the whole food chain.

Traditional detection methods in food and agriculture systems are based on chromatography techniques coupled with mass spectrometry. These methods are time consuming, high cost, and laboratory based. Therefore, implementation of new portable technologies is needed to provide pesticide usage monitoring and regulation.

In this regard, Surface Enhanced Raman Spectroscopy (SERS) has recently attracted a lot of attention as it addresses these requirements. The advantages of SERS include ultrasensitive detection, fast turnover, in-situ sampling, on-site monitoring, low cost, portability of sensors, and the suitability for large-scale screening. Sensors based on SERS have been shown to provide real-time data on soil nutrients [3,4], monitor water run-off and contaminants in water supplies [5], and also detect pesticide residues in food [6,7]. The technique has also found applications in a wide range of sectors including the following: industrial, material, forensic, biological, food safety and electrochemical fields [8–18], with the most common chemical contaminants within environmental and food sectors being pesticides, adulterants, antibiotics and illegal drugs and illegal food dyes.

Photonics **2021**, *8*, 568. https://doi.org/10.3390/photonics8120568

In this paper, we review the main SERS substrate fabrication methods and bring a special focus on applications for the detection of hazardous chemicals in both food and agriculture.

2. Raman Spectroscopy and Surface Enhanced Raman Scattering (SERS)

Raman spectroscopy was first discovered in 1928 by Sir Chandrasekhara Venkata Raman. It was observed that when light excited a molecule, the majority of light was elastically scattered—this is Rayleigh scattering where $E_{incident} = E_{scattered}$, but also a small fraction of photons was inelastically scattered—this corresponds to Raman scattering where $E_{incident} \neq E_{scattered}$. When $E_{scattered} < E_{incident}$, it is called a Stokes shift and when $E_{scattered} > E_{incident}$, it is an anti-Stoke shift. Figure 1 shows an energy diagram for the two types of scattering [19,20]. Each peak on the Raman spectrum corresponds to a vibrational mode of the bonds of the molecule under Investigation. Raman scattering can occur in the near ultraviolet, visible, or near-infrared ranges [21,22]. However, Raman scattering efficiency is very low, with typically only 1 in 10^8 incident photons being Raman scattered. This results in a low signal/noise ratio and can make it impractical and vulnerable to background interferences [23]. Fortunately, the efficiency of this scattering can be increased by using nanostructured plasmonic metals in a technique known as Surface Enhanced Raman Spectroscopy (SERS).

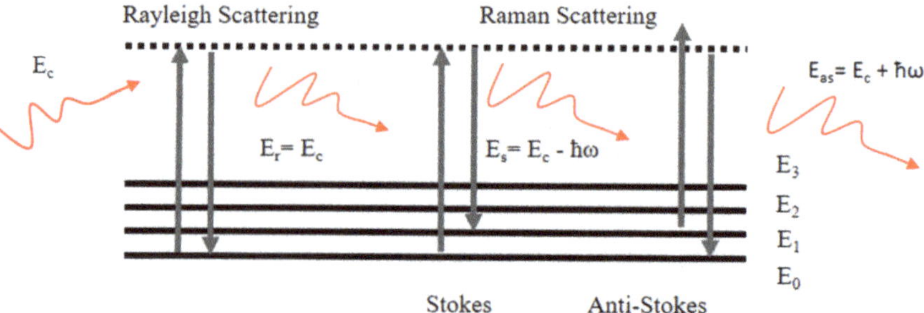

Figure 1. Schematics of energy levels of a molecule for Rayleigh and Raman Scattering [24].

SERS effects were first observed by Fleischman et al. in 1974, when acquiring Raman spectra of pyridine on electrochemically roughened silver [25]. In SERS, lasers are used to excite plasmons in nanostructured metallic surfaces (Au, Ag, Cu, etc.). The SERS enhancement is thought to be two-fold. The first enhancement is based on an electromagnetic enhancement [26,27]. due to localised surface plasmons, while the second originates from chemical resonant energy charge transfer [25,28–30]. The enhancement factor (EF) typically reaches 10^6, which drastically improves the sensitivity of the plasmonic based device.

2.1. Localized Surface Plasmon Resonance (LSPR)—Electromagnetic Enhancement

When laser light excites a metallic nanostructure, the free electrons on its surface oscillate. This collective oscillation is known as Localised Surface Plasmon Resonance (LSPR) [31]. The excited LSPR makes a target molecule highly polarisable and forms a large electric field on the surface. This electric field induces dipole moments in a molecule on the surface of nanostructures, and sequentially produces Raman enhancement. These large, localised electromagnetic fields present around the nanostructures or in nano-gaps between closely-spaced nanostructures are known as "hot spots" [32], see Figure 2. The intensities of the Raman scattered photons are susceptible to enhancement, if their wavelengths are in resonance with the plasmon mode of the nanostructure. LSPR enhancement depends on the size, shape, composition, orientation and local dielectric of the nanostructure. This will be discussed in the next section.

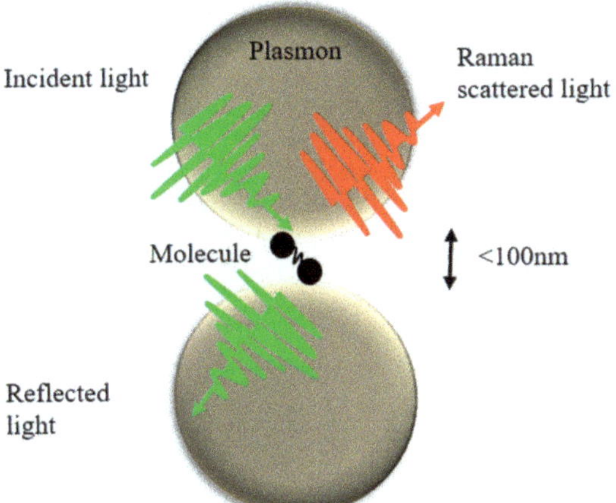

Figure 2. Plasmonic effects: Electric Field around two nanoparticles and the presence of a SERS "hotspot".

2.2. Chemical Enhancement

The origins of the chemical mechanism for SERS enhancement are still under discussion [33]. One hypothesis is that it is based on the change in polarisability of molecule adsorbed to a metal surface. Upon absorption of the incident laser light, charge transfer occurs between the molecule and the metal [34,35]. It is generally estimated that the chemical effect contributes to a factor of 10^2 of the total SERS enhancement [36].

2.3. Enhancement Factor (EF)

The enhancement factor (EF) depends on molecular adsorption on plasmonic surface, as well as the morphology, roughness and homogeneity of this surface and the laser wavelength, etc. The calculation of the enhancement factor is $EF = \frac{I_{SERS}/(\mu_M\,\mu_S\,A_M)}{I_{RS}/(C_{RS}\,H_{eff})}$ [37], where I_{SERS} is the intensity of Surface Enhanced Raman signal, I_{RS} is the intensity of the normal Raman signal, μ_M (m^{-2}) is the surface density of NPs contributing to enhancement, μ_S (m^{-2}) is the surface density of molecules adsorbed to NP, A_M (m^2) is the surface area of metallic NPs, C_{RS} (M) is the concentration of the solution used for non-SERS measurements and H_{eff} (m) is the effective height of the scattering volume.

Experimental values of EFs are typically in the range of 10^4 to 10^6. However, electromagnetic "hotspots" are claimed to have provided massive enhancements of between 10^{11} to 10^{14} orders of magnitude of the SERS signal [38]. This theory is still being largely researched, as it is key to single molecule detection, which may be achieved from selective excitation of single molecules [32,39,40].

3. Fabrication of SERS-Active Substrates

SERS substrates need nanostructured metallic surfaces with well-defined distances in the region of 10–100 nm between nano clusters [41]. By decreasing the distance between the nanostructures, the electric field becomes more localized and concentrated, and the corresponding SERS intensity signal increases accordingly. An example of this was discussed by Lee et al. where the distance between metallic clusters was decreased from 30 nm to 10 nm, and an intensity increase of over 200-fold was observed [42]. There are two fundamentally different approaches to the development of SERS-active nanostructures with "hotspots": bottom-up assembly and top-down fabrication have been used [43].

3.1. Bottom-Up Assembly

Bottom-up approaches refer to the fabrication of nanostructures by chemical synthesis [44,45], colloid aggregation [41,46], electrochemical deposition [47–51], and self-assembly [52–54]. These methods have been used to fabricate a variety of nanostructures ranging from a few nanometers to a few hundred nanometers in size. Metal nanoparticles can be synthesised chemically at low cost with tailored geometries such as nanoparticles [55,56], nanowires [57,58], nanospheres [55,59,60], nanorods [61–64], nanotubes [65,66], nanotriangles [67], nano flower [68], nano-urchins [69], and nanoshells [70,71], see Figure 3. Besides pure metallic nanoparticles, composite materials such as bimetallic or hybrid nanostructures [72], Graphene Oxide/Au nanostars, [73], SiO$_2$@TiO$_2$@Ag [74] etc. and other composite material based on molecular imprint (MIP) have also attracted research attention [68,74–77] and have been reviewed recently [78].

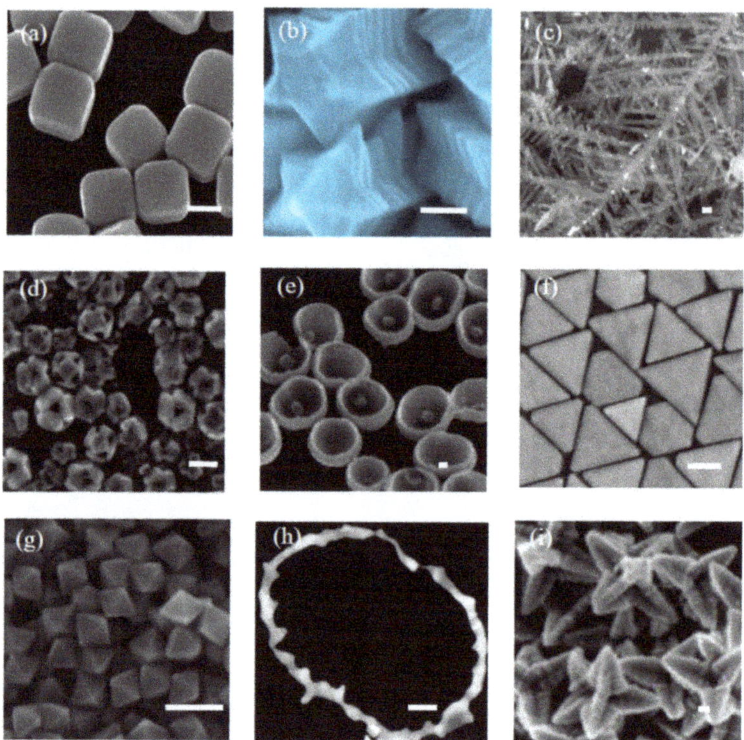

Figure 3. TEM/SEM images of (**a**) Ag/Au nanocubes [79], (**b**) GO/Au nanostars [73], (**c**) Ag nanodendrides [48], (**d**) Au nanocages [80], (**e**) Au nanobowls with Au seed inside [81], (**f**) gold nanotriangles [82], (**g**) Gold octahedrals [83], (**h**) Au nano ring [84], and (**i**) gold hollow stars [85]. Scale bar: 100 nm.

Metal NPs such as gold or silver possess great potential for numerous applications in SERS [86,87]. The most common fabrication of SERS substrates are gold (Au) and silver (Ag) colloids in diameters between 10 and 100 nm, as they yield the greatest enhancements at their "hot spots". These nanoparticles constitute the fundamental SERS "building blocks" and can be assembled in different ways.

For example, they can be presented in a suspension or sol-gel in the presence of the analyte of interest [88–90]. The nanostructures, together with the analyte suspension, can then be drop-casted onto a substrate to create hot spots for Raman enhancement. The disadvantage of this method is that the nanoparticle suspensions must be mixed with the analyte solution for SERS applications [91,92]. However, this drawback was addressed by Yang et al., who grew Ag nanoshells on thiol-modified silica NPs, and deposited them directly on apple skin for analysis [93]. Although there is a greater enhancement observed with these substrates, it is hard to obtain a homogeneous surface to get uniform enhancement. Additionally, they are not suitable for field analysis due to their complex sample preparation steps. In contrast, solid based devices may be more suited for portable and remote sensing, i.e., NPs that are immobilised on a solid substrate [94–97]. For example, Fan et al. fabricated self-assembled Ag NPs onto glass slides by using 3-mercaptopropyltrimethoxysilane (3-MPTMS) [97] This transformation stabilizes the Ag NPs, avoids the usual aggregation process and produces self-sustaining and portable SERS active substrates. Yu et al. fabricated silver colloidal nanoparticles for SERS analysis but alternatively injected them through a filter membrane, thus entrapping them in the filter. The filter therefore was used as the solid portable substrate. It demonstrated 1–2 orders of magnitude better SERS enhancement than the typical approach [98] In addition, Shiohara et al. fabricated gold nanostars and deposited them onto a polydimethylsiloxane (PDMS) platform for SERS evaluation. They used back side illumination for the detection of selected pesticide on fruit skin [99].

Other supports also add additional functionalities to SERS devices. Optical fiber-based SERS sensors have in this regard generated steady interest as a versatile means of extending SERS for portable field applications [63]. There is also a growing interest in the fabrication of flexible SERS substrates. These substrates are ultra-low cost, disposable, easy to use and highly suitable for on-site sensing applications. Polavarapu et al. fabricated SERS substrates by directly writing on paper using a pen filled with plasmonic nanoparticle inks to detect thiabendazole, which is a fungicide and parasiticide [100]. Lee et al. also fabricated SERS paper substrates impregnated with gold nanorods by dip coating [101] Chen et al. combined adhesive tape and SERS activity of Au nanoparticles to fabricate a "SERS tape" substrate. The Au particles were deposited onto the sticky side of the tape that was used to extract pesticides from different kinds of fruit and vegetable peels [102]. These flexible sensors fabricated by different methods dramatically improve the portability and feasibility of SERS detection for pollutants as a promising technique for both laboratory and field-based detection [102].

Overall, bottom up assemblies have shown very high enhancement but they often give inconsistent performance. This is mainly because of a lack of structural uniformity over the entire area of the substrate which could result in poor reproducibility and inhomogeneity, as different size, shape, composition, orientation and local dielectric of the plasmonic structure have different enhancement factors as mentioned before. The arrangements of the aggregates on the nanostructured surface are also hard to control.

3.2. Top-Down Synthesis

Top-down approaches for nanofabrication are scalable and highly reproducible. Top-down approaches include lithography techniques (electron-beam (E-beam) [103–105] and nanoimprint lithography [106,107]), laser etching together with film deposition (sputtering, metal evaporation, atomic layer deposition) [108,109], templating (using anodic aluminium oxide [65], porous polymer (Polyester(PS) [110], masks or molds [57]), inkjet printing [111].

E-beam and nanoimprint lithography are fabrication methods used to create patterns with dimension down to 10 nm. In E-beam lithography, the photon resist is crosslinked after being exposed to the electron beam. The exposed resist can then be washed away, leaving only the unexposed resist on the substrate. Metal layer is then evaporated onto the whole substrate and the unexposed polymer is then removed together with its metal over-layer, leaving only the metal pattern on the substrate. SERS substrates with various geometries such as nanoparticle dimers have been fabricated using E-beam lithography, see Figure 4a–e.

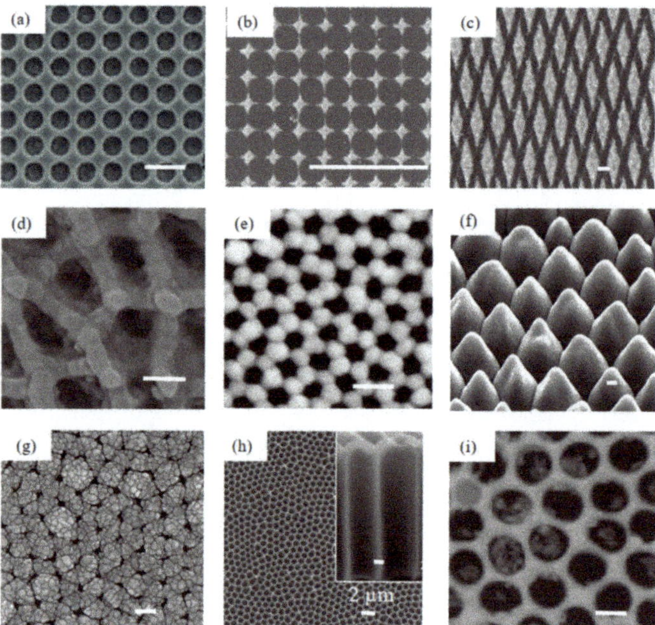

Figure 4. TEM and SEM images of top down approaches including E-beam fabricated (**a**) gold nanodisks [104], (**b**) Au star-like arrays [112], (**c**) Au diamond shaped structures [112], (**d**) Au dimple structures on PEN films treated by ion beam irradiation [113], (**e**) Au NPs distributed on the nanotips of canonical nanopores rims [114]; AAO templated structures: (**f**) Au nanostructure arrays [115], (**g**) Nano-flower like Ag/AAO [116], (**h**) Au nano-island @ Ag-frustum arrays [117], and (**i**) Au nanobipyramids self-assembled onto the AAO substrate [118]. Scale bars are all 100 nm without further indication in graphs.

Besides, Hu et al. fabricated polymer nanofinger structures on Si wafers using nanoimprint lithography, and coated the nanofingers with 70 nm of gold by e-beam evaporation, followed by exposure to solvent, inducing a leaning or self-closing of the nanowires, creating hot spots [119]. The fabricated arrays of electromagnetically coupled Ag nanoparticles on Si, could increase Raman efficiency by controlling the interparticle separation between Ag nanoparticles. These substrates showed high SERS enhancement with good control and reproducibility. However, lithography based methods, although extremely tuneable and scalable, suffer from high cost, slow throughput and are time consuming.

Laser-induced fabrication of SERS substrates has attracted research attention as it is scalabe and cost-effective. The fabrication of laser-induced SERS substrates always involves two steps. Firstly, fabrication of a nano/micro patterned substrate using ultra-fast laser pulses followed by physical vapor deposition to deposit the metal layer on the nano/micro patterned substrates in order to get plasmonic structures. Yang et al. used a nanosecond pulsed laser (1064 nm, pulse duration (τ) (full width at half maximum, FWHM) 5 ns, pulse repetition rate (PRR) 100 kHz, spot size~20 μm and laser ablation speed (v) 100 mm/s) ablation system to create micropatterns and generate different size nanoparticles [108]. The ablated Si surfaces were then deposited with Ag by electron beam evaporation. The enhancement factor of the fabricated substrates was estimated to be ~5.5×10^6. Diebold et al. fabricated Ag SERS substrate using a femtosecond laser (100-fs pulses at a repetition rate of 1 kHz, 800-nm center wavelength) structuring process [120]. This pulse train was frequency-doubled to a center wavelength of 400 nm through a thin $BiBO_3$ crystal. They used an n-type silicon wafer as the substrate. These laser pulses had an average fluence of 10 kJ/m^2 at the surface of a silicon wafer. A thermal deposition at a rate of 0.15 nm/s onto the structured silicon with different thicknesses of 10, 30, 60, 80, 100, and 200 nm was undertaken to get the optimized Ag nanostructure sizes for enhanced SERS performance. Similarly, Indrė Aleknavičienė et al. fabricated fast and scalable SERS substrates at low cost using ultrashort-pulse laser-induced (280 fs, 100 kHz, 350–380 nJ) plasma-assisted ablation (LIPAA) of soda-lime glass. The fabrication speed was as fast as 150 mm/s [121]. After the amorphous nanostructure formation on the glass surface, deposition of a 170 nm silver layer by vacuum deposition (around 100 nm in diameter, forming 1–3 μm size dendrimers) was applied to the glass substrate. This SERS substrate achieved an average enhancement factor (EF) of 3.0 × 10^5 evaluated using thiophenol.

Inkjet printing combined with electrochemical deposition is more convenient than e-beam lithography in terms of time and cost, and allows larger area fabrication at the same time. Inkjet printing method can be used both on solid substrate or on flexible substrate [122–134]. However, the choices of ink for inkjet printing are currently limited.

SERS substrates fabricated by templating methods is another widely used approach and has been demonstrated by a number of groups, see Figure 4f–i. Shanshan Shen et al. have studied substrates based on CdTe quantum dots modified polystyrene (PS) spheres with Ag nanoparticle caps. These substrates showed high enhancement factor (0.71 × 10^6) by using 4-ATP as the model molecule. [135]. Another very popular method is to use Anodised Aluminium Oxide (AAO) as a template to produce nanotubes [136]. Aluminium foil is anodised in acid to create nanopores, which are then used as a template to fabricate SERS substrates. Metals can be pulsed electrodeposited inside the template channels [137], or deposited via electron beam evaporation [138]. Alternatively, polymers can be employed to template the AAO. Lovera et al. fabricated super hydrophobic PS nanotubes by wetting commercial AAO filters and depositing silver onto the resulting PS nanotube structures [65]. Similarly, Zhang et al. patterned Polyethylene terephthalate (PET) on a nanostructured AAO template and deposited Au onto the polymer to create Au ananosturcture arrays. [115]. Other templates used for the fabrication of nanowires and nanotubes for SERS substrates include Polycarbonate membranes (PCM) [139–141], Polystyrene microspheres (PSM) [142–144] and nanochannel glasses. Charconnet et al. fabricated superlattices by templated self-assembly of gold nanoparticles on a flexible support, with tunable lattice-plasmon resonances through macroscopic strain. They found that the highest SERS performance was achieved by matching the lattice plasmon mode to the excitation wavelength, by post-assembly fine-tuning of long-range structural parameters [145]. The above substrates fabricated with "top-down" methods are manufactured reproducibly with high throughput, but often produced weaker signals than 'bottom-up' method due to larger distance between the nano structures and smaller surface density for nano particles that contribute to SERS signal. It involves also the use of expensive equipment and/or complex procedures. Combination of bottom-up and top-down fabricate method could increase the Raman intensity by creating a rougher and more homogeneous plasmonic surface [146].

4. Chemical Functionalization of SERS Substrates

SERS substrates without functionalization have limitations when dealing with real samples, e.g., strong background noise from the environment, or dealing with macro molecular such as DNA/proteins, which would block the SERS signal. For this reason, functionalization of SERS substrates is often undertaken to improve the selectivity or sensitivity in identifying the specific target analyte, which could lower detection time and limit of detection (LOD) [147]. The most widely used functionalization method is the attachment of self-assembled monolayer (SAM) of thiols to the silver or gold metallic nano structures. Functional groups along the thiol are used to phyisorb or chemically bind the analyte of interest to the substrate, thereby pre-concentrating the analyte at the surface leading to a subsequent increased in sensitivity. The thiol attaches to the metal nano structure surface owing to the strong affinity of sulphur with metals. However, the mechanism for the thiol group binding to the metal surface is not totally understood yet. One theory is that the thiol moieties chemisorb to planar gold surfaces, with the loss of hydrogen during the formation of the bond [148]. In 2003, Meirav Cohen-Atiya et al. studied adsorption of thiols on different metal surfaces by potentiometric measurements [149]. This adsorption process involves several complex steps including negative charge transfer and discharge through a reduction process. These steps related to the metal, the surface state of the metal, and the end functional groups in thiols. In 2014, Xue et al. used AFM to study the force between Au NPs and thiols under different experimental condition including oxidized Au surface and reduced Au surface with different pH effects; they stated that the bond between Au and thiol is a covalent bond [150]. In 2019, Inkpen and colleagues suggested that in gold–thiol SAMs prepared from solution deposition of dithiols, the gold–sulfur coupling had a physisorbed character, by both experiment and density functional theory (DFT) caculation [151]. Henrik Grönbeck et al. suggested the role of the thiol chain length must be considered when accessing the stability of monolayer systems on surfaces and clusters [152]. The formation of the Metal-S bond was not always very strong and the attachment of the thiol to the metal surface often happens in a few seconds. In order to gain homogeneous thiol self-assembly layer, longer incubation times, typically 24 h, are required. Thiol concentration affects binding result [153], and spacing between thiols can be adjusted by adding different thiols. Some bio-sensors based on thiolated single stand-DNA [154] or thiolated anti-body [155] have been used to detect DNA/RNA/antigen in environment, while thiol-aptamers are also used for DNA detection, see Figure 5.

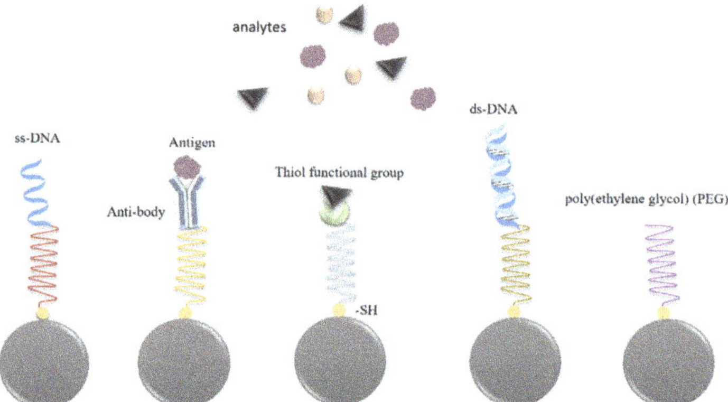

Figure 5. Different functionalisation thiols applied to metal nanoparticles: thiol-ss-DNA, thiol-anti-body, thiol-ds-DNA, thiol-PEG.

The interaction between antibody-antigen is considered strong due to the presence of electrostatic forces such as hydrophobic interactions, hydrogen bonds, van der Waals forces or ionic bonds [156]. Functionalization of SERS substrates with antibodies has gained much interest in recent years because of the significant level of sensitivity and selectivity that can be achieved with them [157]. Aptamers are single-stranded DNA sequences that can be designed to capture specific chemicals. Thiolated aptamers conjugated onto metal nano structures can capture and combine with the targeted chemicals to get surface enhanced Raman spectra. Sensors based on aptamers and SERS have been used for detecting pesticides, DNAs, RNAs, uranyl, biotoxin, pathogen, hazard foodborne etc. [158–163]. Gold/silver nanoparticles modified with polyethylene glycol (thiol-PEG) provide a capping system that stabilizes the antibody and avoid the reticular endothelial system [129,164,165].

5. Application of SERS in Agri-Food

5.1. Detection of Pesticide Residues

According to The United Nations Population Division, it is estimated that in 2050, the global population will reach ~9.7 billion, 30% more people than in 2017. A key challenge therefore, is that food production must keep pace with population growth. To this end, a variety of interventions have been put in place over the years to reduce losses due to disease and pests. Pesticide usage is essential in modern agricultural practices to protect crops and increase yield. There are currently more than 1000 pesticides used commercially around the world to ensure food is not damaged or destroyed by pests. Each pesticide has different properties and toxicological effects. The main drawbacks of pesticides is their potential toxicity to humans and other non-targeted organisms, which can result in significantly reduced biodiversity, through environmental contaminations in soil, water, and other vegetation [166,167]. The widespread use of pesticides therefore needs to be monitored and controlled [168].

Pesticides are classified by (i) the mode of entry, (ii) their function and the pest organism they kill, and (iii) their chemical composition. Based on chemical composition, pesticides are classified into four main groups, namely, organochlorines, organophosphorus, carbamates and pyrethrin/pyrethroids; see Table 1. Organochlorines pesticides (also known as chlorinated hydrocarbons) are organic compounds attached with five or more chlorine atoms. Organophosphates are derivatives of phosphoric acid, while carbamates derived from carbamic acid. Synthetic pyrethroid pesticides are group of organic pesticide that can be synthesized by duplicating the structure of natural pyrethrins.

Table 1. Four categories of pesticides classified by chemical composition.

Pesticides Category	Examples	Degradation in the Environment
Organo-chlorines	DDT, Chlorinated cyclodienes (aldrin, dieldrin, endrin, heptachlor, chlordane and endosulfan), dicofol, mirex, kepone, and pentachlorophenol	Long term residual effect in the environment
Organo-phosphates	Parathion, malathion, methyl parathion chlorpyrifos, diazinon, dichlorvos, phosmet, fenitrothion, tetrachlorvinphos, azamethiphos, azinphos-methyl, terbufos	Biodegradable
Carbamates	Aldicarb, carbofuran, carbaryl, ethienocarb, fenobucarb, oxamyl, and methomyl.	Easily degraded under natural environment with minimum environmental pollution
Pyrethroids	delatmethrin, cyfluthrin, befenthrin, lambda-cyhalothrin, permethrin.	Non-persistent, and break down easily on exposure to light.

Current detection methods of pesticides include high pressure liquid chromatography (HPLC), gas chromatography (GC), liquid chromatography (LG), mass chromatography (MC), spectrofluorimetric techniques as well as electrochemical methods. Han et al., for example, reported 302 targeted contaminants in catfish muscle by fast low-pressure GC–MS/MS and UHPLC-MS/MS methods [169]; Velkoska-Markovska, L. et al. detected malathion using liquid chromatography [170]; Wang et al. detected organic phosphates (OPs) using fluorescent probe [118]. Geto et al. used screen-printed carbon electrodes electrochemical sensors to detect bentazone in water source [171]. Santana et al. detected Carbendazim using electrochemical detection [172].

There are growing demands for the development of novel analytical techniques for a variety of pollutants affecting crops, for example, pesticides. In this respect, methods based on SERS have attracted attention. The first SERS study of pesticides was the detection of organophosphorus pesticides in 1987 by Alak et al. [173]. Since then, the potential toxicity to humans, animals, and the environment has been reported, and tolerance levels were introduced for a large number of harmful pesticides [174,175]. SERS detection methods have developed considerably, resulting in a large number of the more recent reports employing in-situ SERS detection methods on the surface of different foods [93,176]. A search on Web of Science with key words combining the topics of "surface enhanced Raman" and "pesticides" revealed 432 publication results up to 12 November 2021. The majority of SERS and pesticide-related researches demonstrated detection of organophosphate (OP) insecticides, for example, phosmet [176,177], parathion-methyl [178], malathion [170,179,180], chlorpyrifos [181–183]; see Table 2. Other SERS pesticide studies included fungicides (thiram [184], thiabendazol [185,186]), herbicides [179,180,187,188], and neonicotinoids insecticides (imidacloprid [189], thiacloprid [190], acetamiprid [191,192]).

Amongst the pesticides, organophosphates (OP) represent the largest class, making up to 50% of the neurotoxic agents in chemical pesticides [193]. Most OP usage is agricultural, since the Environmental Protection Agency banned their residential use in 2001 [194]. However, their human and animal toxicity still make them a societal health and environmental concern [195–197]. Moreover, pesticides at low concentration have been detected in food and drinking water [182,198–201]. Liu et al. reported on the use of silver-coated gold bimetallic nanoparticles; their Raman Enhancement depends on the silver shell thickness, for in situ detection of a range of pesticides on fruit peels without further sample preparation, with a limit of detection (LOD) below the required maximum residue levels (MRL) [202]. However, as described above, colloidal based solutions are not ideal for portable applications while solid SERS substrates that can be prepared in advance are more suitable. For example, Chen et al. used "SERS tape" to extract OP pesticides (thiram, chlorpyrifos, methyl parathion) from different kinds of fruit and vegetable peels [102]. The tape is placed on to the surface of the produce and peeled off for SERS analysis. This is non-invasive and requires no sample or substrate preparation. Additionally, Li et al. have created a 'smart dust' that easily spreads over a probed surface for in-situ SERS measurements [203]. This method requires no preparation or particle aggregation/concentration on the substrate. Their shell-isolated nanoparticles are used to analyse residues of OP pesticide parathion, on a fresh orange. They present comparable results between a normal Raman and a portable Raman, demonstrating the substrates potential use in-field.

Table 2. Organophosphates (OP) detection by SERS in food industry and environment monitoring.

Organophosphates (Matrix)		SERS Substrate	LOD (Reported)	LOD (Normalised)	Excitation Wavelength
Phosmet	Ooling tea [204]	Ag NPs	0.1 mg/kg	0.1 ppm	633 nm
	fruit [205]	multi-walled carbon nanotubes	0.5 mg/kg	0.5 ppm	785 nm
	paddy water [206]	Au nanorods	0.25 mg/L	0.25 ppm	Portable Raman spectrometer, 785 nm
	fruit skin [207]	polyurethane-Ag NPs	0.6 μg/mL	0.6 ppb	785 nm
	Plant Surfaces [208]	polyurethane micelle/Ag NP	0.08 g/mL	80 ppm	/
Parathion-methyl	fruit or vegetable peels [209]	Snowflake-like Au NPs	0.026 ng/cm	/	638 nm
	solvent [210]	Ag NP decorated ZnO-nanorods	10^{-8} M	2.63 ppb	532 nm
	solvent [211]	nanoporous structure	12 ppb	12 ppb	785 nm
Malathion	solvent [212]	nanostructured Ag	10 nM	3.30 ppb	632.8 nm
Chlorpyrifos	tomato surface [183]	Ag colloid	10^{-9} mol/L	0.35 ppb	638 nm
	soil [213]	Au NP	10 ppm	10 ppm	785 nm
	fruits [214]	Ag NP	10 ng/mL	10 ppb	633 nm

Neonicotinoids are a more recent and relatively powerful class of insecticide, and since the introduction of imidacloprid in 1991, they have been the fastest-growing class of insecticides in modern crop protection [215], representing almost 17% of the global market [216]. Typical detection methods of neonicotinoids are based on enzyme linked immuno-sorbent assays (ELISA) [217,218], HPLC- or GC- mass spectrometry [219,220], surface plasmon resonance [221], and fluorescence spectroscopy [222], none of which are suitable for field analysis. Neonicotinoids are extremely effective against herbivorous insects [223], while having perceived low toxicity to mammals, birds and fish [224]. This has led to their widespread uptake for use on a variety of crops. However, concerns have been raised recently about environmental impact in affecting the homing capacity of honey bees, resulting in global colony collapse of the pollinator population [2,225]. Consequently, the European Union enforced a temporary ban (Dec 2013) [226] reducing the MRL of neonicotinoids to between 0.01 to 3 mg/kg for many fruits and vegetables [227,228].

Studies of detection of neonicotinoids are shown in Table 3. For example, Cao et al. synthesized three types of AuNP/MOF (metal–organic framework) composite to investigate the interaction between acetamiprid and the bridging molecules of the MOFs. Acetamiprid in this case was used to evaluate the characteristics of the SERS substrates. LODs of 0.02 µM, 0.009 µM, and 0.02 µM were achieved for the three composites, which could satisfy the requirement of detection according to the MRLs of acetamiprid. [229] Yang et al. used SERS to evaluate the penetration behaviors of four pesticides (acetamiprid, thiabendazole, ferbam and phosmet) in a variety of fresh produce matrices. They used a pesticide/AgNP complex deposited onto the external surfaces of different fresh produce and measured the penetration depth of the complex using SERS [230]. Although the results are promising, this method requires complex sample preparation with the pesticide and AgNP, including washing steps, and is not ideal for farm-side analysis. On the other hand, Wijaya et al. employed silver dendrites for SERS-based detection of acetamiprid in apple juice and from swabs of the apple surface [231]. The TQ Analyst Software (Thermo Fisher Scientific) was used for SERS spectral data analysis, with second-derivative transformation employed to remove baseline and separate overlapped peaks. Outlier peaks were also removed to gain a more accurate quantification results. Acetamiprid detection was determined using principal component analysis (PCA) and retains the principal components (PCs) that capture the variation between sample treatments. This method does not need pre-treatment for the apple juice samples and the use of the swab is non-invasive to the fruit. This method therefore has the potential to be used for on-site pesticide detection.

Table 3. Neonicotinoids detection by SERS in food industry and environment monitoring.

Neonicotinoids		SERS Substrate	LOD (Reported)	LOD (Normalised)	Excitation Wavelength
Acetamiprid	Solvent [229]	AuNP/MOF (metal–organic framework) composite	0.009 µM	2 ppb	780 nm
	apple juice [232]	Gold nanoparticles (AuNPs) bonded with polyadenine (polyA)-mediated aptamer and Raman tag (MMBN-AuNPs-aptamer)	6.8 nM	1.514 ppb	532 nm
	solvent [233]	co-doped N/Ag carbon dot	0.006 µg/L	6 ppt	633 nm
	green tea [234]	Au NPs	1.76×10^{-8} M	3.91 ppb	micro-Raman spectroscopy, 785 nm

Table 3. *Cont.*

Neonicotinoids		SERS Substrate	LOD (Reported)	LOD (Normalised)	Excitation Wavelength
Imidacloprid	solvent [235]	3-D Ag dendrites on Paper substrate	0.02811 ng/mL	0.02811 ppb	633 nm
	extract solution from apple [214]	Ag NP coated glass	50 ng/mL	50 ppb	633 nm
	green tea [236]	flower shaped Ag nanostructure	10^{-4} µg/mL	10^{-4} ppm	785 nm
	waste water treatment [237]	r-GO supporting Ag meso-flowers and phenyl-modified graphitic carbon nitride	10 mg/mL	10^{4} ppm	632.8 nm
	Solvent [238]	Ag nanostructures on PVDF	1 ng/mL	1 ppb	514 nm
	fresh tea leaves; apple peels [239]	Au NPs	0.5 mg/kg; 0.02 mg/kg	0.5 ppm; 0.02 ppm	780 nm
Thiamethoxam	solvent [240]	Au NPs	0.1 ng/mL	0.1 ppb	785 nm
	solvent [241]	Ag nano structure	g/mL order	10^{6} ppm order	532 nm
Thiacloprid	fruit [242]	Au@Ag NPs	0.1 mg/kg	0.1 ppm	633 n

5.2. Detection of Chemical Additives

5.2.1. Dye Molecules

Dye molecules are used in industries to colour different materials such as silk, wool, cotton and paper. Unfortunately, the wasted water from these industries cause pollution in aquaculture and also cause serious toxic, carcinogenic and mutagenic effects in mammalian cells [243]. Besides, Malachite green and Crystal violet have been used for the treatment of fungal, parasitic and protozoan diseases in fish, and it is found to absorb and metabolise in tissues of fish [244]. The detection of dye molecules such as rhodamine 6G (R6G), malachite green, crystal violet (CV) and 4-aminobenzenthiol are the most reported chemical contaminants due to their ease of detection. These dye molecules are highly Raman active and used indiscriminately as antimicrobials in aquaculture. Moreover, the most significant and influential papers in this field have employed these dye molecules to study single molecular SERS detection [245], enhancement factors [37], and the mechanisms of SERS [246].

5.2.2. Melamine

In 2008, a sanitary scandal involved the intentional contamination of milk powder with melamine to give a false appearance of high protein levels [247]. Monitoring the level of residual melamine has since become important for the dairy industry. Regarding SERS, melamine is probably the most widely documented food adulterant, with Web of Science literature search revealing 309 articles up to 12 November 2021. The majority of melamine detection researches that employ SERS substrates are based on Au and Ag nanoparticle fabrication. Gold substrates include: Au colloids [248], Au NP agglomerates [249–251], 4-mercaptopyridine-modified Au NPs [252], and magnetic Au NPs [253], to name a few. Similarly the silver SERS substrates include: Ag Colloids [254], Ag NP agglomerates [255], Ag NP coated Ag/C nanospheres [256], Ag NP coated polystyrene nanospheres [144,257], cyclodextrin-coated Ag NPs [258], functional graphene/Ag nanocomposite [259].

Peng et al. used self-assembled vertical arrays of nanorods to detect melamine in methanol [260] and similarly, Hu et al. coated Ag nanoparticles on the surfaces of $Fe_3O_4@SiO_2$ composite microspheres to detect a melamine methanol solution [261]. Neither of these articles demonstrated melamine detection in real samples. Zhang et al.

demonstrated melamine detection in milk using silver colloid solution. They reported on an easy pre-treatment for the milk, which, however, still required large instrumentation and, additionally, the colloid NPs required mixing with the diluted and filtered milk [246]. Alternatively, Guo et al. developed self-assembled hollow gold nanospheres to detect melamine in milk on a solid chip platform, which is ideal for remote sensing. However, they employed centrifugation as their only method of sample pre-treatment, which is complex and not suitable for transporting [262].

Another novel method by Betz et al. used a copper tape and a penny coin to fabricate Ag micro- and nanostructures. It was used to analyse infant formula adulterated with melamine [255]. The fact that these substrates form in five minutes on-site without the need for complex equipment, sample pre-treatment, or harsh chemicals enabled the possibility of remote point-of-sampling. However, their LOD (5 ppm) is not sufficient for remote melamine detection.

Finally, Chen et al. reported on the detection of melamine in egg white using fabricated ZnO/Au composite nanoneedle arrays, see Figure 6. The results showed some background interferences from the egg proteins but the characteristic peak for melamine at ~682 cm^{-1} remained detectable and was well resolved [263]. The only sample preparation is a filtering of the egg solution through four layers of gauge, which can easily be employed on-site as it requires no complex instrumentation. Similarly, Kim et al. applied their previously reported gold nanofingers to melamine detection in milk [264]. Although they also require sample pre-treatment, the authors avoid using centrifugation, as it is neither portable nor low cost. Instead, they employ a mini dialysis kit and detect characteristic melamine peaks from the dialysis filtered solutions at 1 ppm. They also demonstrate melamine detection at 100 ppb in infant formula using a solution gel filtration chromatography treatment. These are a few articles that report SERS substrates and methods of sample pre-treatment, that both are fully compatible for field applications in a limited-resource environment.

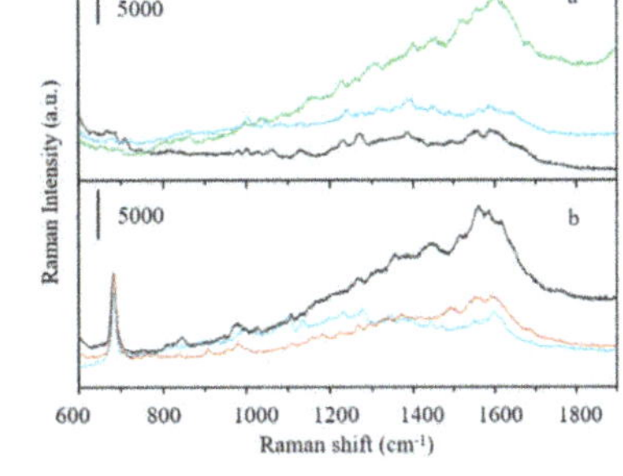

Figure 6. (**a**) SEM image of ZnO/Au composite nanoneedle arrays; (**b**) SERS spectra of egg white solution (6 g/L) (**a**) and the melamine-tainted egg white solution (**b**) on ZnO/Au nanoneedles. The concentration of melamine and egg white in the mixture is 1.0×10^{-5} M and about 6 g/L, by Chen et al. [263].

6. Conclusions and Perspective

In this review paper, SERS as a useful technology for Agri-Food and environmental sensing has been investigated. Compared to other detecting technologies such as chromatography or electrochemical based methods, SERS has the potential to deliver rapid, ultra-sensitive and highly specific detections of a wide range of chemicals and biomolecules. SERS substrates fabrication methods include bottom-up and top-down. Bottom-up approaches refer to the fabrication of nanostructures by chemical synthesis, colloid aggregation, electrochemical deposition and self-assembly; top-down methods include lithography techniques (electron-beam (E-beam) and nanoimprint lithography, laser etching together with metal film deposition (sputtering, metal evaporation, atomic layer deposition), templating (e.g., anodic aluminium oxide), inkjet printing, etc. The combination of top-down and bottom-up methods could improve SERS intensity. The common plasmonic materials are Ag, Cu, Au or their composite material. A wide choice of the support substrate is variable Si, Cu, tissue, paper, leaves and aluminium cans, etc. SERS related devices could widely be used in agri-food and environment monitoring, following the SERS device fabrication method section; this review paper presents SERS applications in detection of pesticides such as organophosphate (OP) insecticides and neonicotinoids insecticides as well as illegal food additive such as dyes and melamine.

However, despite the advancement and strategies presented above, some challenges still need to be overcome for SERS to be widely used in agri-food and environment analytical applications. These include (i) repeatability of SERS substrates and SERS signals for each Raman measurements; (ii) weak interaction—or even repulsion because of the surface energy—between some analytes and SERS surface; (iii) stability of SERS substrate and functional layers that can in some instance react with the targeted analytes or degrade over time (e.g., oxidation) or under continuous laser excitation; (iv) general non suitability for direct detection of heavy metal or macromolecular such as protein; (v) current lack of standardised optical setup or methodologies to compare results obtained by different research groups and (vi) in real sample verification, SERS devices performance could be affected by contamination or interferences from the environment.

Regarding sample reproducibility, SERS substrates could benefit from constant advances in nanofabrication processes and instrumentation as well as in surface chemistry. Regarding quantification, techniques such as isotope labelling [265] or standard addition method have proven promising. Also, new emerging techniques such as electrochemical SERS (EC-SERS), shifted excitation Raman difference spectroscopy (SERDS) or surface enhanced spatially offset resonance Raman spectroscopy (SESORRS) such could help overcome the limitations mentioned above.

All in all, SERS devices benefit from a wide range of choice of plasmonic structures such as Ag/Au NPs, nano rods, nano flowers, nano cubes, and their composite materials etc; supporting substrate, fabrication method and functionalization method. Besides, they also have a numerous of advantages such as rapid response time, high sensitivity and selectivity and possibility to use handheld devices for onsite measurements. However, several challenges still exist in terms of reliability and durability for SERS sensing platform. Ongoing advances in nanofabrication and chemistry have the potential to overcome the current limitations of SERS sensing. As a result, we believe that SERS will soon be a widespread analytical technique for sensitive detection of contaminants in agri-food and environmental applications.

Author Contributions: Conceptualization, Y.Y. and N.C.; data curation, Y.Y. and N.C.; writing—original draft preparation, Y.Y. and N.C.; writing—review and editing, A.O. and P.L.; visualization, Y.Y. and N.C.; supervision, A.O. and P.L.; project administration, A.O. and P.L.; funding acquisition, A.O. and P.L. All authors have read and agreed to the published version of the manuscript.

Funding: This research was funded by the European Union's Horizon 2020 research and innovation programme under the Marie Skłodowska-Curie grant agreement No: H2020-MSCA-ITN-2018-813680 as well as the Irish EPA UisceSense Project (Code: 2015-W-MS-21) and the VistaMilk Center Science Foundation Ireland (SFI) and Department of Agriculture Food and the Marine (DAFM) under Grant Number 16/RC//3835.

Institutional Review Board Statement: Not applicable.

Informed Consent Statement: Not applicable.

Conflicts of Interest: The authors declare no conflict of interest.

References

1. Food Agriculture Organization. *The Future of Food and Agriculture—Trends and Challenges*; United Nations: Rome, Italy, 2017.
2. Henry, M.; Beguin, M.; Requier, F.; Rollin, O.; Odoux, J.-F.; Aupinel, P.; Aptel, J.; Tchamitchian, S.; Decourtye, A. A common pesticide decreases foraging success and survival in honey bees. *Science* **2012**, *336*, 348–350. [CrossRef]

3. Gong, T.X.; Huang, Y.F.; Wei, Z.J.; Huang, W.; Wei, X.B.; Zhang, X.S. Magnetic assembled 3D SERS substrate for sensitive detection of pesticide residue in soil. *Nanotechnology* **2020**, *31*, 205501. [CrossRef]

4. Guselnikova, O.; Postnikov, P.; Elashnikov, R.; Miliutina, E.; Svorcik, V.; Lyutakov, O. Metal-organic framework (MOF-5) coated SERS active gold gratings: A platform for the selective detection of organic contaminants in soil. *Anal. Chim. Acta* **2019**, *1068*, 70–79. [CrossRef]

5. Krafft, B.; Tycova, A.; Urban, R.D.; Dusny, C.; Belder, D. Microfluidic device for concentration and SERS-based detection of bacteria in drinking water. *Electrophoresis* **2021**, *42*, 86–94. [CrossRef] [PubMed]

6. Zhu, C.H.; Zhao, Q.S.; Meng, G.W.; Wang, X.J.; Hu, X.Y.; Han, F.M.; Lei, Y. Silver nanoparticle-assembled micro-bowl arrays for sensitive SERS detection of pesticide residue. *Nanotechnology* **2020**, *31*, 126300. [CrossRef] [PubMed]

7. Wang, K.; Sun, D.-W.; Pu, H.; Wei, Q. Polymer multilayers enabled stable and flexible Au@Ag nanoparticle array for nondestructive SERS detection of pesticide residues. *Talanta* **2021**, *223*, 205303. [CrossRef] [PubMed]

8. Grubisha, D.S.; Lipert, R.J.; Park, H.-Y.; Driskell, J.; Porter, M.D. Femtomolar Detection of Prostate-Specific Antigen: An Immunoassay Based on Surface-Enhanced Raman Scattering and Immunogold Labels. *Anal. Chem.* **2003**, *75*, 5936–5943. [CrossRef] [PubMed]

9. Fan, M.; Andrade, G.F.S.; Brolo, A.G. A review on the fabrication of substrates for surface enhanced Raman spectroscopy and their applications in analytical chemistry. *Anal. Chim. Acta* **2011**, *693*, 7–25. [CrossRef] [PubMed]

10. Justin, L.A.; Jeremy, D.D.; Ralph, A.T.; Yiping, Z. Current Progress on Surface-Enhanced Raman Scattering Chemical/Biological Sensing. In *Functional Nanoparticles for Bioanalysis, Nanomedicine, and Bioelectronic Devices Volume 2*; American Chemical Society: Washington, DC, USA, 2012; Volume 1113, pp. 235–272.

11. Xie, W.; Schlucker, S. Medical applications of surface-enhanced Raman scattering. *Phys. Chem. Chem. Phys.* **2013**, *15*, 5329–5344. [CrossRef]

12. Willets, K.A. Probing nanoscale interfaces with electrochemical surface-enhanced Raman scattering. *Curr. Opin. Electrochem.* **2019**, *13*, 18–24. [CrossRef]

13. Craig, A.P.; Franca, A.S.; Irudayaraj, J. Surface-Enhanced Raman Spectroscopy Applied to Food Safety. *Annu. Rev. Food Sci. Technol.* **2013**, *4*, 369–380. [CrossRef]

14. Fan, C.; Hu, Z.Q.; Mustapha, A.; Lin, M.S. Rapid detection of food- and waterborne bacteria using surface-enhanced Raman spectroscopy coupled with silver nanosubstrates. *Appl. Microbiol. Biotechnol.* **2011**, *92*, 1053–1061. [CrossRef]

15. Zhang, Y.; Huang, Y.; Zhai, F.; Du, R.; Liu, Y.; Lai, K. Analyses of enrofloxacin, furazolidone and malachite green in fish products with surface-enhanced Raman spectroscopy. *Food Chem.* **2012**, *135*, 845–850. [CrossRef] [PubMed]

16. Cialla, D.; Marz, A.; Bohme, R.; Theil, F.; Weber, K.; Schmitt, M.; Popp, J. Surface-enhanced Raman spectroscopy (SERS): Progress and trends. *Anal. Bioanal. Chem.* **2012**, *403*, 27–54. [CrossRef]

17. Granger, J.H.; Schlotter, N.E.; Crawford, A.C.; Porter, M.D. Prospects for point-of-care pathogen diagnostics using surface-enhanced Raman scattering (SERS). *Chem. Soc. Rev.* **2016**, *45*, 3865–3882. [CrossRef] [PubMed]

18. Jarvis, R.M.; Goodacre, R. Characterisation and identification of bacteria using SERS. *Chem. Soc. Rev.* **2008**, *37*, 931–936. [CrossRef] [PubMed]

19. Laserna, J. *Modern Techniques in Raman Spectroscopy*; John Wiley and Sons: New York, NY, USA, 1996.

20. Wen, P.; Yang, F.; Ge, C.; Li, S.; Xu, Y.; Chen, L. Self-assembled nano-Ag/Au@Au film composite SERS substrates show high uniformity and high enhancement factor for creatinine detection. *Nanotechnology* **2021**, *32*, 395502. [CrossRef]

21. Kalantar-zadeh, K.; Fry, B. *Nanotechnology-Enabled Sensors*; Springer Science & Business Media: Berlin/Heidelberg, Germany, 2007.

22. Settle, F.A. *Handbook of Instrumental Techniques for Analytical Chemistry*; Prentice Hall PTR: Hoboken, NJ, USA, 1997.

23. Gardiner, D.J. *Practical Raman Spectroscopy*; Spinger: Berlin/Heidelberg, Germany, 1989.

24. Jablonski, A. Efficiency of anti-Stokes fluorescence in dyes. *Nature* **1933**, *131*, 839–840. [CrossRef]

25. Fleischmann, M.; Hendra, P.J.; McQuillan, A.J. Raman spectra of pyridine adsorbed at a silver electrode. *Chem. Phys. Lett.* **1974**, *26*, 163–166. [CrossRef]

26. Gersten, J.I. The effect of surface roughness on surface enhanced Raman scattering. *J. Chem. Phys.* **1980**, *72*, 5779–5780. [CrossRef]

27. Gersten, J.; Nitzan, A. Electromagnetic theory of enhanced Raman scattering by molecules adsorbed on rough surfaces. *J. Chem. Phys.* **1980**, *73*, 3023–3037. [CrossRef]

28. Moskovits, M. Surface-enhanced spectroscopy. *Rev. Mod. Phys.* **1985**, *57*, 783–826. [CrossRef]

29. Otto, A.; Mrozek, I.; Grabhorn, H.; Akemann, W. Surface-enhanced Raman scattering. *J. Phys. Condens. Matter* **1992**, *4*, 1143. [CrossRef]

30. Stiles, P.L.; Dieringer, J.A.; Shah, N.C.; Duyne, R.P.V. Surface-Enhanced Raman Spectroscopy. *Annu. Rev. Anal. Chem.* **2008**, *1*, 601–626. [CrossRef]

31. Hutter, E.; Fendler, J.H. Exploitation of Localized Surface Plasmon Resonance. *Adv. Mater.* **2004**, *16*, 1685–1706. [CrossRef]

32. Camden, J.P.; Dieringer, J.A.; Wang, Y.; Masiello, D.J.; Marks, L.D.; Schatz, G.C.; Van Duyne, R.P. Probing the Structure of Single-Molecule Surface-Enhanced Raman Scattering Hot Spots. *J. Am. Chem. Soc.* **2008**, *130*, 12616–12617. [CrossRef]

33. Sun, M.; Wan, S.; Liu, Y.; Jia, Y.; Xu, H. Chemical mechanism of surface-enhanced resonance Raman scattering via charge transfer in pyridine–Ag2 complex. *J. Raman Spectrosc.* **2008**, *39*, 402–408. [CrossRef]

34. Adrian, F.J. Charge transfer effects in surface-enhanced Raman scatteringa. *J. Chem. Phys.* **1982**, *77*, 5302–5314. [CrossRef]

35. Schatz, G.C. Theoretical studies of surface enhanced Raman scattering. *Acc. Chem. Res.* **1984**, *17*, 370–376. [CrossRef]

36. Campion, A.; Ivanecky, J., III; Child, C.; Foster, M. On the mechanism of chemical enhancement in surface-enhanced Raman scattering. *J. Am. Chem. Soc.* **1995**, *117*, 11807–11808. [CrossRef]

37. Le Ru, E.C.; Blackie, E.; Meyer, M.; Etchegoin, P.G. Surface Enhanced Raman Scattering Enhancement Factors: A Comprehensive Study. *J. Phys. Chem. C* **2007**, *111*, 13794–13803. [CrossRef]

38. Weaver, M.J.; Zou, S.; Chan, H.Y.H. Peer Reviewed: The New Interfacial Ubiquity of Surface-Enhanced Raman Spectroscopy. *Anal. Chem.* **2000**, *72*, 38A–47A. [CrossRef]

39. Etchegoin, P.G.; Le Ru, E.C. A perspective on single molecule SERS: Current status and future challenges. *Phys. Chem. Chem. Phys.* **2008**, *10*, 6079–6089. [CrossRef] [PubMed]

40. Park, W.H.; Kim, Z.H. Charge Transfer Enhancement in the SERS of a Single Molecule. *Nano Lett.* **2010**, *10*, 4040–4048. [CrossRef]

41. Bell, S.E.; McCourt, M.R. SERS enhancement by aggregated Au colloids: Effect of particle size. *Phys. Chem. Chem. Phys.* **2009**, *11*, 7455–7462. [CrossRef]

42. Lee, S.J.; Guan, Z.; Xu, H.; Moskovits, M. Surface-Enhanced Raman Spectroscopy and Nanogeometry: The Plasmonic Origin of SERS. *J. Phys. Chem. C* **2007**, *111*, 17985–17988. [CrossRef]

43. Lancaster, C.A.; Scholl, W.E.; Ticknor, M.A.; Shumaker-Parry, J.S. Uniting Top-Down and Bottom-Up Strategies Using Fabricated Nanostructures as Hosts for Synthesis of Nanomites. *J. Phys. Chem. C* **2020**, *124*, 6822–6829. [CrossRef]

44. Banholzer, M.J.; Millstone, J.E.; Qin, L.; Mirkin, C.A. Rationally designed nanostructures for surface-enhanced Raman spectroscopy. *Chem. Soc. Rev.* **2008**, *37*, 885–897. [CrossRef]

45. Pan, X.-T.; Liu, Y.-Y.; Qian, S.-Q.; Yang, J.-M.; Li, Y.; Gao, J.; Liu, C.-G.; Wang, K.; Xia, X.-H. Free-Standing Single Ag Nanowires for Multifunctional Optical Probes. *ACS Appl. Mater. Interfaces* **2021**, *13*, 19023–19030. [CrossRef] [PubMed]

46. Xu, K.-X.; Chen, X.; Huang, Z.; Chen, Z.-N.; Chen, J.; Sun, J.-J.; Fang, Y.; Li, J.-F. Ligand-Free Fabrication of Ag Nanoassemblies for Highly Sensitive and Reproducible Surface-Enhanced Raman Scattering Sensing of Antibiotics. *ACS Appl. Mater. Interfaces* **2021**, *13*, 1766–1772. [CrossRef]

47. Choi, S.; Jeong, H.; Choi, K.H.; Song, J.Y.; Kim, J. Electrodeposition of triangular Pd rod nanostructures and their electrocatalytic and SERS activities. *ACS Appl. Mater. Interfaces* **2014**, *6*, 3002–3007. [CrossRef]

48. Ge, D.; Wei, J.; Ding, J.; Zhang, J.; Ma, C.; Wang, M.; Zhang, L.; Zhu, S. Silver Nano-Dendrite-Plated Porous Silicon Substrates Formed by Single-Step Electrochemical Synthesis for Surface-Enhanced Raman Scattering. *ACS Appl. Nano Mater.* **2020**, *3*, 3011–3018. [CrossRef]

49. Jeong, H.; Kim, J. Electrodeposition of nanoflake Pd structures: Structure-dependent wettability and SERS activity. *ACS Appl. Mater. Interfaces* **2015**, *7*, 7129–7135. [CrossRef]

50. Raveendran, J.; Stamplecoskie, K.G.; Docoslis, A. Tunable Fractal Nanostructures for Surface-Enhanced Raman Scattering via Templated Electrodeposition of Silver on Low-Energy Surfaces. *ACS Appl. Nano Mater.* **2020**, *3*, 2665–2679. [CrossRef]

51. Longoni, M.; Zalaffi, M.S.; de Ferri, L.; Stortini, A.M.; Pojana, G.; Ugo, P. Surface Enhanced Raman Spectroscopy With Electrodeposited Copper Ultramicro-Wires With/Without Silver Nanostars Decoration. *Nanomaterials* **2021**, *11*, 518. [CrossRef]

52. Grzelczak, M.; Vermant, J.; Furst, E.M.; Liz-Marzán, L.M. Directed Self-Assembly of Nanoparticles. *ACS Nano* **2010**, *4*, 3591–3605. [CrossRef] [PubMed]

53. Zhu, S.; Fan, C.; Wang, J.; He, J.; Liang, E. Self-assembled Ag nanoparticles for surface enhanced Raman scattering. *Opt. Rev.* **2013**, *20*, 361–366. [CrossRef]

54. Gao, T.; Wang, Y.; Wang, K.; Zhang, X.; Dui, J.; Li, G.; Lou, S.; Zhou, S. Controlled synthesis of homogeneous Ag nanosheet-assembled film for effective SERS substrate. *ACS Appl. Mater. Interfaces* **2013**, *5*, 7308–7314. [CrossRef]

55. Zhang, L. Self-assembly Ag nanoparticle monolayer film as SERS Substrate for pesticide detection. *Appl. Surf. Sci.* **2013**, *270*, 292–294. [CrossRef]

56. Wu, Y.; Dang, H.; Park, S.-G.; Chen, L.; Choo, J. SERS-PCR assays of SARS-CoV-2 target genes using Au nanoparticles-internalized Au nanodimple substrates. *Biosens. Bioelectron.* **2022**, *197*, 113736. [CrossRef] [PubMed]

57. Lee, S.J.; Morrill, A.R.; Moskovits, M. Hot Spots in Silver Nanowire Bundles for Surface-Enhanced Raman Spectroscopy. *J. Am. Chem. Soc.* **2006**, *128*, 2200–2201. [CrossRef]

58. Zheng, G.; Patolsky, F.; Cui, Y.; Wang, W.U.; Lieber, C.M. Multiplexed electrical detection of cancer markers with nanowire sensor arrays. *Nat. Biotechnol.* **2005**, *23*, 1294–1301. [CrossRef] [PubMed]

59. Tang, J.; Zhao, Q.; Zhang, N.; Man, S.-Q. Facile fabrication of large-area and uniform silica nanospheres monolayer for efficient surface-enhanced Raman scattering. *Appl. Surf. Sci.* **2014**, *308*, 247–252. [CrossRef]

60. Li, W.; Camargo, P.H.C.; Lu, X.; Xia, Y. Dimers of Silver Nanospheres: Facile Synthesis and Their Use as Hot Spots for Surface-Enhanced Raman Scattering. *Nano Lett.* **2009**, *9*, 485–490. [CrossRef] [PubMed]

61. Martín, A.; Pescaglini, A.; Schopf, C.; Scardaci, V.; Coull, R.; Byrne, L.; Iacopino, D. Surface-Enhanced Raman Scattering of 4-Aminobenzenethiol on Au Nanorod Ordered Arrays. *J. Phys. Chem. C* **2014**, *118*, 13260–13267. [CrossRef]

62. Martin, A.; Wang, J.J.; Iacopino, D. Flexible SERS active substrates from ordered vertical Au nanorod arrays. *RSC Adv.* **2014**, *4*, 20038–20043. [CrossRef]

63. Xie, Z.; Tao, J.; Lu, Y.; Lin, K.; Yan, J.; Wang, P.; Ming, H. Polymer optical fiber SERS sensor with gold nanorods. *Opt. Commun.* **2009**, *282*, 439–442. [CrossRef]

64. Oh, M.K.; Shin, Y.S.; Lee, C.L.; De, R.; Kang, H.; Yu, N.E.; Kim, B.H.; Kim, J.H.; Yang, J.K. Morphological and SERS Properties of Silver Nanorod Array Films Fabricated by Oblique Thermal Evaporation at Various Substrate Temperatures. *Nanoscale Res. Lett.* **2015**, *10*, 962. [CrossRef]

65. Lovera, P.; Creedon, N.; Alatawi, H.; Mitchell, M.; Burke, M.; Quinn, A.J.; O'Riordan, A. Low-cost silver capped polystyrene nanotube arrays as super-hydrophobic substrates for SERS applications. *Nanotechnology* **2014**, *25*, 175502. [CrossRef]

66. Lovera, P.; Creedon, N.; Alatawi, H.; O'Riordan, A. *Metal Capped Polystyrene Nanotubes Arrays as Super-Hydrophobic Substrates for SERS Applications*; SPIE: Washington, DC, USA, 2014; Volume 9129.

67. Walker, D.A.; Browne, K.P.; Kowalczyk, B.; Grzybowski, B.A. Self-Assembly of Nanotriangle Superlattices Facilitated by Repulsive Electrostatic Interactions. *Angew. Chem. Int. Ed.* **2010**, *49*, 6760–6763. [CrossRef]

68. Kumar-Krishnan, S.; Esparza, R.; Pal, U. Controlled Fabrication of Flower-Shaped Au-Cu Nanostructures Using a Deep Eutectic Solvent and Their Performance in Surface-Enhanced Raman Scattering-Based Molecular Sensing. *ACS Omega* **2020**, *5*, 3699–3708. [CrossRef]

69. Liu, Z.; Yang, Z.; Peng, B.; Cao, C.; Zhang, C.; You, H.; Xiong, Q.; Li, Z.; Fang, J. Highly Sensitive, Uniform, and Reproducible Surface-Enhanced Raman Spectroscopy from Hollow Au-Ag Alloy Nanourchins. *Adv. Mater.* **2014**, *26*, 2431–2439. [CrossRef]

70. Yang, M.; Alvarez-Puebla, R.; Kim, H.-S.; Aldeanueva-Potel, P.; Liz-Marzán, L.M.; Kotov, N.A. SERS-Active Gold Lace Nanoshells with Built-in Hotspots. *Nano Lett.* **2010**, *10*, 4013–4019. [CrossRef]

71. Barbillon, G. Applications of Shell-Isolated Nanoparticle-Enhanced Raman Spectroscopy. *Photonics* **2021**, *8*, 46. [CrossRef]

72. Barbillon, G.; Ivanov, A.; Sarychev, A.K. Hybrid Au/Si Disk-Shaped Nanoresonators on Gold Film for Amplified SERS Chemical Sensing. *Nanomaterials* **2019**, *9*, 1588. [CrossRef]

73. Vigderman, L.; Zubarev, E.R. Starfruit-Shaped Gold Nanorods and Nanowires: Synthesis and SERS Characterization. *Langmuir* **2012**, *28*, 9034–9040. [CrossRef] [PubMed]

74. Li, H.; Wang, Y.; Li, Y.; Zhang, J.; Qiao, Y.; Wang, Q.; Che, G. Fabrication of pollutant-resistance SERS imprinted sensors based on SiO2@TiO2@Ag composites for selective detection of pyrethroids in water. *J. Phys. Chem. Solids* **2020**, *138*, 109254. [CrossRef]

75. Krishnan, S.K.; Chipatecua Godoy, Y. Deep Eutectic Solvent-Assisted Synthesis of Au Nanostars Supported on Graphene Oxide as an Efficient Substrate for SERS-Based Molecular Sensing. *ACS Omega* **2020**, *5*, 1384–1393. [CrossRef] [PubMed]

76. Zhu, C.; Meng, G.; Huang, Q.; Huang, Z.; Chu, Z. Au Hierarchical Micro/Nanotower Arrays and Their Improved SERS Effect by Ag Nanoparticle Decoration. *Cryst. Growth Des.* **2011**, *11*, 748–752. [CrossRef]

77. Subramanian, B.; Theriault, G.; Robichaud, J.; Tchoukanova, N.; Djaoued, Y. Large-area crack-free Au-SiO2 2D inverse opal composite films: Fabrication and SERS applications. *Mater. Chem. Phys.* **2020**, *244*, 122630. [CrossRef]

78. Barbillon, G. Latest Novelties on Plasmonic and Non-Plasmonic Nanomaterials for SERS Sensing. *Nanomaterials* **2020**, *10*, 1200. [CrossRef]

79. Sun, Y.; Xia, Y. Shape-controlled synthesis of gold and silver nanoparticles. *Science* **2002**, *298*, 2176–2179. [CrossRef]

80. Chen, J.; McLellan, J.M.; Siekkinen, A.; Xiong, Y.; Li, Z.-Y.; Xia, Y. Facile Synthesis of Gold-Silver Nanocages with Controllable Pores on the Surface. *J. Am. Chem. Soc.* **2006**, *128*, 14776–14777. [CrossRef] [PubMed]

81. Ye, J.; Van Dorpe, P.; Van Roy, W.; Borghs, G.; Maes, G. Fabrication, Characterization, and Optical Properties of Gold Nanobowl Submonolayer Structures. *Langmuir* **2009**, *25*, 1822–1827. [CrossRef] [PubMed]

82. Liebig, F.; Sarhan, R.M.; Sander, M.; Koopman, W.; Schuetz, R.; Bargheer, M.; Koetz, J. Deposition of Gold Nanotriangles in Large Scale Close-Packed Monolayers for X-ray-Based Temperature Calibration and SERS Monitoring of Plasmon-Driven Catalytic Reactions. *ACS Appl. Mater. Interfaces* **2017**, *9*, 20247–20253. [CrossRef]

83. Li, C.; Shuford, K.L.; Chen, M.; Lee, E.J.; Cho, S.O. A Facile Polyol Route to Uniform Gold Octahedra with Tailorable Size and Their Optical Properties. *ACS Nano* **2008**, *2*, 1760–1769. [CrossRef]

84. Nguyen, P.-D.; Zhang, X.; Su, J. One-Step Controlled Synthesis of Size-Tunable Toroidal Gold Particles for Biochemical Sensing. *ACS Appl. Nano Mater.* **2019**, *2*, 7839–7847. [CrossRef]

85. Qiu, Y.H.; Ding, S.J.; Lin, Y.J.; Chen, K.; Yang, D.J.; Ma, S.; Li, X.; Lin, H.Q.; Wang, J.; Wang, Q.Q. Growth of Au Hollow Stars and Harmonic Excitation Energy Transfer. *ACS Nano* **2020**, *14*, 736–745. [CrossRef] [PubMed]

86. Maier, S.A.; Brongersma, M.L.; Kik, P.G.; Meltzer, S.; Requicha, A.A.; Atwater, H.A. Plasmonics—A route to nanoscale optical devices. *Adv. Mater.* **2001**, *13*, 1501–1505. [CrossRef]

87. Alvarez-Puebla, R.; Liz-Marzán, L.M.; García de Abajo, F.J. Light Concentration at the Nanometer Scale. *J. Phys. Chem. Lett.* **2010**, *1*, 2428–2434. [CrossRef]

88. Aroca, R.; Alvarez-Puebla, R.; Pieczonka, N.; Sanchez-Cortez, S.; Garcia-Ramos, J. Surface-enhanced Raman scattering on colloidal nanostructures. *Adv. Colloid Interface Sci.* **2005**, *116*, 45–61. [CrossRef]

89. Lee, P.C.; Meisel, D. Adsorption and surface-enhanced Raman of dyes on silver and gold sols. *J. Phys. Chem.* **1982**, *86*, 3391–3395. [CrossRef]

90. Rivas, L.; Sanchez-Cortes, S.; García-Ramos, J.V.; Morcillo, G. Mixed Silver/Gold Colloids: A Study of Their Formation, Morphology, and Surface-Enhanced Raman Activity. *Langmuir* **2000**, *16*, 9722–9728. [CrossRef]

91. Faulds, K.; Littleford, R.E.; Graham, D.; Dent, G.; Smith, W.E. Comparison of Surface-Enhanced Resonance Raman Scattering from Unaggregated and Aggregated Nanoparticles. *Anal. Chem.* **2004**, *76*, 592–598. [CrossRef] [PubMed]

92. Schopf, C.; Martín, A.; Burke, M.; Jones, D.; Pescaglini, A.; O'Riordan, A.; Quinn, A.J.; Iacopino, D. Au nanorod plasmonic superstructures obtained by a combined droplet evaporation and stamping method. *J. Mater. Chem. C* **2014**, *2*, 3536–3541. [CrossRef]

93. Yang, J.K.; Kang, H.; Lee, H.; Jo, A.; Jeong, S.; Jeon, S.J.; Kim, H.I.; Lee, H.Y.; Jeong, D.H.; Kim, J.H.; et al. Single-Step and Rapid Growth of Silver Nanoshells as SERS-Active Nanostructures for Label-Free Detection of Pesticides. *ACS Appl. Mater. Interfaces* **2014**, *6*, 12541–12549. [CrossRef] [PubMed]

94. Aroca, R.F.; Goulet, P.J.G.; dos Santos, D.S.; Alvarez-Puebla, R.A.; Oliveira, O.N. Silver Nanowire Layer-by-Layer Films as Substrates for Surface-Enhanced Raman Scattering. *Anal. Chem.* **2005**, *77*, 378–382. [CrossRef]

95. Freeman, R.G.; Grabar, K.C.; Allison, K.J.; Bright, R.M.; Davis, J.A.; Guthrie, A.P.; Hommer, M.B.; Jackson, M.A.; Smith, P.C.; Walter, D.G.; et al. Self-assembled metal colloid monolayers: An approach to SERS substrates. *Science* **1995**, *267*, 1629–1632. [CrossRef] [PubMed]

96. Wang, Z.; Pan, S.; Krauss, T.D.; Du, H.; Rothberg, L.J. The structural basis for giant enhancement enabling single-molecule Raman scattering. *Proc. Natl. Acad. Sci. USA* **2003**, *100*, 8638–8643. [CrossRef]

97. Fan, M.; Brolo, A.G. Silver nanoparticles self assembly as SERS substrates with near single molecule detection limit. *Phys. Chem. Chem. Phys.* **2009**, *11*, 7381–7389. [CrossRef]

98. Yu, W.W.; White, I.M. A simple filter-based approach to surface enhanced Raman spectroscopy for trace chemical detection. *Analyst* **2012**, *137*, 1168–1173. [CrossRef] [PubMed]

99. Shiohara, A.; Langer, J.; Polavarapu, L.; Liz-Marzan, L.M. Solution processed polydimethylsiloxane/gold nanostar flexible substrates for plasmonic sensing. *Nanoscale* **2014**, *6*, 9817–9823. [CrossRef]

100. Polavarapu, L.; La Porta, A.; Novikov, S.M.; Coronado-Puchau, M.; Liz-Marzan, L.M. Pen-on-Paper Approach Toward the Design of Universal Surface Enhanced Raman Scattering Substrates. *Small* **2014**, *10*, 3065–3071. [CrossRef] [PubMed]

101. Lee, C.H.; Tian, L.; Singamaneni, S. Paper-based SERS swab for rapid trace detection on real-world surfaces. *ACS Appl. Mater. Interfaces* **2010**, *2*, 3429–3435. [CrossRef] [PubMed]

102. Chen, J.; Huang, Y.; Kannan, P.; Zhang, L.; Lin, Z.; Zhang, J.; Chen, T.; Guo, L. Flexible and Adhesive Surface Enhance Raman Scattering Active Tape for Rapid Detection of Pesticide Residues in Fruits and Vegetables. *Anal. Chem.* **2016**, *88*, 2149–2155. [CrossRef]

103. Kahl, M.; Voges, E.; Kostrewa, S.; Viets, C.; Hill, W. Periodically structured metallic substrates for SERS. *Sens. Actuators B Chem.* **1998**, *51*, 285–291. [CrossRef]

104. Yu, Q.; Guan, P.; Qin, D.; Golden, G.; Wallace, P.M. Inverted Size-Dependence of Surface-Enhanced Raman Scattering on Gold Nanohole and Nanodisk Arrays. *Nano Lett.* **2008**, *8*, 1923–1928. [CrossRef]

105. Wu, T.; Lin, Y.-W. Surface-enhanced Raman scattering active gold nanoparticle/nanohole arrays fabricated through electron beam lithography. *Appl. Surf. Sci.* **2018**, *435*, 1143–1149. [CrossRef]

106. Krishnamoorthy, S.; Krishnan, S.; Thoniyot, P.; Low, H.Y. Inherently Reproducible Fabrication of Plasmonic Nanoparticle Arrays for SERS by Combining Nanoimprint and Copolymer Lithography. *ACS Appl. Mater. Interfaces* **2011**, *3*, 1033–1040. [CrossRef]

107. Robinson, C.; Justice, J.; Petäjä, J.; Karppinen, M.; Corbett, B.; O'Riordan, A.; Lovera, P. Nanoimprint Lithography–Based Fabrication of Plasmonic Array of Elliptical Nanoholes for Dual-Wavelength, Dual-Polarisation Refractive Index Sensing. *Plasmonics* **2019**, *14*, 951–959. [CrossRef]

108. Yang, J.; Li, J.; Du, Z.; Teng, J.; Hong, M. Laser Hybrid Micro/nano-structuring of Si Surfaces in Air and its Applications for SERS Detection. *Sci. Rep.* **2014**, *4*, 6657. [CrossRef] [PubMed]

109. Sun, W.; Hong, R.; Liu, Q.; Li, Z.; Shi, J.; Tao, C.; Zhang, D. SERS-active Ag–Al alloy nanoparticles with tunable surface plasmon resonance induced by laser ablation. *Opt. Mater.* **2019**, *96*, 109298. [CrossRef]

110. Bian, X.; Xu, J.; Yang, J.; Chiu, K.-L.; Jiang, S. Flexible Ag SERS substrate for non-destructive and rapid detection of toxic materials on irregular surface. *Surf. Interfaces* **2021**, *23*, 100995. [CrossRef]

111. Zhou, Q.; Thokchom, A.K.; Kim, D.-J.; Kim, T. Inkjet-printed Ag micro-/nanostructure clusters on Cu substrates for in-situ pre-concentration and surface-enhanced Raman scattering. *Sens. Actuators B Chem.* **2017**, *243*, 176–183. [CrossRef]

112. Huebner, U.; Boucher, R.; Schneidewind, H.; Cialla, D.; Popp, J. Microfabricated SERS-arrays with sharp-edged metallic nanostructures. *Microelectron. Eng.* **2008**, *85*, 1792–1794. [CrossRef]

113. Yang, J.-Y.; Park, S.-G.; Jung, S.; Byeon, E.-Y.; Kim, D.-G.; Jung, H.S.; Kim, H.J.; Lee, S. SERS substrates based on self-organized dimple nanostructures on polyethylene naphthalate films produced via oxygen ion beam sputtering. *Appl. Surf. Sci.* **2022**, *572*, 151452. [CrossRef]

114. Muhammad, M.; Shao, C.-S.; Huang, Q. Aptamer-functionalized Au nanoparticles array as the effective SERS biosensor for label-free detection of interleukin-6 in serum. *Sens. Actuators B Chem.* **2021**, *334*, 129607. [CrossRef]

115. Zhang, C.; Yi, P.; Peng, L.; Lai, X.; Chen, J.; Huang, M.; Ni, J. Continuous fabrication of nanostructure arrays for flexible surface enhanced Raman scattering substrate. *Sci. Rep.* **2017**, *7*, 39814. [CrossRef] [PubMed]

116. Shi, G.; Wang, M.; Zhu, Y.; Yan, X.; Pan, S.; Zhang, A. Nanoflower-like Ag/AAO SERS platform with quasi-photonic crystal nanostructure for efficient detection of goat serum. *Curr. Appl. Phys.* **2019**, *19*, 1276–1285. [CrossRef]

117. Li, M.; Wu, J.; Wang, C.; Fang, J. The cascade structure of periodic micro/nanoscale Au nano-islands @ Ag-frustum arrays as effective SERS substrates. *Vacuum* **2020**, *175*, 109265. [CrossRef]

118. Lin, B.; Kannan, P.; Qiu, B.; Lin, Z.; Guo, L. On-spot surface enhanced Raman scattering detection of Aflatoxin B1 in peanut extracts using gold nanobipyramids evenly trapped into the AAO nanoholes. *Food Chem.* **2020**, *307*, 125528. [CrossRef]
119. Hu, M.; Ou, F.S.; Wu, W.; Naumov, I.; Li, X.; Bratkovsky, A.M.; Williams, R.S.; Li, Z. Gold Nanofingers for Molecule Trapping and Detection. *J. Am. Chem. Soc.* **2010**, *132*, 12820–12822. [CrossRef]
120. Diebold, E.D.; Mack, N.H.; Doorn, S.K.; Mazur, E. Femtosecond Laser-Nanostructured Substrates for Surface-Enhanced Raman Scattering. *Langmuir* **2009**, *25*, 1790–1794. [CrossRef]
121. Aleknavičienė, I.; Pabrėža, E.; Talaikis, M.; Jankunec, M.; Račiukaitis, G. Low-cost SERS substratefeaturing laser-ablated amorphous nanostructure. *Appl. Surf. Sci.* **2022**, *571*, 151248. [CrossRef]
122. Restaino, S.M.; White, I.M. A critical review of flexible and porous SERS sensors for analytical chemistry at the point-of-sample. *Anal. Chim. Acta* **2019**, *1060*, 17–29. [CrossRef]
123. Li, L.; Yang, S.Y.; Duan, J.L.; Huang, L.; Xiao, G.N. Fabrication and SERS performance of silver nanoarrays by inkjet printing silver nanoparticles ink on the gratings of compact disc recordable. *Spectrochim. Acta A—Mol. Biomol. Spectrosc.* **2020**, *225*, 117598. [CrossRef]
124. Wu, J.; Zhang, L.; Huang, F.; Ji, X.; Dai, H.; Wu, W. Surface enhanced Raman scattering substrate for the detection of explosives: Construction strategy and dimensional effect. *J. Hazard. Mater.* **2020**, *387*, 121714. [CrossRef] [PubMed]
125. Yu, B.; Ge, M.; Li, P.; Xie, Q.; Yang, L. Development of surface-enhanced Raman spectroscopy application for determination of illicit drugs: Towards a practical sensor. *Talanta* **2019**, *191*, 1–10. [PubMed]
126. D'Apuzzo, F.; Sengupta, R.N.; Overbay, M.; Aronoff, J.S.; Rogacs, A.; Barcelo, S.J. A Generalizable Single-Chip Calibration Method for Highly Quantitative SERS via Inkjet Dispense. *Anal. Chem.* **2020**, *92*, 1372–1378. [PubMed]
127. Joshi, P.; Santhanam, V. Inkjet-Based Fabrication Process with Control over the Morphology of SERS-Active Silver Nanostructures. *Ind. Eng. Chem. Res.* **2018**, *57*, 5250–5258. [CrossRef]
128. Kumar, A.; Santhanam, V. Paper swab based SERS detection of non-permitted colourants from dals and vegetables using a portable spectrometer. *Anal. Chim. Acta* **2019**, *1090*, 106–113. [CrossRef]
129. Lan, L.L.; Hou, X.Y.; Gao, Y.M.; Fan, X.C.; Qiu, T. Inkjet-printed paper-based semiconducting substrates for surface-enhanced Raman spectroscopy. *Nanotechnology* **2020**, *31*, 055502. [CrossRef] [PubMed]
130. Li, L.; Xiao, G.N. Research Progress of Preparing Surface-Enhanced Raman Scattering Active Substrates by Printing Technologies. *Spectrosc. Spectr. Anal.* **2019**, *39*, 3326–3332.
131. Micciche, C.; Arrabito, G.; Amato, F.; Buscarino, G.; Agnello, S.; Pignataro, B. Inkjet printing Ag nanoparticles for SERS hot spots. *Anal. Methods* **2018**, *10*, 3215–3223. [CrossRef]
132. Oravec, M.; Sasinkova, V.; Tomanova, K.; Gal, L.; Parciova, S.; Huck, C.W. In-situ surface-enhanced Raman scattering and FT-Raman spectroscopy of black prints. *Vib. Spectrosc.* **2018**, *94*, 16–21. [CrossRef]
133. Weng, G.J.; Yang, Y.; Zhao, J.; Zhu, J.; Li, J.J.; Zhao, J.W. Preparation and SERS performance of Au NP/paper strips based on inkjet printing and seed mediated growth: The effect of silver ions. *Solid State Commun.* **2018**, *272*, 67–73. [CrossRef]
134. Yu, W.W.; White, I.M. Inkjet-printed paper-based SERS dipsticks and swabs for trace chemical detection. *Analyst* **2013**, *138*, 1020–1025. [CrossRef]
135. Shen, S.; Zhao, B.; Wang, H.; Li, Z.; Qu, G.; Guo, Z.; Zhou, T.; Song, W.; Wang, X.; Ruan, W. CdTe quantum dots modified polystyrene spheres with Ag nanoparticle caps: Applications both in fluorescence and in SERS. *Colloids Surf. A Physicochem. Eng. Asp.* **2014**, *443*, 467–472. [CrossRef]
136. Sammi, H.; Nair, R.V.; Sardana, N. Recent advances in nanoporous AAO based substrates for surface-enhanced raman scattering. *Mater. Today Proc.* **2021**, *41*, 843–850. [CrossRef]
137. Huang, Z.; Meng, G.; Huang, Q.; Chen, B.; Zhu, C.; Zhang, Z. Large-area Ag nanorod array substrates for SERS: AAO template-assisted fabrication, functionalization, and application in detection PCBs. *J. Raman Spectrosc.* **2013**, *44*, 240–246. [CrossRef]
138. Ruan, C.; Eres, G.; Wang, W.; Zhang, Z.; Gu, B. Controlled Fabrication of Nanopillar Arrays as Active Substrates for Surface-Enhanced Raman Spectroscopy. *Langmuir* **2007**, *23*, 5757–5760. [CrossRef]
139. Batista, E.A.; dos Santos, D.P.; Andrade, G.F.S.; Sant'Ana, A.C.; Brolo, A.G.; Temperini, M.L.A. Using Polycarbonate Membranes as Templates for the Preparation of Au Nanostructures for Surface-Enhanced Raman Scattering. *J. Nanosci. Nanotechnol.* **2009**, *9*, 3233–3238. [CrossRef] [PubMed]
140. Penn, M.A.; Drake, D.M.; Driskell, J.D. Accelerated Surface-Enhanced Raman Spectroscopy (SERS)-Based Immunoassay on a Gold-Plated Membrane. *Anal. Chem.* **2013**, *85*, 8609–8617. [CrossRef]
141. Wigginton, K.R.; Vikesland, P.J. Gold-coated polycarbonate membrane filter for pathogen concentration and SERS-based detection. *Analyst* **2010**, *135*, 1320–1326. [CrossRef] [PubMed]
142. Piao, L.; Park, S.; Lee, H.B.; Kim, K.; Kim, J.; Chung, T.D. Single Gold Microshell Tailored to Sensitive Surface Enhanced Raman Scattering Probe. *Anal. Chem.* **2010**, *82*, 447–451. [CrossRef]
143. Wang, J.J.; Zhou, F.; Duan, G.T.; Li, Y.; Liu, G.Q.; Su, F.H.; Cai, W.P. A controlled Ag-Au bimetallic nanoshelled microsphere array and its improved surface-enhanced Raman scattering effect. *RSC Adv.* **2014**, *4*, 8758–8763. [CrossRef]
144. Zhao, Y.H.; Luo, W.Q.; Kanda, P.; Cheng, H.W.; Chen, Y.Y.; Wang, S.P.; Huan, S.Y. Silver deposited polystyrene (PS) microspheres for surface-enhanced Raman spectroscopic-encoding and rapid label-free detection of melamine in milk powder. *Talanta* **2013**, *113*, 7–13. [CrossRef]

145. Charconnet, M.; Kuttner, C.; Plou, J.; García-Pomar, J.L.; Mihi, A.; Liz-Marzán, L.M.; Seifert, A. Mechanically Tunable Lattice-Plasmon Resonances by Templated Self-Assembled Superlattices for Multi-Wavelength Surface-Enhanced Raman Spectroscopy. *Small Methods* **2021**, *5*, 2100453. [CrossRef]

146. Wu, L.; Wang, W.; Zhang, W.; Su, H.; Liu, Q.; Gu, J.; Deng, T.; Zhang, D. Highly sensitive, reproducible and uniform SERS substrates with a high density of three-dimensionally distributed hotspots: Gyroid-structured Au periodic metallic materials. *NPG Asia Mater.* **2018**, *10*, e462. [CrossRef]

147. Klutse, C.K.; Mayer, A.; Wittkamper, J.; Cullum, B.M. Applications of Self-Assembled Monolayers in Surface-Enhanced Raman Scattering. *J. Nanotechnol.* **2012**, *2012*, 319038. [CrossRef]

148. Tielens, F.; Santos, E. AuS and SH Bond Formation/Breaking during the Formation of Alkanethiol SAMs on Au(111): A Theoretical Study. *J. Phys. Chem. C* **2010**, *114*, 9444–9452. [CrossRef]

149. Cohen-Atiya, M.; Mandler, D. Studying thiol adsorption on Au, Ag and Hg surfaces by potentiometric measurements. *J. Electroanal. Chem.* **2003**, *550–551*, 267–276. [CrossRef]

150. Xue, Y.; Li, X.; Li, H.; Zhang, W. Quantifying thiol–gold interactions towards the efficient strength control. *Nat. Commun.* **2014**, *5*, 4348. [CrossRef]

151. Inkpen, M.S.; Liu, Z.-F.; Li, H.; Campos, L.M.; Neaton, J.B.; Venkataraman, L. Non-chemisorbed gold–sulfur binding prevails in self-assembled monolayers. *Nat. Chem.* **2019**, *11*, 351–358. [CrossRef]

152. Grönbeck, H.; Curioni, A.; Andreoni, W. Thiols and Disulfides on the Au(111) Surface: The Headgroup-Gold Interaction. *J. Am. Chem. Soc.* **2000**, *122*, 3839–3842. [CrossRef]

153. Levin, C.S.; Bishnoi, S.W.; Grady, N.K.; Halas, N.J. Determining the Conformation of Thiolated Poly(ethylene glycol) on Au Nanoshells by Surface-Enhanced Raman Scattering Spectroscopic Assay. *Anal. Chem.* **2006**, *78*, 3277–3281. [CrossRef]

154. Yang, E.L.; Li, D.; Yin, P.K.; Xie, Q.Y.; Li, Y.; Lin, Q.Y.; Duan, Y.X. A novel surface-enhanced Raman scattering (SERS) strategy for ultrasensitive detection of bacteria based on three-dimensional (3D) DNA walker. *Biosens. Bioelectron.* **2021**, *172*, 112758. [CrossRef]

155. Kamińska, A.; Winkler, K.; Kowalska, A.; Witkowska, E.; Szymborski, T.; Janeczek, A.; Waluk, J. SERS-based Immunoassay in a Microfluidic System for the Multiplexed Recognition of Interleukins from Blood Plasma: Towards Picogram Detection. *Sci. Rep.* **2017**, *7*, 10656. [CrossRef]

156. Oss, C.J.V. Antibody-Antigen Intermolecular Forces. In *Encyclopedia of Immunology*, 2nd ed.; Elsevier: Amsterdam, The Netherlands, 1998; pp. 163–167.

157. Kim, H.; Kang, H.; Kim, H.-N.; Kim, H.; Moon, J.; Guk, K.; Park, H.; Yong, D.; Bae, P.K.; Park, H.G.; et al. Development of 6E3 antibody-mediated SERS immunoassay for drug-resistant influenza virus. *Biosens. Bioelectron.* **2021**, *187*, 113324. [CrossRef]

158. He, D.Y.; Wu, Z.Z.; Cui, B.; Xu, E.B. Aptamer and gold nanorod-based fumonisin B1 assay using both fluorometry and SERS. *Microchim. Acta* **2020**, *187*, 215. [CrossRef]

159. He, X.; Zhou, X.; Liu, Y.; Wang, X.L. Ultrasensitive, recyclable and portable microfluidic surface-enhanced raman scattering (SERS) biosensor for uranyl ions detection. *Sens. Actuators B Chem.* **2020**, *311*, 127676. [CrossRef]

160. Huang, D.D.; Chen, J.M.; Ding, L.; Guo, L.H.; Kannan, P.; Luo, F.; Qiu, B.; Lin, Z.Y. Core-satellite assemblies and exonuclease assisted double amplification strategy for ultrasensitive SERS detection of biotoxin. *Anal. Chim. Acta* **2020**, *1110*, 56–63. [CrossRef] [PubMed]

161. Wang, J.R.; Xia, C.; Yang, L.; Li, Y.F.; Li, C.M.; Huang, C.Z. DNA Nanofirecrackers Assembled through Hybridization Chain Reaction for Ultrasensitive SERS Immunoassay of Prostate Specific Antigen. *Anal. Chem.* **2020**, *92*, 4046–4052. [CrossRef] [PubMed]

162. Wang, Q.; Hu, Y.J.; Jiang, N.J.; Wang, J.J.; Yu, M.; Zhuang, X.M. Preparation of Aptamer Responsive DNA Functionalized Hydrogels for the Sensitive Detection of alpha-Fetoprotein Using SERS Method. *Bioconjugate Chem.* **2020**, *31*, 813–820. [CrossRef] [PubMed]

163. Zhou, S.S.; Lu, C.; Li, Y.Z.; Xue, L.; Zhao, C.Y.; Tian, G.F.; Bao, Y.M.; Tang, L.H.; Lin, J.H.; Zheng, J.K. Gold Nanobones Enhanced Ultrasensitive Surface-Enhanced Raman Scattering Aptasensor for Detecting Escherichia coil O157:H7. *ACS Sens.* **2020**, *5*, 588–596. [CrossRef] [PubMed]

164. Šimáková, P.; Gautier, J.; Procházka, M.; Hervé-Aubert, K.; Chourpa, I. Polyethylene-glycol-Stabilized Ag Nanoparticles for Surface-Enhanced Raman Scattering Spectroscopy: Ag Surface Accessibility Studied Using Metalation of Free-Base Porphyrins. *J. Phys. Chem. C* **2014**, *118*, 7690–7697. [CrossRef]

165. Lin, D.; Hsieh, C.-L.; Hsu, K.-C.; Liao, P.-H.; Qiu, S.; Gong, T.; Yong, K.-T.; Feng, S.; Kong, K.V. Geometrically encoded SERS nanobarcodes for the logical detection of nasopharyngeal carcinoma-related progression biomarkers. *Nat. Commun.* **2021**, *12*, 3430. [CrossRef]

166. Jeyaratnam, J. Acute pesticide poisoning: A major global health problem. *World Health Stat. Q.* **1990**, *43*, 139–144.

167. Aktar, M.W.; Sengupta, D.; Chowdhury, A. Impact of pesticides use in agriculture: Their benefits and hazards. *Interdiscip. Toxicol.* **2009**, *2*, 1–12. [CrossRef]

168. Food Agriculture Organization. *International Code of Conduct on the Distribution and Use of Pesticides*; United Nations: Rome, Italy, 2005.

169. Han, L.J.; Sapozhnikova, Y. Semi-automated high-throughput method for residual analysis of 302 pesticides and environmental contaminants in catfish by fast low-pressure GC-MS/MS and UHPLC-MS/MS. *Food Chem.* **2020**, *319*, 126592. [CrossRef]

170. Velkoska-Markovska, L.; Petanovska-Ilievska, B. Rapid Resolution Liquid Chromatography Method for Determination of Malathion in Pesticide Formulation. *Acta Chromatogr.* **2020**, *32*, 256–259. [CrossRef]
171. Geto, A.; Noori, J.S.; Mortensen, J.; Svendsen, W.E.; Dimaki, M. Electrochemical determination of bentazone using simple screen-printed carbon electrodes. *Environ. Int.* **2019**, *129*, 400–407. [CrossRef]
172. Santana, P.C.A.; Lima, J.B.S.; Santana, T.B.S.; Santos, L.F.S.; Matos, C.R.S.; da Costa, L.P.; Gimenez, I.F.; Sussuchi, E.M. Semiconductor Nanocrystals-Reduced Graphene Composites for the Electrochemical Detection of Carbendazim. *J. Braz. Chem. Soc.* **2019**, *30*, 1302–1308. [CrossRef]
173. Alak, A.M.; Vo-Dinh, T. Surface-enhanced Raman spectrometry of organo phosphorus chemical agents. *Anal. Chem.* **1987**, *59*, 2149–2153. [CrossRef]
174. European Food Safety Authority. The 2010 European Union Report on Pesticide Residues in Food. *EFSA J.* **2013**, *11*, 3130.
175. Song, D.; Yang, R.; Long, F.; Zhu, A. Applications of magnetic nanoparticles in surface-enhanced Raman scattering (SERS) detection of environmental pollutants. *J. Environ. Sci.* **2019**, *80*, 14–34. [CrossRef] [PubMed]
176. Fan, Y.X.; Lai, K.Q.; Rasco, B.A.; Huang, Y.Q. Analyses of phosmet residues in apples with surface-enhanced Raman spectroscopy. *Food Control* **2014**, *37*, 153–157. [CrossRef]
177. Liu, B.; Zhou, P.; Liu, X.M.; Sun, X.; Li, H.; Lin, M.S. Detection of Pesticides in Fruits by Surface-Enhanced Raman Spectroscopy Coupled with Gold Nanostructures. *Food Bioprocess Technol.* **2013**, *6*, 710–718. [CrossRef]
178. Yaseen, T.; Pu, H.B.; Sun, D.W. Rapid detection of multiple organophosphorus pesticides (triazophos and parathion-methyl) residues in peach by SERS based on core-shell bimetallic Au@Ag NPs. *Food Addit. Contam. A—Chem. Anal. Control Expo. Risk Assess.* **2019**, *36*, 762–778. [CrossRef]
179. Benitta, T.A.; Kapoor, S.; Christy, R.S.; Raj, C.I.S.; Kumaran, J.T.T. Surface Enhanced Raman Spectra and Theoretical Study of an Organophosphate Malathion. *Orient. J. Chem.* **2017**, *33*, 760–767. [CrossRef]
180. Nie, Y.H.; Teng, Y.J.; Li, P.; Liu, W.H.; Shi, Q.W.; Zhang, Y.C. Label-free aptamer-based sensor for specific detection of malathion residues by surface-enhanced Raman scattering. *Spectrochim. Acta A—Mol. Biomol. Spectrosc.* **2018**, *191*, 271–276. [CrossRef]
181. Banks, K.E.; Hunter, D.H.; Wachal, D.J. Chlorpyrifos in surface waters before and after a federally mandated ban. *Environ. Int.* **2005**, *31*, 351–356. [CrossRef]
182. Feng, S.L.; Hu, Y.X.; Ma, L.Y.; Lu, X.N. Development of molecularly imprinted polymers-surface-enhanced Raman spectroscopy/colorimetric dual sensor for determination of chlorpyrifos in apple juice. *Sens. Actuators B Chem.* **2017**, *241*, 750–757. [CrossRef]
183. Ma, P.; Wang, L.Y.; Xu, L.; Li, J.Y.; Zhang, X.D.; Chen, H. Rapid quantitative determination of chlorpyrifos pesticide residues in tomatoes by surface-enhanced Raman spectroscopy. *Eur. Food Res. Technol.* **2020**, *246*, 239–251. [CrossRef]
184. Hussain, A.; Sun, D.W.; Pu, H.B. Bimetallic core shelled nanoparticles (Au@AgNPs) for rapid detection of thiram and dicyandiamide contaminants in liquid milk using SERS. *Food Chem.* **2020**, *317*, 126429. [CrossRef]
185. Hu, X.N.; Bian, X.Z.; Yu, S.Z.; Dan, K. Magnetic Fe3O4@SiO2@Ag@COOH NPs/Au Film with Hybrid Localized Surface Plasmon/Surface Plasmon Polariton Modes for Surface-Enhanced Raman Scattering Detection of Thiabendazole. *J. Nanosci. Nanotechnol.* **2020**, *20*, 2079–2086. [CrossRef]
186. Wang, K.Q.; Sun, D.W.; Pu, H.B.; Wei, Q.Y. Two-dimensional Au@Ag nanodot array for sensing dual-fungicides in fruit juices with surface-enhanced Raman spectroscopy technique. *Food Chem.* **2020**, *310*, 125923. [CrossRef] [PubMed]
187. Costa, J.C.S.; Ando, R.A.; Sant'Ana, A.C.; Rossi, L.M.; Santos, P.S.; Temperini, M.L.A.; Corio, P. High performance gold nanorods and silver nanocubes in surface-enhanced Raman spectroscopy of pesticides. *Phys. Chem. Chem. Phys.* **2009**, *11*, 7491–7498. [CrossRef] [PubMed]
188. Dowgiallo, A.M. Trace level pesticide detection utilizing gold nanoparticles and surface enhanced Raman spectroscopy (SERS). In *Synthesis and Photonics of Nanoscale Materials XVI*; Kabashin, A.V., Dubowski, J.J., Geohegan, D.B., Eds.; SPIE: Washington, DC, USA, 2019; Volume 10907.
189. Wang, Q.; Zhao, Y.; Bu, T.; Wang, X.; Xu, Z.; Zhangsun, H.; Wang, L. Semi-sacrificial template growth-assisted self-supporting MOF chip: A versatile and high-performance SERS sensor for food contaminants monitoring. *Sens. Actuators B Chem.* **2022**, *352*, 131025. [CrossRef]
190. Canamares, M.V.; Feis, A. Surface-enhanced Raman spectra of the neonicotinoid pesticide thiacloprid. *J. Raman Spectrosc.* **2013**, *44*, 1126–1135. [CrossRef]
191. Wu, J.; Xi, J.; Chen, H.; Li, S.; Zhang, L.; Li, P.; Wu, W. Flexible 2D nanocellulose-based SERS substrate for pesticide residue detection. *Carbohydr. Polym.* **2022**, *277*, 118890. [CrossRef]
192. Lu, Y.; Tan, Y.; Xiao, Y.; Li, Z.; Sheng, E.; Dai, Z. A silver@gold nanoparticle tetrahedron biosensor for multiple pesticides detection based on surface-enhanced Raman scattering. *Talanta* **2021**, *234*, 122585. [CrossRef]
193. Zaim, M.; Jambulingam, P. *Global Insecticide Use for Vector-Borne Disease Control*; World Health Organization: Geneva, Switzerland, 2007.
194. Kozawa, K.; Aoyama, Y.; Mashimo, S.; Kimura, H. Toxicity and actual regulation of organophosphate pesticides. *Toxin Rev.* **2009**, *28*, 245–254. [CrossRef]
195. Rosenstock, L.; Keifer, M.; Daniell, W.E.; McConnell, R.; Claypoole, K.; The Pesticide Health Effects Study Group. Chronic central nervous system effects of acute organophosphate pesticide intoxication. *Lancet* **1991**, *338*, 223–227. [CrossRef]

196. Eskenazi, B.; Bradman, A.; Castorina, R. Exposures of children to organophosphate pesticides and their potential adverse health effects. *Environ. Health Perspect.* **1999**, *107* (Suppl. 3), 409–419. [CrossRef] [PubMed]

197. Stephens, R.; Spurgeon, A.; Calvert, I.A.; Beach, J.; Levy, L.S.; Harrington, J.; Berry, H. Neuropsychological effects of long-term exposure to organophosphates in sheep dip. *Lancet* **1995**, *345*, 1135–1139. [CrossRef]

198. Fries, E.; Püttmann, W. Occurrence of organophosphate esters in surface water and ground water in Germany. *J. Environ. Monit.* **2001**, *3*, 621–626. [CrossRef]

199. Karalliedde, L.; Eddleston, M.; Murray, V. *The Global Picture of Organophosphate Insecticide Poisoning*; World Scientific: Singapore, 2001; pp. 431–471.

200. Pogačnik, L.; Franko, M. Determination of organophosphate and carbamate pesticides in spiked samples of tap water and fruit juices by a biosensor with photothermal detection. *Biosens. Bioelectron.* **1999**, *14*, 569–578. [CrossRef]

201. Syafrudin, M.; Kristanti, R.A.; Yuniarto, A.; Hadibarata, T.; Rhee, J.; Al-onazi, W.A.; Algarni, T.S.; Almarri, A.H.; Al-Mohaimeed, A.M. Pesticides in Drinking Water—A Review. *Int. J. Environ. Res. Public Health* **2021**, *18*, 468. [CrossRef] [PubMed]

202. Liu, B.; Han, G.; Zhang, Z.; Liu, R.; Jiang, C.; Wang, S.; Han, M.-Y. Shell Thickness-Dependent Raman Enhancement for Rapid Identification and Detection of Pesticide Residues at Fruit Peels. *Anal. Chem.* **2012**, *84*, 255–261. [CrossRef]

203. Li, J.F.; Huang, Y.F.; Ding, Y.; Yang, Z.L.; Li, S.B.; Zhou, X.S.; Fan, F.R.; Zhang, W.; Zhou, Z.Y.; Wu, D.Y.; et al. Shell-isolated nanoparticle-enhanced Raman spectroscopy. *Nature* **2010**, *464*, 392–395. [CrossRef]

204. Chen, X.; Wang, D.H.; Li, J.; Xu, T.T.; Lai, K.Q.; Ding, Q.; Lin, H.T.; Sun, L.; Lin, M.S. A spectroscopic approach to detect and quantify phosmet residues in Oolong tea by surface-enhanced Raman scattering and silver nanoparticle substrate. *Food Chem.* **2020**, *312*, 126016. [CrossRef]

205. Jiang, L.; Gu, K.J.; Liu, R.Y.; Jin, S.Z.; Wang, H.Q.; Pan, C.P. Rapid detection of pesticide residues in fruits by surface-enhanced Raman scattering based on modified QuEChERS pretreatment method with portable Raman instrument. *SN Appl. Sci.* **2019**, *1*, 627. [CrossRef]

206. Weng, S.Z.; Zhu, W.X.; Dong, R.L.; Zheng, L.; Wang, F. Rapid Detection of Pesticide Residues in Paddy Water Using Surface-Enhanced Raman Spectroscopy. *Sensors* **2019**, *19*, 506. [CrossRef]

207. Kang, Y.; Wu, T.; Chen, W.C.; Li, L.; Du, Y.P. A novel metastable state nanoparticle-enhanced Raman spectroscopy coupled with thin layer chromatography for determination of multiple pesticides. *Food Chem.* **2019**, *270*, 494–501. [CrossRef]

208. Kang, Y.; Li, L.; Chen, W.C.; Zhang, F.Y.; Du, Y.P.; Wu, T. Rapid In Situ SERS Analysis of Pesticide Residues on Plant Surfaces Based on Micelle Extraction of Targets and Stabilization of Ag Nanoparticle Aggregates. *Food Anal. Methods* **2018**, *11*, 3161–3169. [CrossRef]

209. Huang, D.D.; Zhao, J.C.; Wang, M.L.; Zhu, S.H. Snowflake-like gold nanoparticles as SERS substrates for the sensitive detection of organophosphorus pesticide residues. *Food Control* **2020**, *108*, 106835. [CrossRef]

210. Tang, F.; Zhang, M.Z.; Li, Z.B.; Du, Z.F.; Chen, B.S.; He, X.; Zhao, S.Y. Hexagonally arranged arrays of urchin-like Ag-nanoparticle decorated ZnO-nanorods grafted on PAN-nanopillars as surface-enhanced Raman scattering substrates. *CrystEngComm* **2018**, *20*, 3550–3558. [CrossRef]

211. Yazdi, S.H.; White, I.M. A nanoporous optofluidic microsystem for highly sensitive and repeatable surface enhanced Raman spectroscopy detection. *Biomicrofluidics* **2012**, *6*, 014105. [CrossRef]

212. Fathi, F.; Lagugne-Labarthet, F.; Pedersen, D.B.; Kraatz, H.B. Studies of the interaction of two organophosphonates with nanostructured silver surfaces. *Analyst* **2012**, *137*, 4448–4453. [CrossRef]

213. Dowgiallo, A.M.; Guenther, D.A. Determination of the Limit of Detection of Multiple Pesticides Utilizing Gold Nanoparticles and Surface-Enhanced Raman Spectroscopy. *J. Agric. Food Chem.* **2019**, *67*, 12642–12651. [CrossRef]

214. Tang, J.S.; Chen, W.W.; Ju, H.X. Rapid detection of pesticide residues using a silver nanoparticles coated glass bead as nonplanar substrate for SERS sensing. *Sens. Actuators B Chem.* **2019**, *287*, 576–583. [CrossRef]

215. Elbert, A.; Haas, M.; Springer, B.; Thielert, W.; Nauen, R. Applied aspects of neonicotinoid uses in crop protection. *Pest Manag. Sci.* **2008**, *64*, 1099–1105. [CrossRef] [PubMed]

216. Jeschke, P.; Nauen, R. Neonicotinoids—From zero to hero in insecticide chemistry. *Pest Manag. Sci.* **2008**, *64*, 1084–1098. [CrossRef]

217. Lee, J.K.; Ahn, K.C.; Park, O.S.; Kang, S.Y.; Hammock, B.D. Development of an ELISA for the Detection of the Residues of the Insecticide Imidacloprid in Agricultural and Environmental Samples. *J. Agric. Food Chem.* **2001**, *49*, 2159–2167. [CrossRef] [PubMed]

218. Watanabe, E.; Baba, K.; Eun, H.; Miyake, S. Application of a commercial immunoassay to the direct determination of insecticide imidacloprid in fruit juices. *Food Chem.* **2007**, *102*, 745–750. [CrossRef]

219. Obana, H.; Okihashi, M.; Akutsu, K.; Kitagawa, Y.; Hori, S. Determination of Acetamiprid, Imidacloprid, and Nitenpyram Residues in Vegetables and Fruits by High-Performance Liquid Chromatography with Diode-Array Detection. *J. Agric. Food Chem.* **2002**, *50*, 4464–4467. [CrossRef]

220. Liu, S.; Zheng, Z.; Wei, F.; Ren, Y.; Gui, W.; Wu, H.; Zhu, G. Simultaneous Determination of Seven Neonicotinoid Pesticide Residues in Food by Ultraperformance Liquid Chromatography Tandem Mass Spectrometry. *J. Agric. Food Chem.* **2010**, *58*, 3271–3278. [CrossRef]

221. Ding, X.; Zhang, W.; Cheng, D.; He, J.; Yang, K.-L. Oligopeptides functionalized surface plasmon resonance biosensors for detecting thiacloprid and imidacloprid. *Biosens. Bioelectron.* **2012**, *35*, 271–276. [CrossRef] [PubMed]

222. Vílchez, J.L.; Valencia, M.C.; Navalón, A.; Molinero-Morales, B.; Capitán-Vallvey, L.F. Flow injection analysis of the insecticide imidacloprid in water samples with photochemically induced fluorescence detection. *Anal. Chim. Acta* **2001**, *439*, 299–305. [CrossRef]

223. Nauen, R.; Bretschneider, T. New modes of action of insecticides. *Pestic. Outlook* **2002**, *13*, 241–245. [CrossRef]

224. Tomizawa, M.; Casida, J.E. Neonicotinoid insecticide toxicology: Mechanisms of selective action. *Annu. Rev. Pharmacol. Toxicol.* **2005**, *45*, 247–268. [CrossRef]

225. Whitehorn, P.R.; O'Connor, S.; Wackers, F.L.; Goulson, D. Neonicotinoid pesticide reduces bumble bee colony growth and queen production. *Science* **2012**, *336*, 351–352. [CrossRef]

226. European Commission. Commission Implementing Regulation (EU) No 485/2013 of 24 May 2013 amending Implementing Regulation (EU) No 540/2011, as regards the conditions of approval of the active substances clothianidin, thiamethoxam and imidacloprid, and prohibiting the use and sale of seeds treated with plant protection products containing those active substances. *Off. J. Eur. Union* **2013**, *139*, 12–14.

227. Ferrer, I.; Thurman, E.M.; Fernández-Alba, A.R. Quantitation and Accurate Mass Analysis of Pesticides in Vegetables by LC/TOF-MS. *Anal. Chem.* **2005**, *77*, 2818–2825. [CrossRef] [PubMed]

228. European Food Safety Authority. The 2014 European Union Report on Pesticide Residues in Food. *EFSA J.* **2016**, *14*, e04611. [CrossRef]

229. Cao, X.L.; Hong, S.H.; Jiang, Z.J.; She, Y.X.; Wang, S.S.; Zhang, C.; Li, H.; Jin, F.; Jin, M.J.; Wang, J. SERS-active metal-organic frameworks with embedded gold nanoparticles. *Analyst* **2017**, *142*, 2640–2647. [CrossRef] [PubMed]

230. Yang, T.; Zhao, B.; Hou, R.; Zhang, Z.; Kinchla, A.J.; Clark, J.M.; He, L. Evaluation of the Penetration of Multiple Classes of Pesticides in Fresh Produce Using Surface-Enhanced Raman Scattering Mapping. *J. Food Sci.* **2016**, *81*, T2891–T2901. [CrossRef]

231. Wijaya, W.; Pang, S.; Labuza, T.P.; He, L. Rapid Detection of Acetamiprid in Foods using Surface-Enhanced Raman Spectroscopy (SERS). *J. Food Sci.* **2014**, *79*, T743–T747. [CrossRef] [PubMed]

232. Sun, Y.; Li, Z.H.; Huang, X.W.; Zhang, D.; Zou, X.B.; Shi, J.Y.; Zhai, X.D.; Jiang, C.P.; Wei, X.O.; Liu, T.T. A nitrile-mediated aptasensor for optical anti-interference detection of acetamiprid in apple juice by surface-enhanced Raman scattering. *Biosens. Bioelectron.* **2019**, *145*, 111672. [CrossRef]

233. Feng, X.Z.; Li, C.N.; Liang, A.H.; Luo, Y.H.; Jiang, Z.L. Doped N/Ag Carbon Dot Catalytic Amplification SERS Strategy for Acetamiprid Coupled Aptamer with 3,3′-Dimethylbiphenyl-4,4′-diamine Oxidizing Reaction. *Nanomaterials* **2019**, *9*, 480. [CrossRef]

234. Li, H.H.; Hu, W.W.; Hassan, M.M.; Zhang, Z.Z.; Chen, Q.S. A facile and sensitive SERS-based biosensor for colormetric detection of acetamiprid in green tea based on unmodified gold nanoparticles. *J. Food Meas. Charact.* **2019**, *13*, 259–268. [CrossRef]

235. Zhao, P.N.; Liu, H.Y.; Zhang, L.N.; Zhu, P.H.; Ge, S.G.; Yu, J.H. Paper-Based SERS Sensing Platform Based on 3D Silver Dendrites and Molecularly Imprinted Identifier Sandwich Hybrid for Neonicotinoid Quantification. *ACS Appl. Mater. Interfaces* **2020**, *12*, 8845–8854. [CrossRef] [PubMed]

236. Chen, Q.S.; Hassan, M.M.; Xu, J.; Zareef, M.; Li, H.H.; Xu, Y.; Wang, P.Y.; Agyekum, A.A.; Kutsanedzie, F.Y.H.; Viswadevarayalu, A. Fast sensing of imidacloprid residue in tea using surface-enhanced Raman scattering by comparative multivariate calibration. *Spectrochim. Acta A—Mol. Biomol. Spectrosc.* **2019**, *211*, 86–93. [CrossRef]

237. Qiu, H.W.; Guo, J.; Wang, M.Q.; Ji, S.D.; Cao, M.H.; Padhiar, M.A.; Bhatti, A.S. Reduced graphene oxide supporting Ag meso-flowers and phenyl-modified graphitic carbon nitride as self-cleaning flexible SERS membrane for molecular trace-detection. *Colloids Surf. A—Physicochem. Eng. Asp.* **2019**, *560*, 9–19. [CrossRef]

238. Creedon, N.; Lovera, P.; Moreno, J.G.; Nolan, M.; O'Riordan, A. Highly Sensitive SERS Detection of Neonicotinoid Pesticides. Complete Raman Spectral Assignment of Clothianidin and Imidacloprid. *J. Phys. Chem. A* **2020**, *124*, 7238–7247. [CrossRef] [PubMed]

239. Hou, R.Y.; Pang, S.; He, L.L. In situ SERS detection of multi-class insecticides on plant surfaces. *Anal. Methods* **2015**, *7*, 6325–6330. [CrossRef]

240. Xu, Y.; Kutsanedzie, F.Y.H.; Hassan, M.; Zhu, J.J.; Ahmad, W.; Li, H.H.; Chen, Q.S. Mesoporous silica supported orderly-spaced gold nanoparticles SERS-based sensor for pesticides detection in food. *Food Chem.* **2020**, *315*, 126300. [CrossRef]

241. Atanasov, P.A.; Nedyalkov, N.N.; Nikov, R.G.; Fukata, N.; Jevasuwan, W.; Subramani, T.; Hirsch, D.; Rauschenbach, B. SERS analyses of thiamethoxam assisted by Ag films and nanostructures produced by laser techniques. *J. Raman Spectrosc.* **2018**, *49*, 397–403. [CrossRef]

242. Atanasov, P.A.; Nedyalkov, N.N.; Fukata, N.; Jevasuwan, W.; Subramani, T. Surface-Enhanced Raman Spectroscopy (SERS) of Neonicotinoid Insecticide Thiacloprid Assisted by Silver and Gold Nanostructures. *Appl. Spectrosc.* **2020**, *74*, 357–364. [CrossRef]

243. Srivastava, S.; Sinha, R.; Roy, D. Toxicological effects of malachite green. *Aquat. Toxicol.* **2004**, *66*, 319–329. [CrossRef]

244. Stammati, A.; Nebbia, C.; de Angelis, I.; Albo, A.G.; Carletti, M.; Rebecchi, C.; Zampaglioni, F.; Dacasto, M. Effects of malachite green (MG) and its major metabolite, leucomalachite green (LMG), in two human cell lines. *Toxicol. Vitr.* **2005**, *19*, 853–858. [CrossRef] [PubMed]

245. Kneipp, K.; Wang, Y.; Kneipp, H.; Perelman, L.T.; Itzkan, I.; Dasari, R.R.; Feld, M.S. Single Molecule Detection Using Surface-Enhanced Raman Scattering (SERS). *Phys. Rev. Lett.* **1997**, *78*, 1667–1670. [CrossRef]

246. Michaels, A.M.; Nirmal, M.; Brus, L. Surface enhanced Raman spectroscopy of individual rhodamine 6G molecules on large Ag nanocrystals. *J. Am. Chem. Soc.* **1999**, *121*, 9932–9939. [CrossRef]

247. Gossner, C.; Schlundt, J.; Embarek, P.; Hird, S.; Lo-Fo-Wong, D.; Beltran, J.; Teoh, K.N.; Tritscher, A. The Melamine Incident: Implications for International Food and Feed Safety. *Environ. Health Perspect.* **2009**, *117*, 1803–1808. [CrossRef] [PubMed]

248. Zhang, X.F.; Zou, M.Q.; Qi, X.H.; Liu, F.; Zhu, X.H.; Zhao, B.H. Detection of melamine in liquid milk using surface-enhanced Raman scattering spectroscopy. *J. Raman Spectrosc.* **2010**, *41*, 1655–1660. [CrossRef]

249. Lee, S.Y.; Ganbold, E.-O.; Choo, J.; Joo, S.-W. Detection of Melamine in Powdered Milk Using Surface-Enhanced Raman Scattering with No Pretreatment. *Anal. Lett.* **2010**, *43*, 2135–2141. [CrossRef]

250. Mecker, L.C.; Tyner, K.M.; Kauffman, J.F.; Arzhantsev, S.; Mans, D.J.; Gryniewicz-Ruzicka, C.M. Selective melamine detection in multiple sample matrices with a portable Raman instrument using surface enhanced Raman spectroscopy-active gold nanoparticles. *Anal. Chim. Acta* **2012**, *733*, 48–55. [CrossRef] [PubMed]

251. Giovannozzi, A.M.; Rolle, F.; Sega, M.; Abete, M.C.; Marchis, D.; Rossi, A.M. Rapid and sensitive detection of melamine in milk with gold nanoparticles by Surface Enhanced Raman Scattering. *Food Chem.* **2014**, *159*, 250–256. [CrossRef]

252. Lou, T.T.; Wang, Y.Q.; Li, J.H.; Peng, H.L.; Xiong, H.; Chen, L.X. Rapid detection of melamine with 4-mercaptopyridine-modified gold nanoparticles by surface-enhanced Raman scattering. *Anal. Bioanal. Chem.* **2011**, *401*, 333–338. [CrossRef]

253. Yazgan, N.N.; Boyaci, I.H.; Topcu, A.; Tamer, U. Detection of melamine in milk by surface-enhanced Raman spectroscopy coupled with magnetic and Raman-labeled nanoparticles. *Anal. Bioanal. Chem.* **2012**, *403*, 2009–2017. [CrossRef] [PubMed]

254. Tang, J.Q.; Tian, C.; Zeng, C.Y.; Man, S.Q. Alkaline Silver Colloid for Surface Enhanced Raman Scattering and Application to Detection of Melamine Doped Milk. *Spectrosc. Spectr. Anal.* **2013**, *33*, 709–713.

255. Betz, J.F.; Cheng, Y.; Rubloff, G.W. Direct SERS detection of contaminants in a complex mixture: Rapid, single step screening for melamine in liquid infant formula. *Analyst* **2012**, *137*, 826–828. [CrossRef] [PubMed]

256. Chen, L.M.; Liu, Y.N. Surface-Enhanced Raman Detection of Melamine on Silver-Nanoparticle-Decorated Silver/Carbon Nanospheres: Effect of Metal Ions. *ACS Appl. Mater. Interfaces* **2011**, *3*, 3091–3096. [CrossRef] [PubMed]

257. Li, J.M.; Ma, W.F.; Wei, C.; You, L.J.; Guo, J.; Hu, J.; Wang, C.C. Detecting Trace Melamine in Solution by SERS Using Ag Nanoparticle Coated Poly(styrene-co-acrylic acid) Nanospheres as Novel Active Substrates. *Langmuir* **2011**, *27*, 14539–14544. [CrossRef]

258. Ma, P.Y.; Liang, F.H.; Sun, Y.; Jin, Y.; Chen, Y.; Wang, X.H.; Zhang, H.Q.; Gao, D.J.; Song, D.Q. Rapid determination of melamine in milk and milk powder by surface-enhanced Raman spectroscopy and using cyclodextrin-decorated silver nanoparticles. *Microchim. Acta* **2013**, *180*, 1173–1180. [CrossRef]

259. Kumar, S.V.; Huang, N.M.; Lim, H.N.; Zainy, M.; Harrison, I.; Chia, C.H. Preparation of highly water dispersible functional graphene/silver nanocomposite for the detection of melamine. *Sens. Actuators B Chem.* **2013**, *181*, 885–893. [CrossRef]

260. Peng, B.; Li, G.Y.; Li, D.H.; Dodson, S.; Zhang, Q.; Zhang, J.; Lee, Y.H.; Demir, H.V.; Ling, X.Y.; Xiong, Q.H. Vertically Aligned Gold Nanorod Monolayer on Arbitrary Substrates: Self-Assembly and Femtomolar Detection of Food Contaminants. *ACS Nano* **2013**, *7*, 5993–6000. [CrossRef]

261. Hu, H.; Wang, Z.; Pan, L.; Zhao, S.; Zhu, S. Ag-Coated Fe3O4@SiO2 Three-Ply Composite Microspheres: Synthesis, Characterization, and Application in Detecting Melamine with Their Surface-Enhanced Raman Scattering. *J. Phys. Chem. C* **2010**, *114*, 7738–7742. [CrossRef]

262. Guo, Z.; Cheng, Z.; Li, R.; Chen, L.; Lv, H.; Zhao, B.; Choo, J. One-step detection of melamine in milk by hollow gold chip based on surface-enhanced Raman scattering. *Talanta* **2014**, *122*, 80–84. [CrossRef]

263. Chen, L.M.; Luo, L.B.; Chen, Z.H.; Zhang, M.L.; Zapien, J.A.; Lee, C.S.; Lee, S.T. ZnO/Au Composite Nanoarrays As Substrates for Surface-Enhanced Raman Scattering Detection. *J. Phys. Chem. C* **2010**, *114*, 93–100. [CrossRef]

264. Kim, A.; Barcelo, S.J.; Williams, R.S.; Li, Z.Y. Melamine Sensing in Milk Products by Using Surface Enhanced Raman Scattering. *Anal. Chem.* **2012**, *84*, 9303–9309. [CrossRef] [PubMed]

265. Subaihi, A.; Xu, Y.; Muhamadali, H.; Mutter, S.T.; Blanch, E.W.; Ellis, D.I.; Goodacre, R. Towards improved quantitative analysis using surface-enhanced Raman scattering incorporating internal isotope labelling. *Anal. Methods* **2017**, *9*, 6636–6644. [CrossRef]

Review

Applications of Shell-Isolated Nanoparticle-Enhanced Raman Spectroscopy

Grégory Barbillon

EPF-Ecole d'Ingénieurs, 92330 Sceaux, France; gregory.barbillon@epf.fr

Abstract: The surface-enhanced Raman scattering (SERS) is mainly used as an analysis or detection tool of biological and chemical molecules. Since the last decade, an alternative branch of the SERS effect has been explored, and named shell-isolated nanoparticle Raman spectroscopy (SHINERS) which was discovered in 2010. In SHINERS, plasmonic cores are used for enhancing the Raman signal of molecules, and a very thin shell of silica is generally employed for improving the thermal and chemical stability of plasmonic cores that is of great interest in the specific case of catalytic reactions under difficult conditions. Moreover, thanks to its great surface sensitivity, SHINERS can enable the investigation at liquid–solid interfaces. In last two years (2019–2020), recent insights in this alternative SERS field were reported. Thus, this mini-review is centered on the applications of shell-isolated nanoparticle Raman spectroscopy to the reactions with CO molecules, other surface catalytic reactions, and the detection of molecules and ions.

Keywords: SHINERS; SERS; core–shell nanoparticles; catalysis; electrochemistry; plasmonics

Citation: Barbillon, G. Applications of Shell-Isolated Nanoparticle-Enhanced Raman Spectroscopy. *Photonics* **2021**, *8*, 46. https://doi.org/10.3390/photonics8020046

Received: 1 January 2021
Accepted: 9 February 2021
Published: 12 February 2021

Publisher's Note: MDPI stays neutral with regard to jurisdictional clai-ms in published maps and institutio-nal affiliations.

1. Introduction

Over the past ten years, the realization of plasmonic structures with a very high sensitivity of detection has significantly increased for application to surface-enhanced spectroscopies [1–10]. Among these enhanced spectroscopies, we find the surface-enhanced Raman scattering (SERS), which uses the plasmonic nanostructures or nanoparticles for amplifying the Raman signal of various molecules. For this amplification, a huge number of geometries has been examined as plasmonic nanodimers [11–15], nanorods [16–20], nanotriangles [21–25] and nanostars [26–30]. Furthermore, plasmonic nanopores have been explored in order to improve the SERS enhancement [31–34]. In addition, another type of SERS substrates has been investigated consisting of a metallic mirror on which plasmonic nanoparticles or nanostructures were deposited or fabricated allowing an enhancement of 1 or 2 magnitude orders due to hybridized modes or a coupling between nanoparticles or nanostructures via surface plasmon polaritons on the metallic mirror (film) [35–39]. Another way to improve the Raman signal was the use of hybrid nanomaterials based on zinc oxide or silicon coupled to a plasmonic layer or plasmonic nanoparticles [40–49], and also based on bimetallic nanoparticles [50–55] or other materials as metal oxides [56–61]. Another branches of the SERS field have been also explored, such as the photo-induced enhanced Raman spectroscopy [62–64], the SERS effect generated by high pressure [65,66], and the shell-isolated nanoparticle-enhanced Raman spectroscopy (SHINERS) [67–69]. Concerning SHINERS, this technique has been discovered in 2010 in order to overcome the limitations of SERS regarding the accurate characterization of different surface morphologies, materials, and biological samples [67]. The base concept of SHINERS consists of plasmonic cores that are employed for enhancing the Raman signal of molecules, and a very thin shell of silica improving the thermal and chemical stability of the plasmonic cores, being of significant interest in the specific case of catalytic reactions under difficult conditions [70,71]. By using SHINERS, several groups have already studied catalytic reactions [72–74], applications in electrochemistry [75], and also reported the detection of different chemical molecules [76–81].

The goal of this mini-review is to present the recent advances on the most used applications of SHINERS, such as the catalytic reaction-monitoring processes and the detection of molecules, over the period 2019–2020. Firstly, we will explore the SHINERS applications to the reactions with CO molecules, which are well-known model reactions, then other surface catalytic reactions, and finally the detection of molecules and ions in order to examine the potential of this SHINERS technique. In the final section, we will discuss points to be improved and advantages of the SHINERS technique, and we will address the future directions of this latter.

2. What Is Shell-Isolated Nanoparticle-Enhanced Raman Spectroscopy?

The shell-isolated nanoparticle-enhanced Raman spectroscopy belongs to the SERS field, and is based on the enhancement of the Raman signal obtained with strong electric fields coming from plasmonic core–shell nanoparticles. In SHINERS, each core-shell nanoparticle plays the role of a metallic tip as for the tip-enhanced Raman spectroscopy (TERS), and this technique allows to obtain a couple of thousand of "TERS tips" on the substrate surface to be analyzed. Thus, the enhanced Raman signal can be jointly obtained from all these plasmonic core-shell nanoparticles ("TERS tips"), allowing a gain of two to three magnitude orders compared to a single TERS tip. Furthermore, the use of the metallic nanoparticles coated with a chemically inert shell can enable the protection of the plasmonic core (SERS-active part) from the substrate surface to be analyzed and the environment. This inert shell can conform to different morphologies of substrates, and also prevent the agglomeration of these core-shell nanoparticles and the oxidation of their plasmonic core. The principal merits of such a technique are a more significant detection sensitivity and a great number of practical applications in life and materials sciences, as well as in food science and environmental pollution.

3. Applications of Shell-Isolated Nanoparticle-Enhanced Raman Spectroscopy

3.1. SHINERS Application to the Reactions with CO Molecules

In this section, we present a couple of investigations on the reactions with CO molecules (see Table 1) [82–86]. These reactions with CO molecules are well-known model reactions.

Table 1. Shell-isolated nanoparticle Raman spectroscopy (SHINERS) application to the reactions with CO molecules (NPs = nanoparticles; G = Graphene).

Samples	Core/Shell Size (nm)	Reaction	Refs
PtFe/Au@SiO$_2$ NPs	120/2	CO oxidation	[82]
Pt/Au@SiO$_2$ NPs	55/2	CO electrooxidation	[83]
Pt/Au@SiO$_2$ NPs	101/2	CO adsorption	[84]
Ni/Au@SiO$_2$ NPs	90/2	CO adsorption	[85]
Au@SiO$_2$ NPs, Au@TiO$_2$ NPs	90/3–105/2	Hydrogenation/dissociation of CO	[86]
Pt/Au@G NPs	500/5 layers of G	CO adsorption	[87]
Au@SiO$_2$ NPs on Cu foil	50/2	CO adsorption	[88,89]

Wang et al., have reported on the CO oxidation probed SHINERS technique based on the use of Au@SiO$_2$ nanoparticles, on which are deposited Pt or PtFe nanocatalysts (see Figure 1a) [82]. The authors demonstrated that PtFe catalysts were more active and stable than Pt ones in the CO oxidation. In the CO oxidation, two Raman peaks have been observed at 397 and 485 cm^{-1} for Pt catalysts and at 389 and 480 cm^{-1} for PtFe catalysts, both corresponding to the adsorption of Pt-CO (see Figure 1b). The redshift of these two Raman peaks recorded for the case with PtFe catalysts has indicated that the CO adsorption was lower on PtFe than on Pt (see the inset of Figure 1b). Moreover, three Raman peaks of oxygen species were detected with PtFe nanocatalysts, indicating that the Pt-C binding was weakened by the presence of the ferrous center thus producing oxygen species (see Figure 1b) [82].

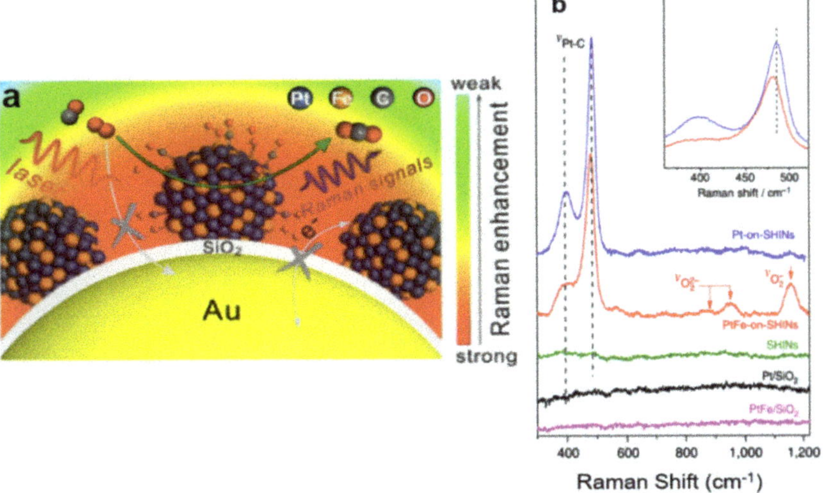

Figure 1. (**a**) Principle scheme of a SHINERS catalyst (PtFe/Au@SiO$_2$) for CO oxidation. (**b**) SHINERS spectra for CO oxidation on Pt (blue line) and PtFe (red line) catalysts. The inset displays a zoom of the two SHINERS spectra (blue and red lines) in the range 350–550 cm^{-1} in order to better observe the redshift of Raman peaks. All the figures are reprinted (adapted) with permission from [82], Copyright 2019 American Chemical Society.

Su et al., have reported on the detection of OH and COOH species during the CO electrooxidation process on three Pt surfaces (Pt(100), Pt(110) and Pt(111)) with the SHINERS technique by using Au@SiO$_2$ nanoparticles [83]. The authors have observed that the activity of CO electrooxidation was higher for Pt(111)/Pt(100) surfaces than Pt (110) surface. This increased activity of CO electrooxidation for Pt(111)/Pt(100) surfaces was due to the presence of OH and COOH species. For Pt(110) surface, this activity was weaker due to its high adsorption and coverage of CO on this surface [83]. Furthermore, Wondergem et al. have demonstrated the CO adsorption on Pt nanoparticles which are themselves deposited on Au@SiO$_2$ nanoparticles (see Figure 2a,b) [84]. This investigation was realized by employing the SHINERS technique. From the SHINERS spectra, two characteristic Raman peaks of CO adsorption were recorded in the two ranges of wavenumbers: 350–600 cm^{-1} and 1900–2150 cm^{-1}. The first Raman peak corresponds to the CO adsorbed on Pt in a bridge configuration located at 430 and 2010 cm^{-1} in these two ranges of wavenumbers (see Figure 2c,d), then the second one corresponds to the CO adsorbed on Pt in a linear configuration located at 505 and 2070 cm^{-1} in these two wavenumber ranges (see Figure 2c,d) [84]. Next, the same group has studied the effect of the fabrication method of nickel (Ni) catalysts on Au@SiO$_2$ nanoparticles for SHINERS investigations [85]. Three methods have been tested: spark ablation (SA), colloidal deposition (Col) and precursor (Pr) method. The authors have studied the CO adsorption on these three types of Ni catalysts, and concluded that Pr and Col methods were not suitable for the SHINERS technique due to the use of a high-temperature treatment of reduction. Finally, the SA technique is the most efficient for direct deposition of the nickel catalyst on Au@SiO$_2$ nanoparticles [85].

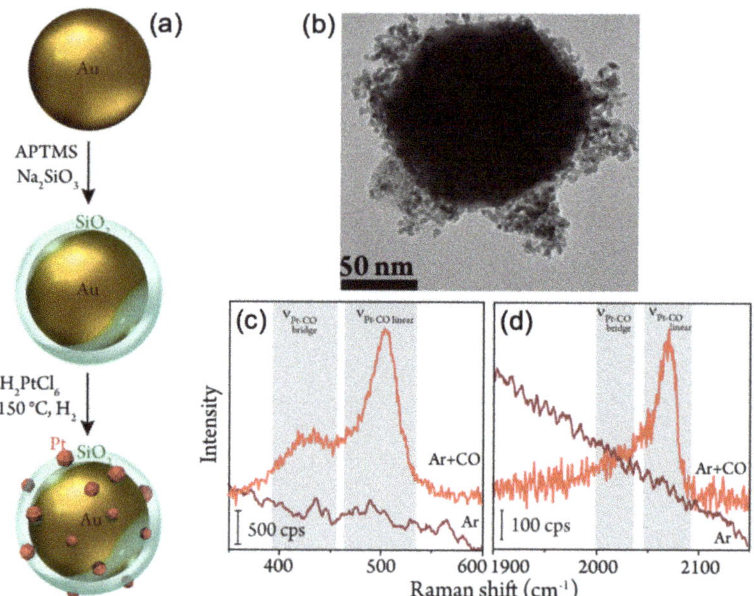

Figure 2. (**a**) Fabrication process of a SHINERS catalyst (Pt/Au@SiO$_2$). (**b**) TEM picture of a SHINERS catalyst (Pt/Au@SiO$_2$). SHINERS spectra for CO adsorption on Pt for (**c**) the low range of wavenumbers and (**d**) the high range of wavenumbers. The grey zones correspond to the Raman peaks of Pt-CO bridge at 430 and 2010 cm^{-1}, and Pt-CO linear at 505 and 2070 cm^{-1}. All the figures are reprinted (adapted) with permission from [84] (https://pubs.acs.org/doi/10.1021/acscatal.9b03010 (accessed on 1 January 2021)), Copyright 2019 American Chemical Society (for all further reuses related to the excerpted material, further permissions should be directed to the American Chemical Society).

Hartman et al. have investigated the support effect on the interaction of rhodium (Rh) with CO molecules probed by SHINERS technique [86]. Two types of extrudate support have been tested by introducing alternatively CO and H$_2$: the first one was Rh/SiO$_2$ on which were deposited Au@SiO$_2$ nanoparticles and the second one was Rh/TiO$_2$ on which were also deposited Au@TiO$_2$ nanoparticles. Under the same conditions of CO then H$_2$, the shifts of a Raman peak (named "unknown" by the authors) for the CO hydrogenation were of 70 cm^{-1} and 25 cm^{-1} for Rh/TiO$_2$ and Rh/SiO$_2$ extrudate supports, respectively. From these Raman shifts for the CO hydrogenation, the authors have deduced that the Rh/TiO$_2$ extrudate support had the strongest interaction with CO molecules during the catalytic process compared to the Rh/SiO$_2$ extrudate support. Thus, this powerful interaction has resulted in a catalyst with higher efficiency for the CO dissociation [86]. Zhang et al., have reported on the adsorption of CO molecules on Pt nanocatalysts, which were deposited on novel core-shell nanoparticles composed of a gold core covered by graphene layers (Au@G nanoparticles). The authors have demonstrated via SHINERS measurements that the adsorption of CO molecules on these Pt/Au@G nanoparticles has occurred in a linear configuration [87]. To finish this section, two groups have reported on the CO reduction catalysis on Cu foil by using the SHINERS technique [88,89]. The authors have employed Au@SiO$_2$ nanoparticles, and allowed them to deduce that the CuO$_x$/(OH)$_y$ species were detected on the Cu foil during the CO reduction. Thus, the authors have concluded that the oxygenated species of Cu were unlikely to be the active sites easing the formation of C$_{2+}$ oxygenates during the process of CO reduction [88,89].

3.2. SHINERS Application to Other Surface Catalytic Reactions

Here, we address a couple of studies on other surface catalytic reactions (see Table 2) [90–100].

Table 2. SHINERS application to other surface catalytic reactions (NPs = nanoparticles; IrO_x = Iridium oxide; MBT = 2-mercaptobenzothiazole; SnO_2 = Tin oxide; pNTP = para-nitrothiophenol; RhB = Rhodamine B; CNNDs = g-C_3N_4 nanodots; EGLs = Electrochemical exfoliated graphene layers; ITO = Indium tin oxide).

Samples	Core/Shell Size (nm)	Reaction	Refs
Au@SiO$_2$ NPs/Pt	90/2	Adsorption of Propargyl Alcohol	[90]
Au@SiO$_2$ NPs/IrO$_x$ surface	55/2	Water oxidation	[91]
Au@SiO$_2$ NPs/Au(111)	55/2	Configuration of interfacial water	[92]
Au@SiO$_2$ NPs/pyrite	55/2	Adsorption of MBT	[93]
Au@SiO$_2$ NPs/Cu surface	55/2	Oxidation of Cu surfaces	[94]
Au@SiO$_2$ NPs/Pt(*hkl*)	55/2–3	Oxygen reduction reaction	[95]
Au@SiO$_2$ NPs/Pt(*hkl*)	55/2	Oxygen reduction reaction	[96]
Au@SiO$_2$ NPs/Pt(*hkl*)	55/2	Oxygen reduction reaction	[97]
Au@SnO$_2$ NPs/steel surface	35/4	Steel surface corrosion	[98]
Pt/Au@SiO$_2$ NPs	120/2	Hydrogenation of pNTP	[99]
Ag@SiO$_2$/RhB/CNND/EGL/ITO	75/3	Photodegradation of RhB	[100]

Guan et al. have demonstrated that the adsorption of propargyl alcohol (PA) on Pt(*hkl*) surfaces by using the SHINERS technique [90]. The authors have employed Au@SiO$_2$ nanoparticles for SHINERS experiments, and they obtained adsorption of PA-privileged on Pt(100) and Pt(110) surfaces than on Pt(111) surface. The better surface reactivity for Pt(100) compared to two other Pt surfaces (Pt(100) > Pt(110) > Pt(111)) was due to the more important presence of the primary alcohol group [90]. In the next two examples, the studies of the water oxidation and the configuration of the interfacial water are addressed. Firstly, Saeed et al., have reported on the employment of the SHINERS technique to analyze in real-time the mechanisms of water oxidation with iridium oxides (IrO$_x$) as electrocatalyst [91]. To do that, Au@SiO$_2$ nanoparticles were used and deposited on IrO$_x$ surface for the Raman characterization (SHINERS). Thus, the authors demonstrated that SHINERS allowed to observe in real-time the chemical changes on the IrO$_x$ surface during the oxidation of water [91]. Secondly, Li et al., have reported on the configuration of the interfacial water at Au(111) surface, probed by the SHINERS technique using Au@SiO$_2$ nanoparticles for this study [92]. The authors have observed redshifts of the Raman peak corresponding to the $O - H$ stretching mode of the interfacial water when the potential went towards more negative values, indicating a configuration variation of interfacial water. Thus, the authors have shown three configurations of the interfacial water named parallel, one-H-down and two-H-down, respectively, when the potential shifted to more negative values [92]. In addition, Guo et al., have studied the adsorption of 2-mercaptobenzothiazole (MBT) on pyrite by SHINERS [93]. Au@SiO$_2$ nanoparticles were employed and deposited on pyrite for SHINERS experiments. From the SHINERS spectra recorded with an MBT concentration of 0.01 mM, a Raman peak at 1406 cm^{-1} was observed and corresponded to NCS ring stretch mode (see Figure 3a). This Raman peak suggested that double "minerophilic" groups of MBT were bound to pyrite surfaces in the configuration displayed in Figure 3b on the left. From the SHINERS spectra recorded with an MBT concentration of 0.1 mM, two Raman peaks at 1389 and 1409 cm^{-1} were observed and also corresponded to NCS ring stretch mode (see Figure 3c). The Raman peak at 1409 cm^{-1} has indicated that double "minerophilic" groups of MBT were bound to pyrite surface. The Raman peak at 1389 cm^{-1} (starting to appear at −200 mV, see Figure 3c) has also indicated that the MBT molecule was bound to pyrite with the exocyclic sulfur atom without the presence of any nitrogen–metal bond in the configuration displayed in Figure 3b on the right. In summary, the authors have concluded that the configuration of MBT molecules was preferentially the one in Figure 3b on the left for weaker concentrations of MBT and negative potentials and the one in Figure 3b on the right for higher concentrations of MBT and positive potentials [93].

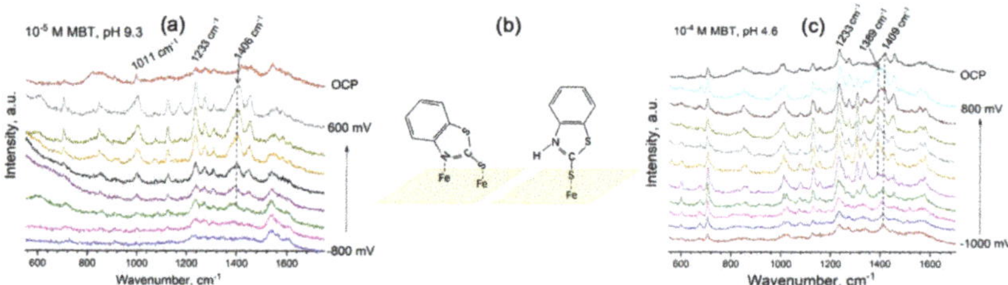

Figure 3. (**a**) SHINERS spectra of a MBT solution (0.01 mM) recorded at pH 9.3 for different potentials. The black dotted line indicates the Raman peak at 1406 cm^{-1}. (**b**) Potential configurations of the MBT adsorption on pyrite. (**c**) SHINERS spectra of a MBT solution (0.1 mM) recorded at pH 4.6 for different potentials. The black dotted lines indicate the Raman peaks at 1389 and 1409 cm^{-1}. All the figures are reprinted from [93], Copyright 2020, with permission from Elsevier.

Bodappa et al., have investigated the electrochemical oxidation of Cu(111) and polycrystalline Cu (Cu(poly)) surfaces by using the SHINERS technique [94]. Au@SiO$_2$ nanoparticles were employed for studying the oxidation mechanism of Cu(111) and Cu(poly) surfaces (see Figure 4a). From the SHINERS spectra for Cu(111) oxidation, intermediate species Cu-OH, Cu-O$_{ad}$, and (Cu$_2$O)$_{surf}$ were observed during the oxidation process (i.e., when the potential increased, see Figure 4b). For Cu(poly) oxidation, only Cu-OH and (Cu$_2$O)$_{surf}$ were spotted during the oxidation on the SHINERS spectra (see Figure 4c). Thus, the authors have remarked a difference in the presence of the intermediate species during the oxidation process [94].

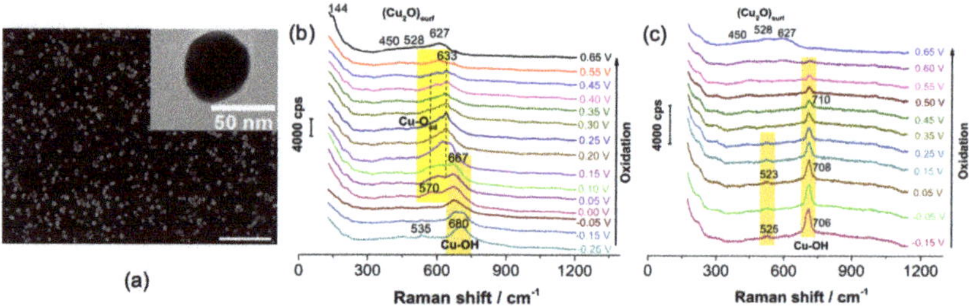

Figure 4. (**a**) SEM picture of Au@SiO$_2$ nanoparticles on Cu surface (scale bar = 1 μm). The inset displays a TEM picture of a Au@SiO$_2$ nanoparticle. (**b**) SHINERS spectra for Cu(111) oxidation in a 0.01 M KOH solution. (**c**) SHINERS spectra for Cu(poly) oxidation in a 0.01 M KOH solution. All the figures are reprinted (adapted) with permission from [94], Copyright 2019 American Chemical Society.

The next three works have addressed the topic of oxygen reduction reaction (ORR) on Pt(*hkl*) surfaces by employing SHINERS spectroscopy. At first, Galloway et al., have demonstrated the surface specificity of the ORR on Pt(*hkl*) surfaces by SHINERS in sodium–oxygen electrochemistry [95]. The reduction of NaO$_2$ to Na$_2$O$_2$ was favored on Pt(111) and Pt(110) surfaces in 0.1 M NaClO$_4$ dissolved in dimethyl sulfoxide (DMSO), whereas this reduction was not detected on Pt(100) and Pt(poly) surfaces (no characteristic Raman peak of Na$_2$O$_2$ recorded) due to the restricted interactions with adsorbed oxygens [95]. Next, the second work realized by Dong et al. dealt with the observation of intermediate species for ORR on different Pt(*hkl*) surfaces examined by SHINERS. The authors have spectroscopically evidenced the fact that ORRs on Pt(111) surface was obtained by the generation of OOH species, while for Pt(110) and Pt(100) surfaces by the formation of OH species [96]. Finally, the same group has demonstrated the presence of intermediate species during ORR on high-index Pt(*hkl*) surfaces by SHINERS spectroscopy [97]. Au@SiO$_2$ nanoparticles were used for the study of the ORR activity of these Pt surfaces (see Figure 5a). The authors have observed intermediate species for the two Pt surfaces (Pt(311) and Pt(211)) studied for ORR. From SHINERS spectra recorded for different values of potential, two characteristic Raman peaks at ~765 and ~1041 cm^{-1} were observed and corresponded to OOH and OH species, respectively (see Figure 5b,c). Moreover, they concluded that the Pt(211) surface had a better reactivity than the Pt(311) surface due to the greater adsorption energy for OOH species with the Pt(311) surface [97].

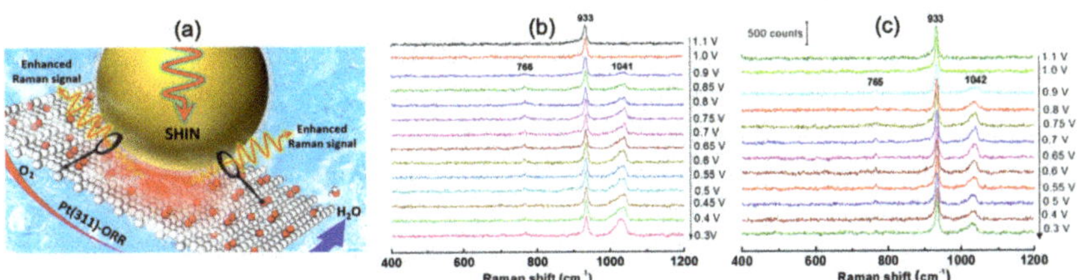

Figure 5. (**a**) Principle scheme of a SHINERS (Au@SiO$_2$) measurement for oxygen reduction reaction on Pt(*hkl*) surfaces. For different values of potential, SHINERS spectra recorded in HClO$_4$ solution (0.1 M) saturated in O$_2$ are shown for the surfaces of (**b**) Pt(311) and (**c**) Pt(211). All the figures are reprinted (adapted) with permission from [97], Copyright 2020 American Chemical Society.

Barlow et al. have investigated the corrosion of 304 stainless steel by using the SHINERS technique. Au@SnO$_2$ nanoparticles were employed for examining this possible corrosion [98]. For the 304 stainless steel, the authors have evidenced a characteristic Raman (SHINERS) peak corresponding to amorphous Fe(OH)$_2$, and also another Raman peak attributed to Cr(VI)–O bindings from a blended oxide based on Cr(VI). When KCl is present in the electrolyte, a Raman peak attributed to γ-FeOOH was observed. Finally, the authors have evidenced no green rust, i.e., no intermediate species during the conversion from Fe(OH)$_2$ to γ-FeOOH [98]. Besides, Wang et al., have reported on the effects of the size and the nature of nanocatalysts on the hydrogenation of para-nitrothiophenol (pNTP) by employing the SHINERS spectroscopy [99]. Au@SiO$_2$ nanoparticles were used for SHINERS experiments on which Pt, PtCu, PtNi nanocatalysts have been assembled via electrostatic interactions. At first, the authors have studied the size effect of Pt nanocatalysts on the pNTP hydrogenation, and have reported on an optimal size of 6.8 nm for Pt nanoparticles. Then, the authors have studied the kinetics of reaction for Pt, PtCu and PtNi nanocatalysts. They observed that PtCu and PtNi nanocatalysts have shown a quicker and quasi-complete conversion of pNTP than for Pt ones [99]. To finish this section on SHINERS applications to other surface catalytic reactions, Qiu et al. have investigated the effect of the presence of g-C$_3$N$_4$ nanodots (CNNDs) and electrochemical exfoliated graphene layers (EGLs) on the photodegradation of Rhodamine B (RhB) molecules probed by SHINERS technique [100]. For SHINERS experiments, Ag@SiO$_2$ nanoparticles and an illumination wavelength of 632.8 nm were employed for examining this photocatalytic process of degradation of RhB molecules (see Figure 6a). In the absence of CNNDs and EGLs on the ITO substrate, the authors have noted no significant degradation of the intensity of the Raman peaks of RhB molecules. In contrast, the authors have observed a complete degradation of the intensity of these Raman peaks for an illumination time of 20 min with the presence of CNNDs and EGLs (see Figure 6b) [100].

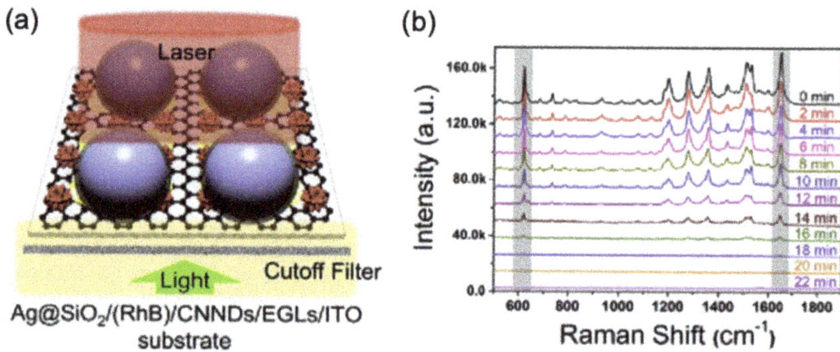

Figure 6. (**a**) Principle scheme for the photocatalytic process of degradation of RhB molecules (RhB = red shapes) probed by SHINERS (Ag@SiO$_2$ NPs = blue spheres with a grey shell). The yellow shapes correspond to the g-C$_3$N$_4$ nanodots (CNNDs), and the EGLs/ITO substrate is represented by the white rectangle on which the hexagonal lattice of graphene (EGLs) is depicted. (**b**) SHINERS spectra of RhB molecules recorded for various illumination times in the range 0–22 min. All the figures are reprinted from [100], Copyright 2019, with permission from Elsevier.

3.3. SHINERS Application for the Detection of Molecules and Ions

In this final section for the SHINERS applications, we report on a couple of works on the detection of molecules and ions (see Table 3) [101–107].

Table 3. SHINERS application for the detection of molecules and ions (NPs = nanoparticles; cc-Au = concave cubic gold; PPy = Polypyrrole).

Samples	Core/Shell Size (nm)	Detection	Refs
(Au/SiO$_2$)@SiO$_2$ NPs on Si	(61/900)/1–2	Rhodamine 6G	[101]
Au@SiO$_2$ NPs on TiO$_2$(*hkl*)	55/2	Photoinduced behavior of dyes	[102]
Ag@SiO$_2$ NPs on filter paper	45/3	Thiram	[103]
Ag@TiO$_2$ NPs	10-30/2–10	Copper ions	[104]
Ag@TiO$_2$ NPs	20/2–10	Copper oxidation states	[105]
cc-Au@Ag@SiO$_2$ NPs	50/4/2–5	4-mercaptobenzoic acid	[106]
Au@PPy bipyramids	100 (length)/1	γ-aminobutyric acid	[108]
Au@SiO$_2$ NPs	40/5	*Metschnikowia pulcherrima* yeast cells	[109]
Au@SiO$_2$ NPs	55/2	Atypical hyperplasia	[110]
Ag@SiO$_2$ NPs	53/2–5	Tumor cells in blood	[107]

Wondergem et al. have reported the detection of Rhodamine 6G (R6G) molecules by SHINERS spectroscopy [101]. For SHINERS experiments, (Au/SiO$_2$)@SiO$_2$ nanoparticles were used in order to avoid contact between gold nanoparticles and the liquid medium. The authors have obtained a detection limit of 10^{-12} M for R6G molecules with these (Au/SiO$_2$)@SiO$_2$ plasmonic superstructures. Moreover, these plasmonic superstructures can enable the study of catalytic reactions in liquids by using SHINERS [101]. In addition, Zhang et al. have investigated the photoinduced behavior of dyes (N719) molecules on three rutile TiO$_2$(*hkl*) surfaces [102]. Au@SiO$_2$ nanoparticles were employed for SHINERS experiments. The authors have remarked that the SCN group of N719 molecules was the group that primarily adsorbed on these three rutile TiO$_2$(*hkl*) surfaces. The authors have evidenced a shift of the Raman peak corresponding to the SCN group after an illumination time of 36 min on TiO$_2$(110) and TiO$_2$(001) surfaces, whereas no shift of this Raman peak was observed for the TiO$_2$(111) surface. They concluded that the N719 molecules adsorbed on TiO$_2$(111) surface were very stable in the time, whereas the N719 molecules had desorbed on TiO$_2$(110) and TiO$_2$(001) surfaces caused by the cleavage of the S = C binding [102]. Furthermore, Sun et al., have reported a detection limit of 10^{-9} M for thiram molecules (pesticides) probed by the SHINERS spectroscopy [103]. Ag@SiO$_2$ nanoparticles on filter paper were employed as well as a miniaturized portable Raman analyzer based on smart-phone for SHINERS experiments. The authors have recorded SHINERS spectra for each concentration of thiram molecules (concentration range = 10^{-9}–10^{-3} M), where four characteristic Raman peaks of thiram molecules located at 440, 559, 1145 and 1379 cm^{-1} are displayed (see Figure 7). By using the Raman peak at 1379 cm^{-1}, the authors have deduced the detection limit (see Figure 7).

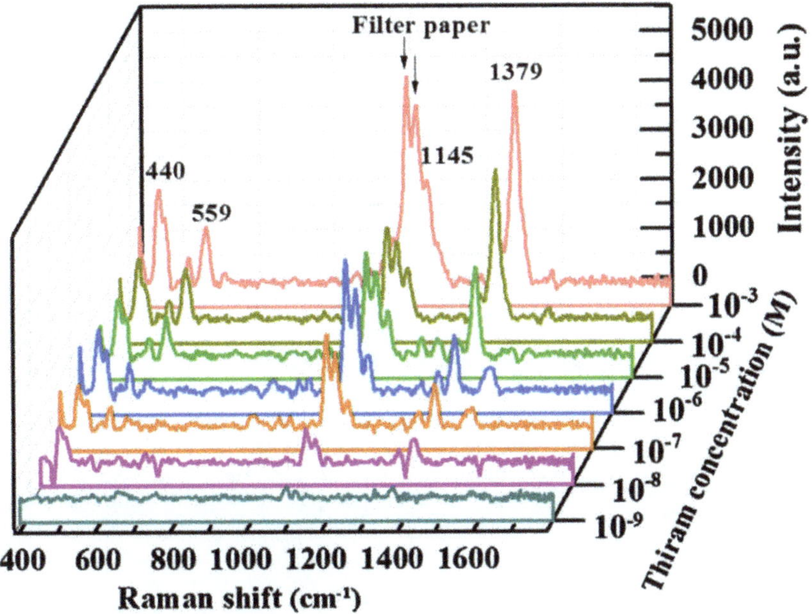

Figure 7. SHINERS spectra of thiram recorded for various concentrations in the range 10^{-9}–10^{-3} M, where are indicated the four characteristic Raman peaks as well as the two peaks associated to the filter paper. The figure is reprinted from [103], Copyright 2019, with permission from Elsevier.

In the next two works realized by the same group [104,105], the detection of copper ions and their oxidation states by SHINERS is addressed. Firstly, Forato et al., have investigated the detection of Cu(II) ions by SHINERS spectroscopy. Ag@TiO$_2$ nanoparticles and three excitation wavelengths (514, 633, and 785 nm) were used for SHINERS experiments. The authors have demonstrated an optimal efficiency for the detection of the Raman peak of the Cu–N binding at an excitation wavelength of 633 nm [104]. Then, Quéffelec et al., have reported on the distinctness of Cu(I) and Cu(II) ions by SHINERS measurements [105]. The authors have used the same Ag@TiO$_2$ nanoparticles functionalized with 2,2′-bipyridine phosphonate (bpy-PA) and an excitation wavelength of 633 nm where the efficiency was optimal [104]. From the SHINERS spectra recorded for Ag@TiO$_2$@bpy-PA, Ag@TiO$_2$@bpy-PA-Cu(I) and Ag@TiO$_2$@bpy-PA-Cu(II), the authors have detected the characteristic Raman peaks of N–Cu(II) and N-Cu(I) vibrational modes at 230 and 290 cm^{-1}, respectively (see Figure 8).

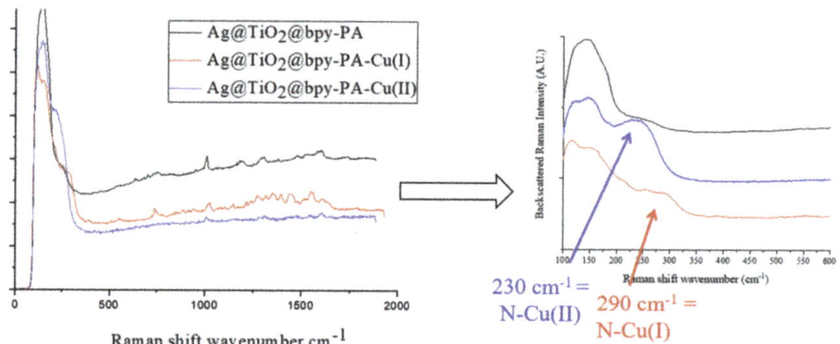

Figure 8. SHINERS spectra for Ag@TiO$_2$@bpy-PA (in black), Ag@TiO$_2$@bpy-PA-Cu(I) (in red) and Ag@TiO$_2$@bpy-PA-Cu(II) (in blue), where the Raman peaks of N-Cu(II) and N-Cu(I) vibrational modes are depicted. The figure is reproduced from [105] with permission from the Royal Society of Chemistry, Copyright 2020.

Krajczewski et al., have reported the detection of four-mercaptobenzoic acid (p-MBA) by SHINERS measurements [106]. Au@Ag concave cubic nanoparticles (noted: cc-Au@Ag NPs) were employed for the p-MBA detection probed by SHINERS measurements. The authors have demonstrated that these cc-Au@Ag NPs have improved by 35% the enhancement factor of the Raman signal of p-MBA molecules compared to the cc-Au NPs without the thin Ag shell. Moreover, the authors have added a thin layer of SiO$_2$ (<5 nm) on cc-Au@Ag NPs, and they recorded a reduction by 50% of the Raman signal of p-MBA molecules [106]. El-Said et al., have demonstrated the detection of neurotransmitters, such as γ-aminobutyric acid (GABA), by using the SHINERS technique [108]. Au@PPy nanobipyramids were used for this investigation, and allowed the detection of GABA with high sensitivity (detection limit of 116 nM). Moreover, these Au@PPy nanobipyramids have also enabled the detection of GABA in the presence of human serum, representing a real sample [108]. Zdaniauskienė et al., have used the SHINERS technique in order to study *Metschnikowia pulcherrima* yeast cells [109]. Au@SiO$_2$ nanoparticles were employed for this investigation, and allowed to obtain SHINERS spectra more enhanced than SERS spectra. Moreover, the Au@SiO$_2$ nanoparticles also allowed to suppress the appearance of supplementary bands due to potential interactions with the gold nanoparticles, and to identify the wall of yeast cells and their functional elements [109]. To conclude this section and also this review, two works related to cancer research, focusing on the identification of the atypical hyperplasia (AH) of the breast and the detection of tumor cells, are addressed. Zheng et al., have explored the identification of the breast AH by employing the SHINERS technique, which can provide a non-invasive diagnosis and study the cancer mechanisms at a molecular level [110]. The authors have remarked via changes in the Raman band intensities from SHINERS spectra that DNA strands have begun to snap in breast AH, and the presence of amino acid residues was more important than in normal breast tissues [110]. Nicinski et al., have reported on the improvement of the detection sensitivity of tumor cells such as renal cell carcinoma, and blood cells. This improvement was achieved by using Ag@SiO$_2$ nanoparticles via SHINERS measurements. The authors have observed variations in the intensities of Raman bands between cancer and healthy cells due to changes in the structure and quantity of molecules present during the formation of cancer cells [107].

4. Discussions and Future Directions

The shell-isolated nanoparticle-enhanced Raman spectroscopy has been generally employed with core-shell nanoparticles composed of gold cores (or silver) and silica shells, because the gold or silver cores presented a strong SERS activity. It would be interesting to use other well-known plasmonic materials, such as aluminum, copper, palladium, and other alternative plasmonic materials (e.g., transparent conductive oxides and transition-metal nitrides) in order to study the influence of the nature of plasmonic material on the efficiency of the SHINERS technique to be analyzed different surface reactions at normal and high temperatures. Other materials for the shell fabrication, such as polymers, can be used to investigate the effect of shell material on this same efficiency cited previously. Through the three sections presented above, we have observed that the shell-isolated nanoparticle-enhanced Raman spectroscopy was non-invasive thanks to the catalytically inactive dielectric shell, and also had other advantages, such as the study of different surface catalytic reactions or adsorption of reactants on several surfaces of different natures and morphologies, and the detection of different molecules. Both these studies [84] were realized under normal conditions of temperature (T = 20–150 °C) and pressure. However, a great number of catalytic reactions are produced at high temperatures (T = 300–1000 °C) typically in industry. Thus, the thermal stability of SHINERS substrates should be improved for industrial applications. Moreover, the influence of high pressures on the stability of SHINERS substrates is still a research issue to be solved. Another challenging improvement to be addressed is to reduce the shell thickness (<1 nm; without pinhole in shell) in order to achieve a better Raman enhancement. Besides, another advantage would be to use the shell-isolated nanoparticle-enhanced Raman spectroscopy in catalysis as local nanosensors of molecules during the catalytic reactions in order to have a deep understanding of these catalytic reactions at the subnanometer scale [111]. Furthermore, the shell-isolated nanoparticle-enhanced Raman spectroscopy can be extended to other enhanced spectroscopies as tip-enhanced Raman spectroscopy (named shell-isolated TERS) [112] and sum-frequency generation spectroscopy (named SHINE-SFG) [113]. The shell-isolated TERS can allow the exclusion of interferences which are due to the presence of contaminants, and also the investigation of catalytic reactions at the level of a solid–liquid interface [112]. To finish, the SHINE-SFG spectroscopy can enable a novel type of enhancement coming from the non-linear coupling of SHINE-SFG with difference frequency generation. Thus, alternative substrates based on this type of coupling can be designed in order to enhance different signals [113].

5. Conclusions

In this mini-review, we addressed the applications of the shell-isolated nanoparticle-enhanced Raman spectroscopy. We started with the SHINERS application to the reactions with CO molecules. The reactions of oxidation, hydrogenation, and adsorption of CO molecules with various catalysts have been presented. Next, we explored SHINERS studies on other surface catalytic reactions. Among these reactions, we presented a couple of works on oxygen reduction reactions realized on Pt(*hkl*) surfaces. Then, oxidation reactions of water and Cu surfaces have been exposed. Hydrogenation and photodegradation reactions, molecule adsorption, and corrosion have been also addressed. Finally, we reported on the detection of molecules and ions by SHINERS spectroscopy. The SHINERS experiments have enabled to improve the detection sensitivity of pesticides (thiram), tumor cells, and to distinguish the copper oxidation states. In conclusion, the shell-isolated nanoparticle-enhanced Raman spectroscopy can be very useful for obtaining various information on surface catalytic reactions, such as their mechanism and the intermediate species present during these reactions. Moreover, the SHINERS substrates based on core–shell nanoparticles can be employed as very sensitive nanosensors of molecules and ions.

Funding: This research received no external funding.

Conflicts of Interest: The author declares no conflict of interest.

References

1. Castro-Grijalba, A.; Montes-Garcia, V.; Cordero-Ferradas, M.J.; Coronado, E.; Perez-Juste, J.; Pastoriza-Santos, I. SERS-Based Molecularly Imprinted Plasmonic Sensor for Highly Sensitive PAH Detection. *ACS Sens.* **2020**, *5*, 693–702. [CrossRef]
2. Blanco-Formoso, M.; Pazos-Perez, N.; Alvarez-Puebla, R.A. Fabrication and SERS properties of complex and organized nanoparticle plasmonic clusters stable in solution. *Nanoscale* **2020**, *12*, 14948–14956. [CrossRef]
3. Lan, L.; Fan, X.; Gao, Y.; Li, G.; Hao, Q.; Qiu, T. Plasmonic metal carbide SERS chips. *J. Mater. Chem. C* **2020**, *8*, 14523–14530. [CrossRef]
4. Huang, J.A.; Mousavi, M.Z.; Zhao, Y.Q.; Hubarevich, A.; Omeis, F.; Giovannini, G.; Schutte, M.; Garoli, D.; De Angelis, F. SERS discrimination of single DNA bases in single oligonucleotides by electro-plasmonic trapping. *Nat. Commun.* **2019**, *10*, 5321. [CrossRef] [PubMed]
5. Yang, L.T.; Lee, J.H.; Rathnam, C.; Hou, Y.N.; Choi, J.W.; Lee, K.B. Dual-Enhanced Raman Scattering-Based Characterization of Stem Cell Differentiation Using Graphene-Plasmonic Hybrid Nanoarray. *Nano Lett.* **2019**, *19*, 8138–8148. [CrossRef] [PubMed]
6. Zhong, J.H.; Vogelsang, J.; Yi, J.M.; Wang, D.; Wittenbecher, L.; Mikaelsson, S.; Korte, A.; Chimeh, A.; Arnold, C.L.; Schaaf, P.; et al. Nonlinear plasmon-exciton coupling enhances sum-frequency generation from a hybrid metal/semiconductor nanostructure. *Nat. Commun.* **2020**, *11*, 1464. [CrossRef] [PubMed]
7. Gao, M.; He, Y.H.; Chen, Y.; Shih, T.M.; Yang, W.M.; Chen, H.Y.; Yang, Z.L.; Wang, Z.H. Enhanced sum frequency generation for ultrasensitive characterization of plasmonic modes. *Nanophotonics* **2020**, *9*, 815–822. [CrossRef]
8. Dalstein, L.; Humbert, C.; Ben Haddada, M.; Boujday, S.; Barbillon, G.; Busson, B. The Prevailing Role of Hotspots in Plasmon-Enhanced Sum-Frequency Generation Spectroscopy. *J. Phys. Chem. Lett.* **2019**, *10*, 7706–7711. [CrossRef] [PubMed]
9. Barbillon, G.; Noblet, T.; Busson, B.; Tadjeddine, A.; Humbert, C. Localised detection of thiophenol with gold nanotriangles highly structured as honeycombs by nonlinear sum frequency generation spectroscopy. *J. Mater. Sci.* **2018**, *53*, 4554–4562. [CrossRef]
10. Dalstein, L.; Ben Haddada, M.; Barbillon, G.; Humbert, C.; Tadjeddine, A.; Boujday, S.; Busson, B. Revealing the Interplay between Adsorbed Molecular Layers and Gold Nanoparticles by Linear and Nonlinear Optical Properties. *J. Phys. Chem. C* **2015**, *115*, 17146–17155. [CrossRef]
11. Ma, W.; Sun, M.; Xu, L.; Wang, L.; Kuang, H.; Xu, C. A SERS active gold nanostar dimer for mercury ion detection. *Chem. Commun.* **2013**, *49*, 4989–4991. [CrossRef] [PubMed]
12. Kessentini, S.; Barchiesi, D.; D'Andrea, C.; Toma, A.; Guillot, N.; Di Fabrizio, E.; Fazio, B.; Marago, O.M.; Gucciardi, P.G.; Lamy de la Chapelle, M. Gold Dimer Nanoantenna with Slanted Gap for Tunable LSPR and Improved SERS. *J. Phys. Chem. C* **2014**, *118*, 3209–3219. [CrossRef]
13. Hakonen, A.; Svedendahl, M.; Ogier, R.; Yang, Z.-Y.; Lodewijks, K.; Verre, R.; Shegai, T.; Andersson, P.O.; Käll, M. Dimer-on-mirror SERS substrates with attogram sensitivity fabricated by colloidal lithography. *Nanoscale* **2015**, *7*, 9405–9410. [CrossRef] [PubMed]
14. Prinz, J.; Heck, C.; Ellerik, L.; Merk, V.; Bald, I. DNA origami based Au–Ag-core–shell nanoparticle dimers with single-molecule SERS sensitivity. *Nanoscale* **2016**, *8*, 5612–5620. [CrossRef]
15. Ghosh, P.; Paria, D.; Balasubramanian, K.; Ghosh, A.; Narayanan, R.; Raghavan, S. Directed Microwave-Assisted Self-Assembly of Au–Graphene–Au Plasmonic Dimers for SERS Applications. *Adv. Mater. Interfaces* **2019**, *6*, 1900629. [CrossRef]
16. Ma, L.; Wang, J.; Ren, C.; Ju, P.; Huang, Y.; Zhang, F.; Zhao, F.; Zhang, Z.; Zhang, D. Detection of corrosion inhibitor adsorption via a surface-enhanced Raman spectroscopy (SERS) silver nanorods tape sensor. *Sens. Actuators B* **2020**, *321*, 128617. [CrossRef]
17. Kuttner, C.; Höller, R.P.M.; Quintanilla, M.; Schnepf, M.J.; Dulle, M.; Fery, A.; Liz-Marzan, L.M. SERS and plasmonic heating efficiency from anisotropic core/satellite superstructures. *Nanoscale* **2019**, *11*, 19561–19570. [CrossRef]
18. Reguera, J.; Langer, J.; Jimenez de Aberasturi, D.; Liz-Marzan, L.M. Anistropic metal nanoparticles for surface enhanced Raman scattering. *Chem. Soc. Rev.* **2017**, *46*, 3866–3885. [CrossRef]
19. Tang, L.; Li, S.; Han, F.; Liu, L.; Xu, L.; Ma, W.; Kuang, H.; Li, A.; Wang, L.; Xu, C. SERS-active Au@Ag nanorod dimers for ultrasensitive dopamine detection. *Biosens. Bioelectron.* **2015**, *71*, 7–12. [CrossRef]
20. Alexander, K.D.; Skinner, K.; Zhang, S.; Wei, H.; Lopez, R. Tunable SERS in Gold Nanorod Dimers through Strain Control on an Elastomeric Substrate. *Nano Lett.* **2010**, *10*, 4488–4493. [CrossRef] [PubMed]
21. Höller, R.P.M.; Kuttner, C.; Mayer, M.; Wang, R.; Dulle, M.; Contreras-Caceres, R.; Fery, A.; Liz-Marzan, L.M. Colloidal Superstructures with Triangular Cores: Size Effects on SERS Efficiency. *ACS Photonics* **2020**, *7*, 1839–1848. [CrossRef]
22. Koetz, J. The Effect of Surface Modification of Gold Nanotriangles for Surface-Enhanced Raman Scattering Performance. *Nanomaterials* **2020**, *10*, 2187. [CrossRef] [PubMed]
23. Liebig, F.; Sarhan, R.M.; Bargheer, M.; Schmitt, C.N.Z.; Poghasyan, A.H.; Shahinsyan, A.A.; Koetz, J. Spiked gold nanotriangles: Formation, characterization and applications in surface-enhanced Raman spectroscopy and plasmon-enhanced catalysis. *RSC Adv.* **2020**, *10*, 8152–8160. [CrossRef]
24. Kuttner, C.; Mayer, M.; Dulle, M.; Moscoso, A.; Lopez-Romero, J.M.; Forster, S.; Fery, A.; Perez-Juste, J.; Contreras-Caceres, R. Seeded Growth Synthesis of Gold Nanotriangles: Size Control, SAXS Analysis, and SERS Performance. *ACS Appl. Mater. Interfaces* **2018**, *10*, 11152–11163. [CrossRef] [PubMed]
25. Bryche, J.-F.; Tsigara, A.; Bélier, B.; Lamy de la Chapelle, M.; Canva, M.; Bartenlian, B.; Barbillon, G. Surface enhanced Raman scattering improvement of gold triangular nanoprisms by a gold reflective underlayer for chemical sensing. *Sens. Actuators B* **2016**, *228*, 31–35. [CrossRef]

26. Su, Q.; Ma, X.; Dong, J.; Jiang, C.; Qian, W. A Reproducible SERS Substrate Based on Electrostatically Assisted APTES-Functionalized Self-Assembly of Gold Nanostars. *ACS Appl. Mater. Interfaces* **2011**, *3*, 1873–1879. [CrossRef]

27. Wang, Y.; Polavarapu, L.; Liz-Marzan, L.M. Reduced Graphene Oxide-Supported Gold Nanostars for Improved SERS Sensing and Drug Delivery. *ACS Appl. Mater. Interfaces* **2014**, *6*, 21798–21805. [CrossRef]

28. Indrasekara, A.S.D.S.; Meyers, S.; Shubeita, S.; Feldman, L.C.; Gustafsson, T.; Fabris, L. Gold nanostar substrates for SERS-based chemical sensing in the femtomolar regime. *Nanoscale* **2014**, *6*, 8891–8899. [CrossRef]

29. Jimenez de Aberasturi, D.; Serrano-Montes, A.B.; Langer, J.; Henriksen-Lacey, M.; Parak, W.J.; Liz-Marzan, L.M. Surface enhanced Raman scattering encoded gold nanostars for multiplexed cell discrimination. *Chem. Mater.* **2016**, *28*, 6779–6790. [CrossRef]

30. Serrano-Montes, A.B.; Langer, J.; Henriksen-Lacey, M.; Jimenez de Aberasturi, D.; Solis, D.M.; Taboada, J.M.; Obelleiro, F.; Sentosun, K.; Bals, S.; Bekdemir, A.; et al. Gold nanostar-coated polystyrene beads as multifunctional nanoprobes for SERS bioimaging. *J. Phys. Chem. C* **2016**, *120*, 20860–20868. [CrossRef]

31. Zhang, L.; Lang, X.; Hirata, A.; Chen, M. Wrinkled Nanoporous Gold Films with Ultrahigh Surface-Enhanced Raman Scattering Enhancement. *ACS Nano* **2011**, *5*, 4407–4413. [CrossRef] [PubMed]

32. Zhang, X.; Zheng, Y.; Liu, X.; Lu, W.; Dai, J.; Lei, D.; MacFarlane, D.R. Hierarchical Porous Plasmonic Metamaterials for Reproducible Ultrasensitive Surface-Enhanced Raman Spectroscopy. *Adv. Mater.* **2015**, *27*, 1090–1096. [CrossRef]

33. Hubarevich, A.; Huang, J.-A.; Giovanni, G.; Schirato, A.; Zhao, Y.; Maccaferri, N.; De Angelis, F.; Alabastri, A.; Garoli, D. λ-DNA through Porous Materials–Surface-Enhanced Raman Scattering in a Single Plasmonic Nanopore. *J. Phys. Chem. C* **2020**, *124*, 22663–22670. [CrossRef]

34. Cao, J.; Liu, H.-L.; Yang, J.-M.; Li, Z.-Q.; Yang, D.-R.; Ji, L.-N.; Wang, K.; Xia, X.-H. SERS Detection of Nucleobases in Single Silver Plasmonic Nanopores. *ACS Sens.* **2020**, *5*, 2198–2204. [CrossRef]

35. Lum, W.; Bruzas, I.; Gorunmez, Z.; Unser, S.; Beck, T.; Sagle, L. Novel Liposome-Based Surface-Enhanced Raman Spectroscopy (SERS) Substrate. *J. Phys. Chem. Lett.* **2017**, *8*, 2639–2646. [CrossRef]

36. Yue, W.; Wang, Z.; Whittaker, J.; Lopez-Royo, F.; Yang, Y.; Zayats, A.V. Amplification of surface-enhanced Raman scattering due to substrate-mediated localized surface plasmons in gold nanodimers. *J. Mater. Chem. C* **2017**, *5*, 4075–4084. [CrossRef]

37. Benz, F.; Chikkaraddy, R.; Salmon, A.; Ohadi, H.; de Nijs, B.; Mertens, J.; Carnegie, C.; Bowman, R.W.; Baumberg, J.J. SERS of Individual Nanoparticles on a Mirror: Size Does Matter, but so Does Shape. *J. Phys. Chem. Lett.* **2016**, *7*, 2264–2269. [CrossRef] [PubMed]

38. Huang, Y.; Ma, L.; Hou, M.; Li, J.; Xie, Z.; Zhang, Z. Hybridized plasmon modes and near-field enhancement of metallic nanoparticle-dimer on a mirror. *Sci. Rep.* **2016**, *6*, 30011. [CrossRef]

39. Sobhani, A.; Manjavacas, A.; Cao, Y.; McClain, M.J.; Javier Garcia de Abajo, F.; Nordlander, P.; Halas, N.J. Pronounced Linewidth Narrowing of an Aluminum Nanoparticle Plasmon Resonance by Interaction with an Aluminum Metallic Film. *Nano Lett.* **2015**, *15*, 6946–6951. [CrossRef] [PubMed]

40. Convertino, A.; Mussi, V.; Maiolo, L. Disordered array of Au covered silicon nanowires for SERS biosensing combined with electrochemical detection. *Sci. Rep.* **2016**, *6*, 25099. [CrossRef]

41. Wang, H.; Jiang, X.; Lee, S.T.; He, Y. Silicon nanohybrid-based surface-enhanced Raman scattering sensors. *Small* **2014**, *10*, 4455–4468. [CrossRef]

42. Akin, M.S.; Yilmaz, M.; Babur, E.; Ozdemur, B.; Erdogan, H.; Tamer, U.; Demirel, G. Large area uniform deposition of silver nanoparticles through bio-inspired polydopamine coating on silicon nanowire arrays for practical SERS applications. *J. Mater. Chem. B* **2014**, *10*, 4455–4468. [CrossRef] [PubMed]

43. Bryche, J.-F.; Bélier, B.; Bartenlian, B.; Barbillon, G. Low-cost SERS substrates composed of hybrid nanoskittles for a highly sensitive sensing of chemical molecules. *Sens. Actuators B* **2017**, *239*, 795–799. [CrossRef]

44. Barbillon, G.; Ivanov, A.; Sarychev, A.K. Hybrid Au/Si Disk-Shaped Nanoresonators on Gold Film for Amplified SERS Chemical Sensing. *Nanomaterials* **2019**, *9*, 1588. [CrossRef]

45. Graniel, O.; Iatsunskyi, I.; Coy, E.; Humbert, C.; Barbillon, G.; Michel, T.; Maurin, D.; Balme, S.; Miele, P.; Bechelany, M. Au-covered hollow urchin-like ZnO nanostructures for surface-enhanced Raman scattering sensing. *J. Mater. Chem. C* **2019**, *7*, 15066–15073. [CrossRef]

46. Cui, S.; Dai, Z.; Tian, Q.; Liu, J.; Xiao, X.; Jiang, C.; Wu, W.; Roy, V.A.L. Wetting properties and SERS applications of ZnO/Ag nanowire arrays patterned by a screen printing method. *J. Mater. Chem. C* **2016**, *4*, 6371–6379. [CrossRef]

47. Lee, Y.; Lee, J.; Lee, T.K.; Park, J.; Ha, M.; Kwak, S.K.; Ko, H. Particle-on-film gap plasmons on antireflective ZnO nanocone arrays for molecular-level surface-enhanced Raman scattering sensors. *ACS Appl. Mater. Interfaces* **2015**, *7*, 26421–26429. [CrossRef]

48. Song, W.; Ji, W.; Vantasin, S.; Tanabe, I.; Zhao, B.; Ozaki, Y. Fabrication of a highly sensitive surface-enhanced Raman scattering substrate for monitoring the catalytic degradation of organic pollutants. *J. Mater. Chem. A* **2015**, *3*, 13556–13562. [CrossRef]

49. He, X.; Yue, C.; Zang, Y.; Yin, J.; Sun, S.; Li, J.; Kang, J. Multi-hotspot configuration on urchin-like Ag nanoparticle/ZnO hollow nanosphere arrays for highly sensitive SERS. *J. Mater. Chem. A* **2013**, *1*, 15010–15015. [CrossRef]

50. Liu, X.W.; Wang, D.S.; Li, Y.D. Synthesis and catalytic properties of bimetallic nanomaterials with various architectures. *Nano Today* **2012**, *7*, 448–466. [CrossRef]

51. Su, Y.; Xu, S.; Zhang, J.; Chen, X.; Jiang, L.-P.; Zheng, T.; Zhu, J.-J. Plasmon Near-Field Coupling of Bimetallic Nanostars and a Hierarchical Bimetallic SERS "Hot Field": Toward Ultrasensitive Simultaneous Detection of Multiple Cardiorenal Syndrome Biomarkers. *Anal. Chem.* **2019**, *91*, 864–872. [CrossRef] [PubMed]

52. Joseph, D.; Kwak, C.H.; Huh, Y.S.; Han, Y.-K. Synthesis of AuAg@Ag core@shell hollow cubic nanostructures as SERS substrates for attomolar chemical sensing. *Sens. Actuators B* **2019**, *281*, 471–477. [CrossRef]

53. Vu, T.D.; Duy, P.K.; Chung, H. Nickel foam-caged Ag-Au bimetallic nanostructure as a highly rugged and durable SERS substrate. *Sens. Actuators B* **2019**, *282*, 535–540. [CrossRef]

54. Barbillon, G. Latest Novelties on Plasmonic and Non-Plasmonic Nanomaterials for SERS Sensing. *Nanomaterials* **2020**, *10*, 1200. [CrossRef]

55. Hussain, A.; Sun, D.-W.; Pu, H. Bimetallic core shelled nanoparticles (Au@AgNPs) for rapid detection of thiram and dicyandiamide contaminants in liquid milk using SERS. *Food Chem.* **2020**, *317*, 126429. [CrossRef]

56. Cong, S.; Yuan, Y.; Chen, Z.; Hou, J.; Yang, M.; Su, Y.; Zhang, Y.; Li, L.; Li, Q.; Geng, F.; et al. Noble metal-comparable SERS enhancement from semiconducting metal oxides by making oxygen vacancies. *Nat. Commun.* **2015**, *6*, 7800. [CrossRef]

57. Liu, W.; Bai, H.; Li, X.; Li, W.; Zhai, J.; Li, J.; Xi, G. Improved Surface-Enhanced Raman Spectroscopy Sensitivity on Metallic Tungsten Oxide by the Synergistic Effect of Surface Plasmon Resonance Coupling and Charge Transfer. *J. Phys. Chem. Lett.* **2018**, *9*, 4096–4100. [CrossRef] [PubMed]

58. Hou, X.; Fan, X.; Wei, P.; Qiu, T. Planar transition metal oxides SERS chips: A general strategy. *J. Mater. Chem. C* **2019**, *7*, 11134–11141. [CrossRef]

59. Wei, W.; Yao, Y.; Zhao, Q.; Xu, Z.; Wang, Q.; Zhang, Z.; Gao, Y. Oxygen defect-induced localized surface plasmon resonance at the WO_{3-x} quantum dot/silver nanowire interface: SERS and photocatalysis. *Nanoscale* **2019**, *11*, 5535–5547. [CrossRef]

60. He, R.; Lai, H.; Wang, S.; Chen, T.; Xie, F.; Chen, Q.; Liu, P.; Chen, J.; Xie, W. Few-layered vdW MoO_3 for sensitive, uniform and stable SERS applications. *Appl. Surf. Sci.* **2020**, *507*, 145116. [CrossRef]

61. Wang, X.; Li, J.; Shen, Y.; Xie, A. An assembled ordered $W_{18}O_{49}$ nanowire film with high SERS sensitivity and stability for the detection of RB. *Appl. Surf. Sci.* **2020**, *504*, 144073. [CrossRef]

62. Ben-Jaber, S.; Peveler, W.J.; Quesada-Cabrera, R.; Cortés, E.; Sotelo-Vazquez, C.; Abdul-Karim, N.; Maier, S.A.; Parkin, I.P. Photo-induced enhanced Raman spectroscopy for universal ultra-trace detection of explosives, pollutants and biomolecules. *Nat. Commun.* **2016**, *7*, 12189. [CrossRef] [PubMed]

63. Glass, D.; Cortés, E.; Ben-Jaber, S.; Brick, T.; Peveler, W.J.; Blackman, C.S.; Howle, C.R.; Quesada-Cabrera, R.; Parkin, I.P.; Maier, S.A. Dynamics of Photo-Induced Surface Oxygen Vacancies in Metal-Oxide Semiconductors Studied Under Ambient Conditions. *Adv. Sci.* **2019**, *6*, 1901841. [CrossRef] [PubMed]

64. Barbillon, G.; Noblet, T.; Humbert, C. Highly crystalline ZnO film decorated with gold nanospheres for PIERS chemical sensing. *Phys. Chem. Chem. Phys.* **2020**, *22*, 21000–21004. [CrossRef]

65. Sun, H.H.; Yao, M.G.; Song, Y.P.; Zhu, L.Y.; Dong, J.J.; Liu, R.; Li, P.; Zhao, B.; Liu, B.B. Pressure-induced SERS enhancement in a MoS_2/Au/R6G system by a two-step charge transfer process. *Nanoscale* **2019**, *11*, 21493–21501. [CrossRef]

66. Barbillon, G. Nanoplasmonics in High Pressure Environment. *Photonics* **2020**, *7*, 53. [CrossRef]

67. Li, J.-F.; Huang, Y.F.; Ding, Y.; Yang, Z.L.; Li, S.B.; Zhou, X.S.; Fan, F.R.; Zhang, W.; Zhou, Z.Y.; Wu, D.Y.; et al. Shell-isolated nanoparticle-enhanced Raman spectroscopy. *Nature* **2010**, *464*, 392–395. [CrossRef]

68. Ding, S.-Y.; Yi, J.; Li, J.-F.; Ren, B.; Wu, D.-Y.; Panneerselvam, R.; Tian, Z.-Q. Nanostructure-based plasmon-enhanced Raman spectroscopy for surface analysis of materials. *Nat. Rev. Mater.* **2016**, *1*, 16021. [CrossRef]

69. Li, J.-F.; Zhang, Y.-J.; Ding, S.-Y.; Panneerselvam, R.; Tian, Z.-Q. Core-Shell Nanoparticle-Enhanced Raman Spectroscopy. *Chem. Rev.* **2017**, *111*, 5002–5069. [CrossRef]

70. Li, C.-Y.; Meng, M.; Huang, S.-C.; Li, L.; Huang, S.-R.; Chen, S.; Meng, L.-Y.; Panneerselvam, R.; Zhang, S.-J.; Ren, B.; et al. "Smart" Ag Nanostructures for Plasmon-Enhanced Spectroscopies. *J. Am. Chem. Soc.* **2015**, *137*, 13784–13787. [CrossRef] [PubMed]

71. Hartman, T.; Weckhuysen, B.M. Thermally Stable TiO_2- and SiO_2-Shell-Isolated Au Nanoparticles for In Situ Plasmon-Enhanced Raman Spectroscopy of Hydrogenation Catalysts. *Chem. Eur. J.* **2018**, *24*, 3733–3741. [CrossRef] [PubMed]

72. Hartman, T.; Wondergem, C.S.; Weckhuysen, B.M. Pratical Guidelines for Shell-Isolated Nanoparticle-Enhanced Raman Spectroscopy of Heterogeneous Catalysts. *ChemPhysChem* **2018**, *19*, 2461–2467. [CrossRef]

73. Zhang, H.; Zhang, X.G.; Wei, J.; Wang, C.; Chen, S.; Sun, H.L.; Wang, Y.H.; Chen, B.H.; Yang, Z.L.; Wu, D.Y.; et al. Revealing the Role of Interfacial Properties on Catalytic Behaviors by In-Situ Surface-Enhanced Raman Spectroscopy. *J. Am. Chem. Soc.* **2017**, *139*, 10339–10346. [CrossRef]

74. Zhang, H.; Wang, C.; Sun, H.L.; Fu, G.; Chen, S.; Zhang, Y.J.; Chen, B.H.; Anema, J.R.; Yang, Z.L.; Li, J.F.; et al. In Situ Dynamic Tracking of Heterogeneous Nanocatalytic Processes by Shell-Isolated Nanoparticle-Enhanced Raman Spectroscopy. *Nat. Commun.* **2017**, *8*, 15447. [CrossRef]

75. Lin, X.-D.; Li, J.-F.; Huang, Y.-F.; Tian, X.-D.; Uzayisenga, V.; Li, S.-B.; Ren, B.; Tian, Z.-Q. Shell-isolated nanoparticle-enhanced Raman spectroscopy: Nanoparticle synthesis, characterization and applications in electrochemistry. *J. Electroanal. Chem.* **2013**, *688*, 5–11. [CrossRef]

76. Liu, Y.; Hu, Y.; Zhang, J. Few-Layer Graphene-Encapsulated Metal Nanoparticles for Surface-Enhanced Raman Spectroscopy. *J. Phys. Chem. C* **2014**, *118*, 8993–8998. [CrossRef]

77. Chen, S.N.; Li, X.; Zhao, Y.Y.; Chang, L.M.; Qi, J.Y. Graphene oxide shell-isolated Ag nanoparticles for surface-enhanced Raman scattering. *Carbon* **2015**, *81*, 767–772. [CrossRef]

78. Yang, C.; Zhang, C.; Huo, Y.Y.; Jiang, S.Z.; Qiu, H.W.; Xu, Y.Y.; Li, X.H.; Man, B.Y. Shell-isolated graphene@Cu nanoparticles on graphene@Cu substrates for the application in SERS. *Carbon* **2016**, *98*, 526–533. [CrossRef]

79. Lin, X.-D.; Uzayisenga, V.; Li, J.-F.; Fang, P.-P.; Wu, D.-Y.; Ren, B.; Tian, Z.-Q. Synthesis of ultrathin and compact Au@MnO$_2$ nanoparticles for shell-isolated nanoparticle-enhanced Raman spectroscopy. *J. Raman Spectrosc.* **2012**, *43*, 40–45. [CrossRef]

80. Honesty, N.R.; Gewirth, A.A. Shell-isolated nanoparticle enhanced Raman spectroscopy (SHINERS) investigation of benzotriazole film formation on Cu(100), Cu(111), and Cu(poly). *J. Raman Spectrosc.* **2012**, *43*, 46–50. [CrossRef]

81. Li, S.-B.; Li, L.-M.; Anema, J.R.; Li, J.-F.; Yang, Z.-L.; Ren, B.; Sun, J.-J.; Tian, Z.-Q. Shell-Isolated Nanoparticle-Enhanced Raman Spectroscopy (SHINERS) Based on Gold-Core Silica-Shell Nanorods. *Z. Phys. Chem.* **2011**, *225*, 775–783. [CrossRef]

82. Wang, Y.-H.; Wei, J.; Radjenovic, P.; Tian, Z.-Q.; Li, J.-F. In Situ Analysis of Surface Catalytic Reactions Using Shell-Isolated Nanoparticle-Enhanced Raman Spectroscopy. *Anal. Chem.* **2019**, *91*, 1675–1685. [CrossRef]

83. Su, M.; Dong, J.-C.; Le, J.-B.; Zhao, Y.; Yang, W.-M.; Yang, Z.-L.; Attard, G.; Liu, G.-K.; Cheng, J.; Wei, Y.-M.; et al. In Situ Raman Study of CO Electrooxidation Pt(*Hkl*) Single-Cryst. Surfaces Acidic Solution. *Angew. Chem. Int. Ed.* **2020**, *59*, 1–6. [CrossRef]

84. Wondergem, C.S.; Hartman, T.; Weckhuysen, B.M. In Situ Shell-Isolated Nanoparticle-Enhanced Raman Spectroscopy to Unravel Sequential Hydrogenation of Phenylacetylene over Platinum Nanoparticles. *ACS Catal.* **2019**, *9*, 10794–10802. [CrossRef]

85. Wondergem, C.S.; Kromwijk, J.J.G.; Slagter, M.; Vrijburg, W.L.; Hensen, E.J.M.; Monai, M.; Vogt, C.; Weckhuysen, B.M. In Situ Shell-Isolated Nanoparticle-Enhanced Raman Spectroscopy of Nickel-Catalyzed Hydrogenation Reactions. *ChemPhysChem* **2020**, *21*, 625–632. [CrossRef]

86. Hartman, T.; Geitenbeek, R.G.; Whiting, G.T.; Weckhuysen, B.M. Operando monitoring of temperature and active species at the single catalyst particle level. *Nat. Catal.* **2019**, *2*, 986–996. [CrossRef]

87. Zhang, Y.-J.; Chen, Q.-Q.; Chen, X.; Wang, A.; Tian, Z.-Q.; Li, J.-F. Graphene-coated Au nanoparticle-enhanced Raman spectroscopy. *J. Raman Spectrosc.* **2020**, 1–7. [CrossRef]

88. Kuruvinashetti, K.; Zhang, Y.; Li, J.; Kornienko, N. Shell isolated nanoparticle enhanced Raman spectroscopy for renewable energy electorcatalysis. *New J. Chem.* **2020**, *44*, 19953–19960. [CrossRef]

89. Zhao, Y.; Chang, X.; Malkani, A.S.; Yang, X.; Thompson, L.; Jiao, F.; Xu, B. Speciation of Cu Surfaces During the Electrochemical CO Reduction Reaction. *J. Am. Chem. Soc.* **2020**, *142*, 9735–9743. [CrossRef]

90. Guan, S.; Attard, G.A.; Wain, A.J. Observation of Substituent Effects in the Electrochemical Adsorption and Hydrogenation of Alkynes on Pt{*Hkl*} Using SHINERS. *ACS Catal.* **2020**, *10*, 10999–11010. [CrossRef]

91. Saeed, K.H.; Forster, M.; Li, J.-F.; Hardwick, L.J.; Cowan, A.J. Water oxidation intermediates on iridium oxide electrodes probed by in situ electrochemical SHINERS. *Chem. Commun.* **2020**, *56*, 1129–1132. [CrossRef] [PubMed]

92. Li, C.-Y.; Le, J.-B.; Wang, Y.-H.; Chen, S.; Yang, Z.-L.; Li, J.-F.; Cheng, J.; Tian, Z.-Q. In situ probing electrified interfacial water structures at atomically flat surfaces. *Nat. Mater.* **2019**, *18*, 697–701. [CrossRef]

93. Guo, B.; Lin, X.; Burgess, I.J.; Yu, C. Electrochemical SHINERS investigation of the adsorption of butyl xanthate and 2-mercaptobenzothiazole on pyrite. *Appl. Surf. Sci.* **2020**, *529*, 147118. [CrossRef]

94. Bodappa, N.; Su, M.; Zhao, Y.; Le, J.-B.; Yang, W.-M.; Radjenovic, P.; Dong, J.-C.; Cheng, J.; Tian, Z.-Q.; Li, J.-F. Early Stages of Electrochemical Oxydation of Cu(111) and Polycrystalline Cu Surfaces Revealed by in Situ Raman Spectroscopy. *J. Am. Chem. Soc.* **2019**, *141*, 12192–12196. [CrossRef]

95. Galloway, T.A.; Dong, J.-C.; Li, J.-F.; Attard, G.; Hardwick, L.J. Oxygen reactions on Pthkl in a non-aqueous Na+ electrolyte: Site selective stabilisation of a sodium peroxy species. *Chem. Sci.* **2019**, *10*, 2956–2964. [CrossRef]

96. Dong, J.-C.; Zhang, X.-G.; Briega-Martos, V.; Jin, X.; Yang, J.; Chen, S.; Yang, Z.-L.; Wu, D.-Y.; Feliu, J.M.; Williams, C.T.; et al. In situ Raman spectroscopic evidence for oxygen reduction reaction intermadiates at platinum single-crystal surfaces. *Nat. Energy* **2019**, *4*, 60–67. [CrossRef]

97. Dong, J.-C.; Su, M.; Briega-Martos, V.; Li, L.; Le, J.-B.; Radjenovic, P.; Zhou, X.-S.; Feliu, J.M.; Tian, Z.-Q.; Li, J.-F. Direct in Situ Raman Spectroscopic Evidence of Oxygen Reduction Reaction Intermediates at High-Index Pt(hkl) Surfaces. *J. Am. Chem. Soc.* **2020**, *142*, 715–719. [CrossRef]

98. Barlow, B.C.; Guo, B.; Situm, A.; Grosvenor, A.P.; Burgess, I.J. Shell isolated nanoparticle enhanced Raman spectroscopy (SHINERS) studies of steel surface corrosion. *J. Electroanal. Chem.* **2019**, *853*, 113559. [CrossRef]

99. Wang, C.; Chen, X.; Chen, T.-M.; Wei, J.; Qin, S.-N.; Zheng, J.-F.; Zhang, H.; Tian, Z.-Q.; Li, J.-F. In-situ SHINERS Study of the Size and Composition Effect of Pt-Based Nanocatalysts in Catalytic Hydrogenation. *ChemCatChem* **2020**, *12*, 75–79. [CrossRef]

100. Qiu, H.; Wang, M.; Zhang, L.; Cao, M.; Yang, Z.; Dou, J.; Ji, S.; Ji, Y.; Kou, S.; Bhatti, A.S. Insights into the role of graphene in hybrid photocatalytic system by in-situ shell-isolated nanoparticle-enhanced Raman Spectroscopy. *Carbon* **2019**, *152*, 305–315. [CrossRef]

101. Wondergem, C.S.; van Swieten, T.P.; Geitenbeek, R.G.; Erné, B.H.; Weckhuysen, B.M. Extented Surface-Enhanced Raman Spectroscopy to Liquids Using Shell-Isolated Plasmonic Superstructures. *Chem. Eur. J.* **2019**, *25*, 15772–15778. [CrossRef]

102. Zhang, S.-P.; Lin, J.-S.; Lin, R.-K.; Radjenovic, P.M.; Yang, W.-M.; Xu, J.; Dong, J.-C.; Yang, Z.-L.; Hang, W.; Tian, Z.-Q.; et al. In situ Raman study of the photoinduced behavior of dye molecules on TiO$_2$(*hkl*) single crystal surfaces. *Chem. Sci.* **2020**, *11*, 6431–6435. [CrossRef]

103. Sun, M.; Li, B.; Liu, X.; Chen, J.; Mu, T.; Zhu, L.; Guo, J.; Ma, X. Performance enhancement of paper-based SERS chips by shell-isolated nanoparticle-enhanced Raman spectroscopy. *J. Mater. Sci. Technol.* **2019**, *35*, 2207–2212. [CrossRef]

104. Forato, F.; Talebzadeh, S.; Rousseau, N.; Mevellec, J.-Y.; Bujoli, B.; Knight, D.A.; Queffélec, C.; Humbert, B. Functionalized core–shell Ag@TiO$_2$ nanoparticles for enhanced Raman spectroscopy: A sensitive detection method for Cu(II) ions. *Phys. Chem. Chem. Phys.* **2019**, *21*, 3066–3072. [CrossRef] [PubMed]

105. Queffélec, C.; Forato, F.; Bujoli, B.; Knight, D.A.; Fonda, E.; Humbert, B. Investigation of copper oxidation states in plasmonic nanomaterials by XAS and Raman spectroscopy. *Phys. Chem. Chem. Phys.* **2020**, *22*, 2193–2199. [CrossRef]
106. Krajczewski, J.; Kedziora, M.; Kołataj, K.; Kudelski, A. Improved synthesis of concave cubic gold nanoparticles and their applications for Raman analysis of surfaces. *RSC Adv.* **2019**, *9*, 18609–18618. [CrossRef]
107. Nicinski, K.; Krajczewski, J.; Kudelski, A.; Witkowska, E.; Trzcinska-Danielewicz, J.; Girstun, A.; Kaminska, A. Detection of circulating tumor cells in blood by shell-isolated nanoparticle—Enhanced Raman spectroscopy (SHINERS) in microfluidic device. *Sci. Rep.* **2019**, *9*, 9267. [CrossRef]
108. El-Said, W.A.; Alshitari, W.; Choi, J.-W. Controlled fabrication of gold nanobipyramids/polypyrrole for shell-isolated nanoparticle-enhanced Raman spectroscopy to detect γ-aminobutyric acid. *Spectrochim. Acta Part A Mol. Biomol. Spectrosc.* **2020**, *229*, 117890. [CrossRef] [PubMed]
109. Zdaniauskienė, A.; Charkova, T.; Ignatjev, I.; Melvydas, V.; Garjonytė, R.; Matulaitienė, I.; Talaikis, M.; Niaura, G. Shell-isolated nanoparticle-enhanced Raman spectroscopy for characterization of living yeast cells. *Spectrochim. Acta Part A Mol. Biomol. Spectrosc.* **2020**, *240*, 118560. [CrossRef]
110. Zheng, C.; Jia, H.Y.; Liu, L.Y.; Wang, Q.; Jiang, H.C.; Teng, L.S.; Geng, C.Z.; Jin, F.; Tang, L.L.; Zhang, J.G.; et al. Molecular fingerprint of precancerous lesions in breast atypical hyperplasia. *J. Int. Med. Res.* **2020**, *48*. [CrossRef]
111. Hartman, T.; Geitenbeek, R.G.; Wondergem, C.S.; van der Stam, W.; Weckhuysen, B.M. Operando Nanoscale Sensors in Catalysis: All Eyes on Catalyst Particles. *ACS Nano* **2020**, *14*, 3725–3735. [CrossRef] [PubMed]
112. Zhang, H.; Duan, S.; Radjenovic, P.M.; Tian, Z.-Q.; Li, J.-F. Core–Shell Nanostructure-Enhanced Raman Spectroscopy for Surface Catalysis. *Acc. Chem. Res.* **2020**, *53*, 729–739. [CrossRef] [PubMed]
113. He, Y.; Ren, H.; You, E.-M.; Radjenovic, P.M.; Sun, S.-G.; Tian, Z.-Q.; Li, J.-F.; Wang, Z. Polarization- and Wavelength-Dependent Shell-Isolated-Nanoparticle-Enhanced Sum-Frequency Generation with High Sensitivity. *Phys. Rev. Lett.* **2020**, *125*, 047401. [CrossRef] [PubMed]

Review

Nanoplasmonics in High Pressure Environment

Grégory Barbillon

EPF-Ecole d'Ingénieurs, 3 bis rue Lakanal, 92330 Sceaux, France; gregory.barbillon@epf.fr

Received: 7 July 2020; Accepted: 25 July 2020; Published: 28 July 2020

Abstract: An explosion in the interest for nanoplasmonics has occurred in order to realize optical devices, biosensors, and photovoltaic devices. The plasmonic nanostructures are used for enhancing and confining the electric field. In the specific case of biosensing, this electric field confinement can induce the enhancement of the Raman signal of different molecules, or the localized surface plasmon resonance shift after the detection of analytes on plasmonic nanostructures. A major part of studies concerning to plasmonic modes and their application to sensing of analytes is realized in ambient environment. However, over the past decade, an emerging subject of nanoplasmonics has appeared, which is nanoplasmonics in high pressure environment. In last five years (2015–2020), the latest advances in this emerging field and its application to sensing were carried out. This short review is focused on the pressure effect on localized surface plasmon resonance of gold nanosystems, the supercrystal formation of plasmonic nanoparticles stimulated by high pressure, and the detection of molecules and phase transitions with plasmonic nanostructures in high pressure environment.

Keywords: plasmonics; localized surface plasmon resonance; high pressure; sensing; SERS

---

1. Introduction

During the past decade, nanoplasmonics was employed for the production of photovoltaic devices [1–6], optical devices [7–15], and biosensors [16–20]. Additionally, nanoplasmonics enabled the enhancement of photocatalysis [21–23], the luminescence upconversion enhancement [24,25], and the optical tuning of luminescence and upconversion luminescence [26,27]. In addition, nanoplasmonics can also enhance the sum-frequency generation signal [28–32] and the Förster resonance energy transfer (FRET) [33–37]. Gold and silver were largely used for the production of plasmonic nanostructures, and other alternative plasmonic materials were also employed, such as aluminum [38,39], copper [40,41], palladium [42,43], transition-metal nitrides [44,45], and transparent conductive oxides [46,47]. The plasmonic nanostructures allowed confining the electromagnetic (EM) field into subwavelength-size zones. Concerning to the application to plasmonic biosensing, this EM field confinement allowed inducing an enhancement of Raman signal of analytes named surface enhanced Raman scattering (SERS) [48–53] or the localized surface plasmon resonance (LSPR) shift after detection of analytes on plasmonic nanostructures [54–58]. The confinement of the EM field can be controlled by adjusting the geometry and spatial organization of plasmonic nanosystems, for instance, which can be realized with various techniques of lithography [45,59–66]. In addition, various plasmonic modes can be used for biosensing based on SERS effect or LSPR shifting as dipolar and multipolar resonances [67,68], surface lattice resonances [69,70], and hybridized resonances [71,72]. In the majority of studies cited previously concerning the plasmonic sensing of analytes, the LSPR shifting and SERS measurements were realized in ambient environment (e.g., pressure). However, a relevant subject of nanoplasmonics has emerged over the past decade. This latter concerns nanoplasmonics in high pressure environment [73,74], and its potential application to sensing of molecules. In the nanoplasmonics in high pressure environment, the mechanisms for the LSPR shifts of metallic nanoparticles induced by high pressure are generally based on variations of the refractive index or the phase transitions of the surrounding medium, or deformation of metallic nanoparticles [73,74].

For instance, a study reporting on the effect of high pressure on the LSPR shift of colloidal gold nanoparticles demonstrated that the LSPR redshift of Au nanoparticles in water was due to the linear increasing of the refractive index of the water with pressure [73].

The aim of this short review is to discuss the latest advances on nanoplasmonics in high pressure environment over the period 2015–2020. Firstly, we will present the pressure effect on localized surface plasmon resonance of gold nanosystems, then the use of high pressures for the supercrystal formation of gold nanoparticles, and finish the detection of molecules and phase transitions with plasmonic nanostructures in high pressure environment.

2. Nanoplasmonics in High Pressure Environment

2.1. Effect of High Pressure on Localized Surface Plasmon Resonance of Metallic Nanoparticles

In this first section, we discuss the effect of high pressure on LSPR modes of gold nanoparticles with different shapes (see Table 1).

Table 1. Effect of high pressure on localized surface plasmon resonance (LSPR) of metallic nanoparticles.

Samples	Study	References
Au spheroidal nanoparticles	LSPR shifts	[75]
Au nanocrystals	LSPR shifts	[76]
Au nanospheres and Au nanorods	LSPR shifts for both modes	[77]

Bao et al. reported the effects of high pressure and the thickness of the gasket in a diamond anvil cell (DAC) on the localized surface plasmon resonance of a colloidal solution of Au spheroidal nanoparticles (AuSNPs). Authors have measured the LSPR of AuSNPs (size = 80 nm) by varying the pressure from 2 to 12 GPa for two gaskets pre-indented to 140 µm (called GPI140) and 317 µm (called GPI317). For the GPI140, the authors have recorded the absorption spectra of AuSNPs for pressures from 2.24 GPa to 11.8 GPa (see Figure 1a).

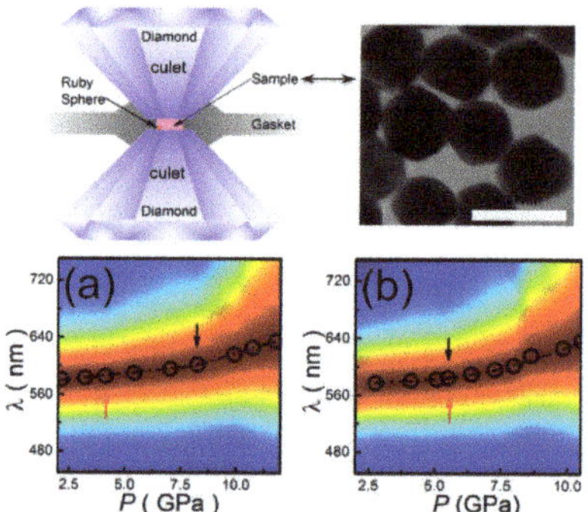

Figure 1. At top, principle scheme of a DAC with a TEM picture of sample (AuSNP size = 80 nm; scale bar = 100 nm). The ruby sphere is employed in order to measure the pressure in chamber by fluorescence. At bottom, absorption spectra of AuSNPs are displayed as function of the wavelength and the pressure with a gasket pre-indented to (**a**) 140 µm and (**b**) 317 µm. Black circles correspond to experimental measurements of the absorption maximum. Red arrows represent the starting of the broadening of absorption peak. Black arrows represent the brutal change in the LSPR shift magnitude for AuSNPs. All of the figures are reprinted from [75], with the permission of AIP Publishing.

A broadening of the absorption peak of AuSNPs has occurred from the pressure of 4.13 GPa (indicated by the red arrow in Figure 1a). Subsequently, a sudden variation in the magnitude of LSPR shift for AuSNPs has occurred at 8.24 GPa (indicated by the black arrow in Figure 1a). This sudden variation is attributed to the deformation of the AuSNP shape. For the second gasket GPI317, they have recorded the absorption spectra of AuSNPs for pressures from 2.75 GPa to 10.47 GPa (see Figure 1b). The broadening of the absorption spectrum and the sudden variation in the LSPR shift magnitude for AuSNPs have occurred at the same pressure of 5.5 GPa (indicated by the red and black arrow, respectively, in Figure 1b). Furthermore, the authors remarked that the sudden variation in the LSPR shift magnitude for AuSNPs was achieved at a higher pressure for the thinnest gasket (GPI140). This was due to a better support of the part of the thinnest gasket located outside the culets in order to do a sharp expansion or contraction of the chamber where the sample is located, emerging at a higher pressure [75]. Gu et al. investigated the effect of quasihydrostatic and non-hydrostatic high pressures on the LSPR of gold nanocrystals (size = 3.9 nm) in a DAC [76]. The used quasihydrostatic and non-hydrostatic pressure media were ethylcyclohexane [78] and toluene [79], respectively. The authors have recorded no variation in the LSPR wavelength of Au nanocrystals for quasihydrostatic high pressures. On contrary, for non-hydrostatic high pressures, they observed a redshift of the LSPR of Au nanocrystals achieving 68 nm, and this latter was reversible when the pressure was decreased. This redshift was due to the deformation of Au nanocrystals (deformed shape with an aspect ratio of ∼2). When the non-hydrostatic pressure was decreased down to ambient pressure, the shape of Au nanocrystal came back its original shape [76]. Martin-Sanchez et al. demonstrated the effects of the hydrostatic pressure on LSPR of gold nanospheres and nanorods [77]. Firstly, the authors reported on the changes in the absorbance spectra of gold nanospheres (AuNS; diameter = 20 nm) in parrafin with pressure. Paraffin was used as solvent due to its easibility of stabilizing gold nanospheres in non-polar media. Authors observed a redshift of the localized surface plasmon resonance of AuNS when the pressure was increased from 0 to 17 GPa, and the redshift magnitude was around 3% of the LSPR wavelength for AuNS (see Figure 2a).

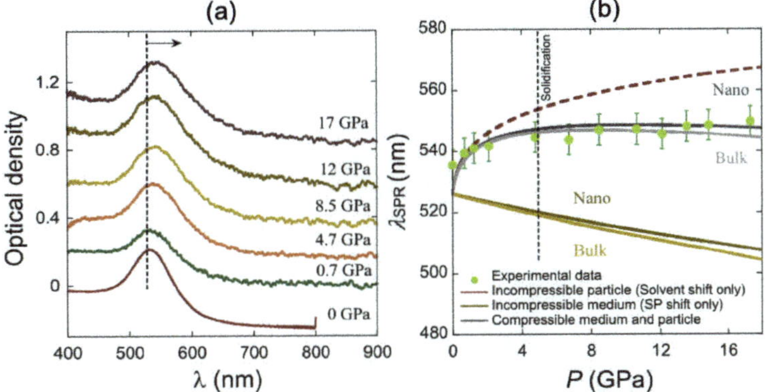

Figure 2. (a) Absorbance (optical density) spectra of Au nanospheres in paraffin at different pressures. (b) Localized surface plasmon resonance wavelength versus pressure. The experimental data are displayed as green points. All of the lines correspond to a fit with the Mie–Gans model where the parameters vary according the considered case. The red dashed line corresponds to the case of an imcompressible particle. The dark and light gray lines correspond to the case where medium and particle are compressible with K_0 = 190 GPa (called Nano) and K_0 = 167 GPa (called Bulk), respectively. The brown lines correspond to the case of an incompressible medium for a gold nanoparticle (called Nano; in dark brown) and for bulk gold material (called Bulk; in light brown). All of the figures are reprinted (adapted) with permission from [77], Copyright 2019 American Chemical Society.

These weaker LSPR variations with pressure were caused by the higher shape factor (L = 1/3), which decreased the solvent impact. Furthermore, the authors have determined from the experimental measurements the bulk modulus (K_0) of the gold nanoparticles (called Nano) by using the Mie–Gans model [77]. They found a value of K_0 equal to 190 GPa, which is bigger as compared to this obtained in the work of Heinz et al. [80] which is equal to 167 GPa for bulk gold (called bulk, see Figure 2b). The Mie–Gans model enables to express the wavelength of the localized surface plasmon resonance at a given pressure (P) as follows (for more details, see reference [77]):

$$\lambda_{LSPR}(P) = \lambda_p(0)\sqrt{\frac{V(P)}{V_0}}\sqrt{\epsilon(0) + \frac{1-L}{L}\epsilon_m(P)} \tag{1}$$

where $\lambda_p(0)$ corresponds to the bulk plasma wavelength at the ambient pressure (corresponding to the pressure $P = 0$), $V(P)$ and V_0 are the particle volume at the pressure P and at the ambient pressure, respectively. The ratio $V(P)/V_0$ depends on the bulk modulus of gold (K_0) and the first derivative K_0'. This ratio and K_0' express, as follows:

$$\frac{V(P)}{V_0} = \left(\frac{PK_0'}{K_0} + 1\right)^{-1/K_0'}, K_0' = \left(\frac{\partial K}{\partial P}\right)_{P=0} \tag{2}$$

where $K = K_0 + K_0'P$ is the bulk modulus of a material (here gold) at a given pressure P. K_0 is the bulk modulus of a material (here gold) at the ambient pressure. The value of K_0' is fixed at 6 [80,81] for all of the studies presented here. $\epsilon_m(P)$ and $\epsilon(0)$ are the solvent dielectric function at the pressure P and the dielectric constant of gold in the short wavelength limit $\lambda \to 0$ (or $\omega \to \infty$, which is commonly noted ϵ_∞), respectively. $\epsilon_m(P)$ depends on the ratio $V_0/V(P)$. L is the shape factor of the nanoparticle.

Secondly, the authors investigated the pressure effect on the LSPR of gold nanorods in hydrostatic regime and beyond this latter for two mixtures of a methanol–ethanol solution. For the first methanol–ethanol (1:4) solution with gold nanorods whose the aspect ratio (AR) is 3.7 (dimensions: 21.7 nm × 5.6 nm, see Figure 3a), they have experimentally observed a redshift of longitudinal plasmonic mode in hydrostatic ($P = 1$–4 GPa) and non-hydrostatic ($P > 4$ GPa) regimes (see Figure 3b). However, in the non-hydrostatic regime (after solution solidification), the optical density decreased abruptly (see Figure 3b).

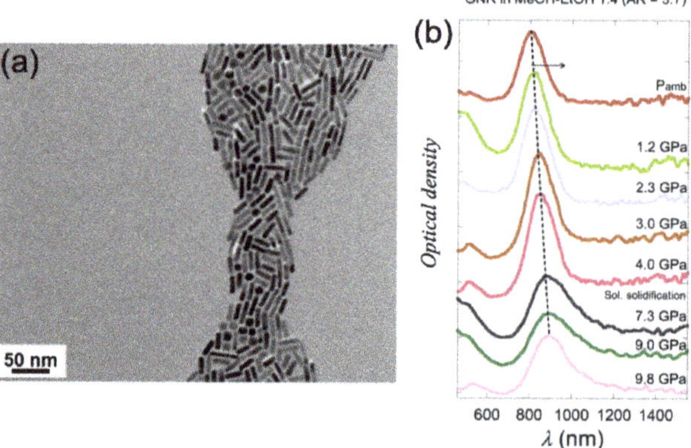

Figure 3. (**a**) TEM picture of the Au nanorods with AR = 3.7 (dimensions: 21.7 nm × 5.6 nm). (**b**) Absorbance (optical density) spectra of Au nanorods (AR = 3.7) in methanol–ethanol (1:4) solution at different pressures. All of the figures are reprinted (adapted) with permission from [77], Copyright 2019 American Chemical Society.

Finally, the authors studied gold nanorods (AR = 3.7, see Figure 3a) in a methanol–ethanol (4:1) solution. They have experimentally observed a redshift of longitudinal plasmonic mode in hydrostatic (1–10 GPa) regime, i.e., up to solution solidification (see Figure 4a). Then, a blueshift in the LSPR wavelength of the longitudinal mode was observed after switching from hydrostatic to non-hydrostatic regime (see Figure 4). In this non-hydrostatic (P > 10 GPa) regime, the LSPR wavelength of the longitudinal mode was again redshifted when the pressure was increased (see Figure 4). However, the optical density decreased more abruptly than in the case of the methanol–ethanol (1:4) solution (see Figure 4a). Besides, weaker LSPR blueshifts for the transversal mode were also observed (see Figure 4b). These blueshifts of the transversal mode were due to the compression of Au nanorods and a higher electron density [77]. The Mie–Gans theory was in agreement with the experimental results for the measurement of the position of the plasmon peak in the hydrostatic regime for both longitudinal and transversal modes. For the non-hydrostatic regime, a difference between experiments and the Mie–Gans theory was observed.

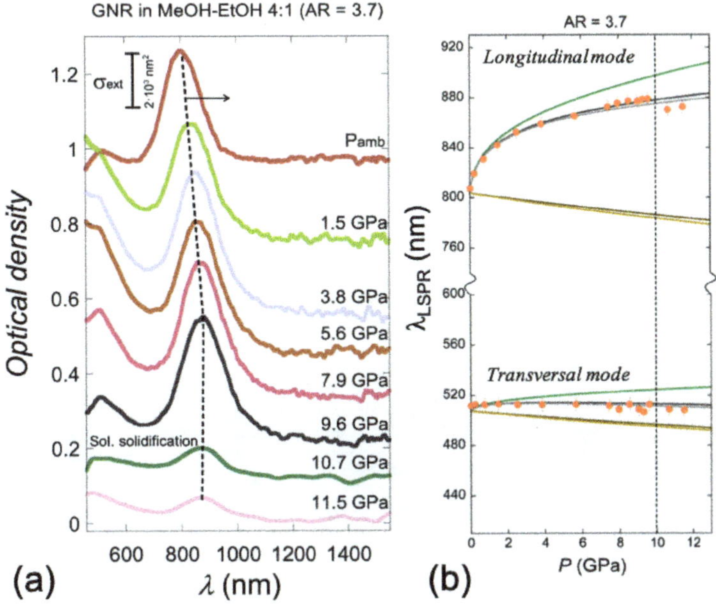

Figure 4. (**a**) Absorbance (optical density) spectra of Au nanorods (AR = 3.7) in methanol–ethanol (4:1) solution recorded at different pressures. (**b**) LSPR wavelength versus hydrostatic pressure for the longitudinal and transversal modes of Au nanorods (AR = 3.7) in methanol–ethanol (4:1) solution. Orange points correspond to experimental data. All of the lines correspond to a fit with the Mie–Gans model where the parameters vary according the considered case. For both plasmonic modes, the green line corresponds to the case of an incompressible particle. The gray lines correspond to the case where particle and solvent are compressible (in dark gray = the bulk modulus of gold called Nano; in light gray = the bulk modulus of gold called Bulk). The brown lines correspond to the case of an incompressible medium (in dark brown = the bulk modulus of gold called Nano; in light brown = the bulk modulus of gold called Bulk). The vertical dashed line corresponds to the solution solidification. All of the figures are reprinted (adapted) with permission from [77], Copyright 2019 American Chemical Society.

To conclude this section, a dramatic decrease of the optical density at the LSPR peak was recorded after the solution solidification for both methanol–ethanol solutions with gold nanorods (AR = 3.7). When the pressure of the solution solidification was higher, the optical density decay was more significant (see Figures 3b and 4a, and reference [77]).

2.2. Use of High Pressures for the Supercrystal Formation of Gold Nanoparticles

In this section, we present studies regarding the pressure effect on the supercrystal formation with gold nanoparticles (see Table 2).

Table 2. Studies for the supercrystal formation of metallic nanoparticles stimulated by high pressure.

Samples	Study	References
Au spherical nanoparticles	Supercrystal formation	[82]
Au spherical nanoparticles	Kinetics of nanocrystal superlattice formation	[83]
Au nanorods	Supercrystal formation	[84]

Schroer et al. investigated the pressure effect on reversibility of the supercrystal formation with a gold nanoparticle suspension. The authors have used Au nanoparticles (AuNPs) functionalized with a shell of poly(ethyleneglycol) (PEG). The radius of AuNPs is around 6 nm, and two lengths of PEG were used (2 and 5 kDa), and these Au nanoparticles coated with PEG were called AuNP@PEG2k and AuNP@PEG5k, respectively. First, the authors have recorded patterns of small-angle X-ray scattering (SAXS) for Au@PEG5k in a CsCl solution of 2 M for two pressures: 1 bar and 4000 bar. For the SAXS pattern obtained for the pressure of 1 bar, they observed a strong forward scattering corresponding to a liquid state of the AuNP@PEG5k solution (see at left in Figure 5a,b). Then, for SAXS pattern recorded for the pressure of 4000 bar, Debye–Scherrer rings were observed, indicating the formation of supercrystals under the form of a face-centered cubic (fcc) superlattice (see at middle in Figure 5a,b). Finally, they observed a reversibility of the state of the AuNP@PEG5k solution after the pressure reduction down to 1 bar (see at right in Figure 5a,b) [82].

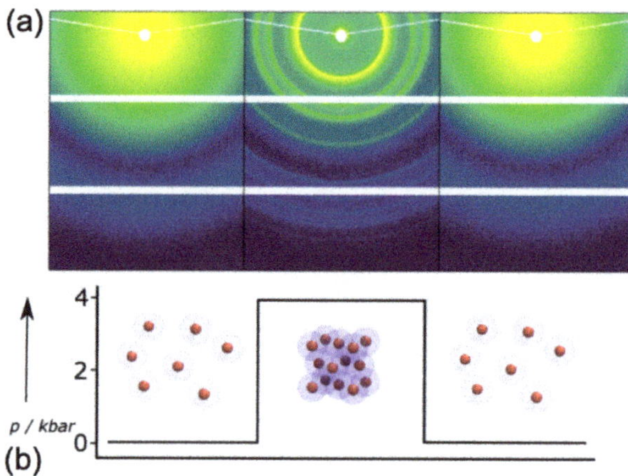

Figure 5. (**a**) Small-angle X-ray scattering (SAXS) patterns of the AuNP@PEG5k in a CsCl solution of 2 M recorded for a pressure of 1 bar (at left), 4000 bar (at middle), again 1 bar (at right). (**b**) Corresponding scheme of the structural assembly of AuNP@PEG5k: at left, liquid state; at middle, face-centered cubic crystallites; at right, return to liquid state. All of the figures are reprinted (adapted) with permission from [82], Copyright 2018 American Chemical Society.

Subsequently, the authors studied the pressure effect on the supercrystal formation with the two types of AuNPs (AuNP@PEG2k and AuNP@PEG5k) in four chloride salts (CsCl, KCl, NaCl, RbCl) at a concentration of 2 M. They remarked that the constant a of the fcc superlattice of the AuNP@PEG2k and AuNP@PEG5k had decreased when the pressure had increased (see Figure 6a,b). This decreasing was dependent on the cation of the chloride salt solution. Moreover, this pressure effect on the constant a of the fcc lattice enabled the authors to calculate the effective compressibility κ_{eff} of the superlattice at the pressure of 4000 bar and at the fixed concentration of 2 M for each chloride salt solution. They observed higher values of κ_{eff} for the KCl solution for two types of AuNPs: 17.4×10^{-5} bar^{-1} for AuNP@PEG2k and 39.5×10^{-5} bar^{-1} for AuNP@PEG5k. They concluded that the decreasing of the lattice constant is primarily due to the compression of the PEG layer, because the Au core shape was not modified [82].

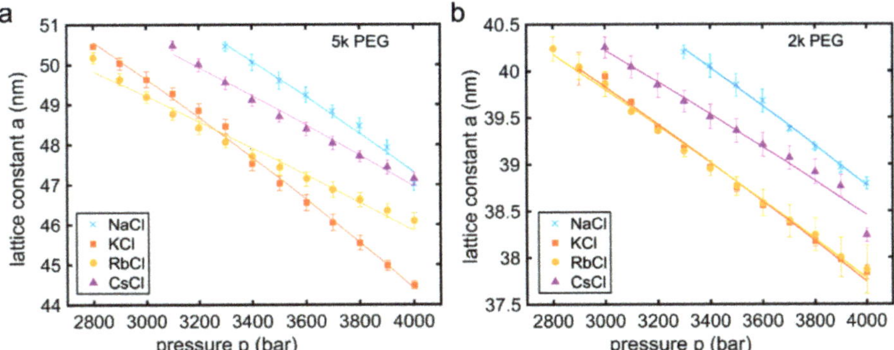

Figure 6. Pressure effect on the lattice constant *a* for (**a**) AuNP@PEG5k and (**b**) AuNP@PEG2k in each chloride salt solution of 2 M (blue crosses for NaCl, red squares for KCl, orange disks for RbCl, and purple triangles for CsCl). All of the figures are reprinted (adapted) with permission from [82], Copyright 2018 American Chemical Society.

In the same research group, Lehmkühler et al. studied the kinetics of the supercrystal formation induced by pressure [83]. The authors have taken the same radius of 6 nm than previously for the gold spherical nanoparticles (AuNPs) functionalized with α-methoxypoly(ethylene glycol)-ω-(11-mercaptoundecanoate) ligands (PEGMUA). The molecular weight of PEGMUA is 5000 g.mol^{-1}. These PEGMUA-coated AuNPs were disseminated in an 2 M chloride salt solution (RbCl). Authors observed that the time of the supercrystal formation has decreased when the jump from initial pressure (below the crystallisation pressure) to final pressure (beyond the crystallisation pressure) was more important. The time scale of this supercrystal formation has varied from 25 s to 0.3 s with the increasing of the pressure jump. This effect is linked to an improvement of the crystal quality caused by a larger speed of supercrystal formation [83].

Finally, Schroer et al. (same research group) also reported on the supercrystal formation of Au nanorods (AuNRs) stimulated by high pressure. The Au nanorods were functionalized with α-methoxypoly(ethylene glycol)-ω-(11-mercaptoundecanoate) ligands (PEGMUA2k). The dimensions of AuNRs (see Figure 7a) were 75 nm for the length and 22 nm for the width, and PEGMUA2k had a molecular weight of 2000 g·mol^{-1}. First, authors have recorded SAXS patterns for AuNR@PEGMUA2k in a RbCl solution of 2 M for two pressures: 1 bar and 4000 bar. For SAXS pattern that was obtained for the pressure of 1 bar, they observed a same behavior than in the case of Au spherical nanoparticles seen previously, i.e., the AuNR@PEGMUA2k solution was in a liquid state (see Figure 7b). For the second SAXS pattern at the pressure of 4000 bar, Debye–Scherrer rings were distinguished showing the formation of supercrystals under the form of a 2D hexagonal superlattice (see Figure 7b). They also observed that the formation was very fast (a few seconds) and also reversible [84]. The authors also calculated the effective compressibilities κ_{eff} from the dependence of the interparticle distance to the pressure (see Figure 7c). They found $\kappa_{eff,liquid} = 10.6 \times 10^{-5}$ bar^{-1} and $\kappa_{eff,supercrystal} = 6.8 \times 10^{-5}$ bar^{-1}.

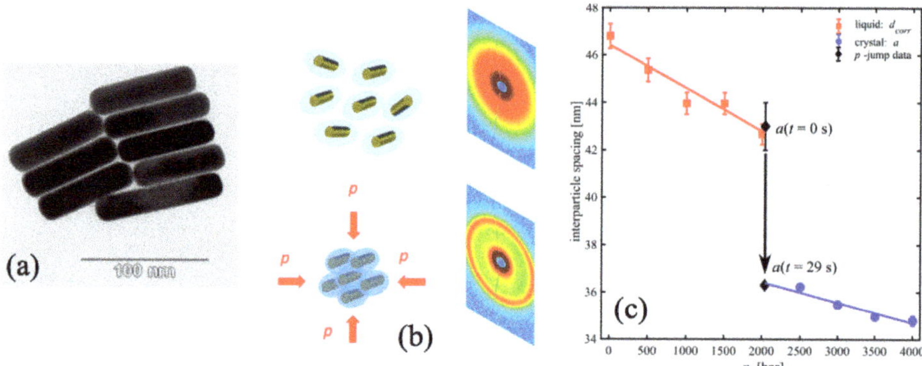

Figure 7. (a) TEM picture of an AuNR@PEGMUA2k assembly. (b) SAXS patterns of the AuNR@PEGMUA2k in a RbCl solution of 2 M recorded for a pressure of 1 bar (at **top**), 4000 bar (at **bottom**) with the corresponding scheme of the structural assembly (liquid state and supercrystal with a two-dimensional (2D) hexagonal superlattice, respectively). (c) Interparticle distance for AuNR@PEGMUA2k versus pressure. The red and blue data correspond to the liquid state and supercrystal formation, respectively. The black data correspond to the switching from liquid state to supercrystals. All the figures are reprinted (adapted) with permission from [84], Copyright 2019 American Chemical Society.

To finish this section, the same authors demonstrated that, by changing the shape of gold nanoparticles, they obtained supercrystals under the form of a different superlattice as a face-centered cubic lattice with the spherical Au nanoparticles [82], and a 2D hexagonal lattice with the Au nanorods [84].

2.3. Detection of Molecules and Phase Transitions in High Pressure Environment

In this section, we present studies regarding the detection of molecules and phase transitions in high pressure environment (see Table 3).

Table 3. Detection of molecules and phase transitions in high pressure environment (RI = Refractive Index, AuNPs = Gold nanoparticles; MoS$_2$ NFs = Molybdenum disulphide nanoflowers).

Samples	Detection	References
Au spherical nanoparticles	RI of Methanol-ethanol mixture	[85]
Au nanorods	Phases of water	[86]
Au nanorods	Phases of water and urea	[87]
AuNPs/MoS$_2$ NFs	Rhodamine 6G	[88]

Martin-Sanchez et al. demonstrated the detection of the refractive index of a methanol-ethanol (4:1) mixture in high pressure environment with gold spherical nanoparticles (AuNPs) of 20-nm diameter by following their LSPR shift [85]. First, the authors observed the LSPR shift of AuNPs in the methanol-ethanol (4:1) solution as a function of pressure (see Figure 8a–c). In the hydrostatic regime (from 0 to 10 GPa), they recorded a redshift of the AuNP LSPR due to a larger compressibility of solvent when compared to this of gold. In the non-hydrostatic regime (from 10 to 60 GPa), a blueshift of the AuNP LSPR was observed caused by the plasmon compression, which is more important than this of solvent in this case.

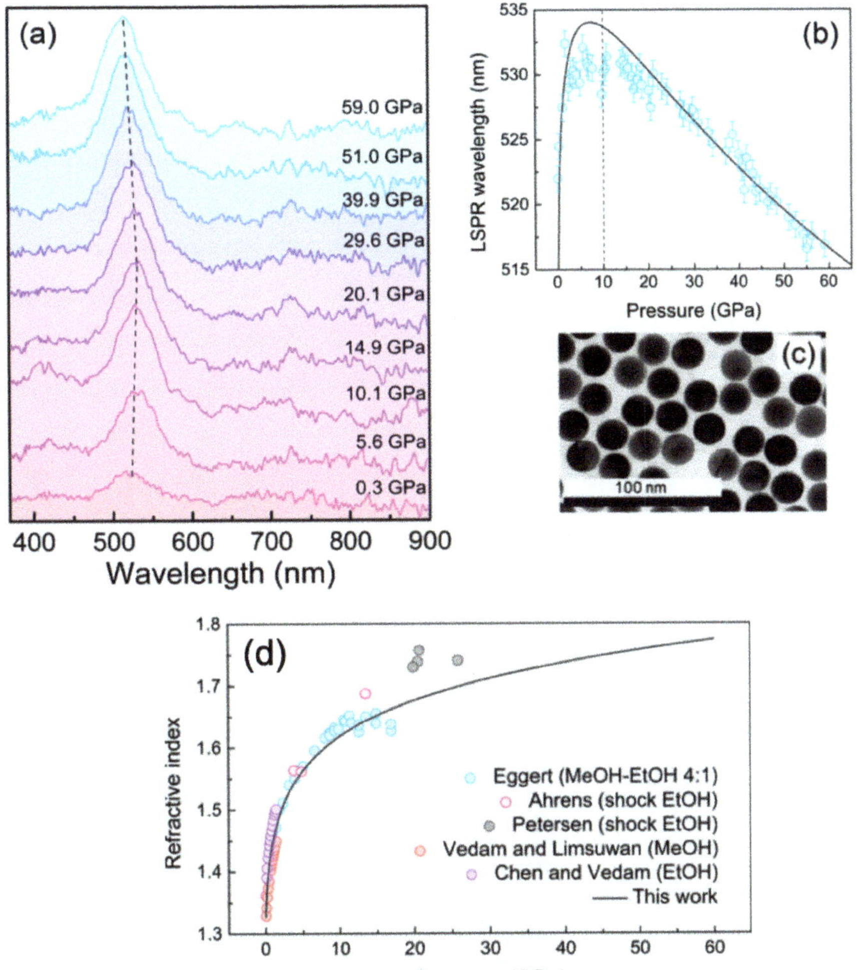

Figure 8. (**a**) Extinction spectra of AuNPs in the methanol-ethanol (4:1) solution at different pressures. (**b**) LSPR wavelength versus pressure. The dashed line represents the limit between the hydrostatic and non-hydrostatic regimes. (**c**) TEM picture of Au spherical nanoparticles with a diameter of 20 nm. (**d**) Refractive index of the methanol-ethanol (4:1) solution versus pressure. All of the colored points correspond to data found in the literature. The gray line corresponds to experimental RI measurements obtained with the expression (3). All the figures are reprinted (adapted) with permission from [85], Copyright 2020 American Chemical Society.

Subsequently, they studied the variations of the refractive index (RI) of the methanol-ethanol (4:1) solution with pressure. The authors described these RI variations with pressure by using the expression of Murnaghan type:

$$n = n_0 \left(\frac{P\alpha}{\beta} + 1 \right)^{1/\alpha} \tag{3}$$

where n_0 corresponds to the RI of the methanol-ethanol (4:1) solution taken at ambient pressure ($P = 0$). α and β correspond to parameters of fit. These parameters were obtained by fitting the expression (1) of the LSPR wavelength at the pressure P by employing the expressions (2) and (3) in order to depict the variations of electron density of gold and dielectric function of solvent, respectively. They obtained $\alpha = 19.3$ and $\beta = 4.3$ in the whole range of pressure (0–60 GPa) by taking $K_0 = 190$ GPa and $K_0' = 6$ (see Section 2.1 and references [85,86]). Afterwards, the authors compared their RI values as function of pressure to the literature [89–93], and these latter were generally in good agreement with this literature (see Figure 8d). Furthermore, the authors of this same research group investigated the detection of the refractive index of water in its liquid, ice VI, and ice VII phases by measuring the LSPR shift of aqueous solutions of Au nanorods at different high pressures [86] (see Figure 9). The dimensions of Au nanorods were 45.7 nm for the length and 13.4 nm for the width. By using the expressions (1)–(3), the RI values for each water state were calculated from the LSPR wavelength of Au nanorods. For the liquid phase ($P = 0$–1.8 GPa), the authors have taken $n_0 = 1.33$, $\alpha = 26$, and $\beta = 6$. For the ice VI phase ($P = 1.5$–2.2 GPa), the values of n_0, α, and β were equal to 1.40, 34, and 14, respectively. For the ice VII phase ($P = 2.2$–9 GPa), these values of n_0, α, and β were equal to 1.43, 13.7, and 30, respectively. For all of the water states, α and β were obtained with the same method as described previously (with $K_0 = 190$ GPa and $K_0' = 6$), and n_0 corresponds to the RI of each water state at ambient pressure. Subsequently, these RI values of each water state were compared to the literature [92,94–96], and a good agreement between them was obtained [86].

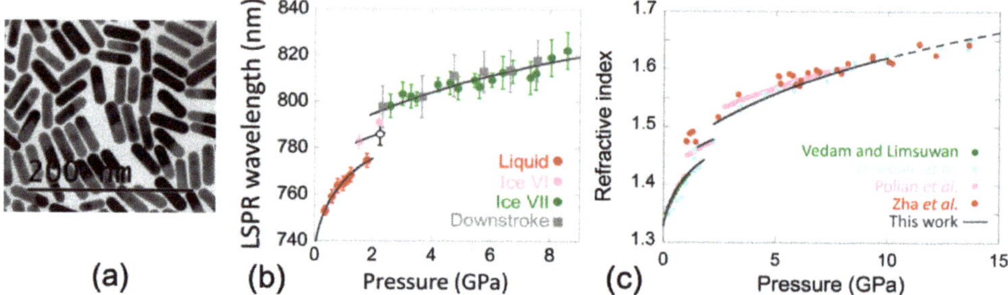

Figure 9. (**a**) TEM picture of Au nanorods employed for experiments (scale bar = 200 nm). (**b**) LSPR wavelength versus pressure. The red, pink and green points correspond to experimental values for the liquid state, ice VI phase, and ice VII phase, respectively. The gray line represents the values determined with the Mie–Gans model. (**c**) Refractive index of the water versus pressure. All of the colored points correspond to data referenced in the literature. The gray line corresponds to experimental RI measurements obtained with the expression (3), and the dashed gray line corresponds to the values extrapolated with this model. All of the figures are reprinted (adapted) with permission from [86], Copyright 2019 American Chemical Society.

In addition, Runowski et al. demonstrated the detection of phase transitions of different media by measuring the LSPR shift of gold nanorods with pressure [87]. The dimensions of Au nanorods were 100 nm for the length and 40 nm for the width. Authors have found the phase transition from liquid water to ice VI for a pressure of 1 GPa due to an abrupt redshift of the LSPR of both longitudinal and transversal modes of Au nanorods caused by a large jump of the refractive index of water (from liquid state to ice VI). Subsequently, they observed the transition from ice VI to ice VII at a pressure of 2.2 GPa due to a short blueshift for both plasmonic modes caused by the presence of both ice VI and ice VII. Then, authors investigated the detection phase transitions of urea [87]. A transition pressure was observed at 0.5 GPa for the phase from urea I to urea III measured by an abrupt redshift in the LSPR of Au nanorods due to the crystal lattice change from the tetragonal structure (phase I) to the orthorhombic one (phase III) corresponding to a significant deformation between these phases [97]. A second transition pressure was observed at 2.8 GPa for the transition from urea III to urea IV characterized by a smaller redshift of the Au nanorod LSPR due to the crystal lattice change from the orthorhombic phase III to the orthorhombic phase IV corresponding to a weaker deformation between these phases [97].

In the work of Sun et al., the authors reported on the surface enhanced Raman scattering (SERS) enhancement induced by high pressure with semiconducting nanoflowers (MoS_2 NFs) decorated with gold nanoparticles (AuNPs) [88]. The diameters of the MoS_2 NFs and AuNPs were 700 nm and 10 nm, respectively (see Figure 10a). They used the rhodamine 6G (R6G) molecules as SERS probe in the experimental measurements. The authors recorded the SERS spectra of R6G molecules on AuNPs/Mo_2 NFs at the excitation wavelength of 532 nm for pressures varying from 0 to 8.38 GPa. The highest enhancement of the SERS signal was recorded for the pressure of 2.39 GPa (see Figure 10b). At this pressure, the SERS enhancement was due to a better alignment of the energy levels between MoS_2 NFs, Au, and R6G molecules. This better alignment was obtained by the reduction of the energy of the band gap of MoS_2 NFs, and gap of the highest occupied molecular orbital (HOMO) and lowest unoccupied molecular orbital (LUMO) levels for R6G molecules, when the applied pressure increased. The Fermi energy level of gold was kept almost constant when the pressure increased [88]. Thus, at this pressure of 2.39 GPa, two charge transfers (CTs) have occurred for enhancing the SERS signal (see Figure 10c).

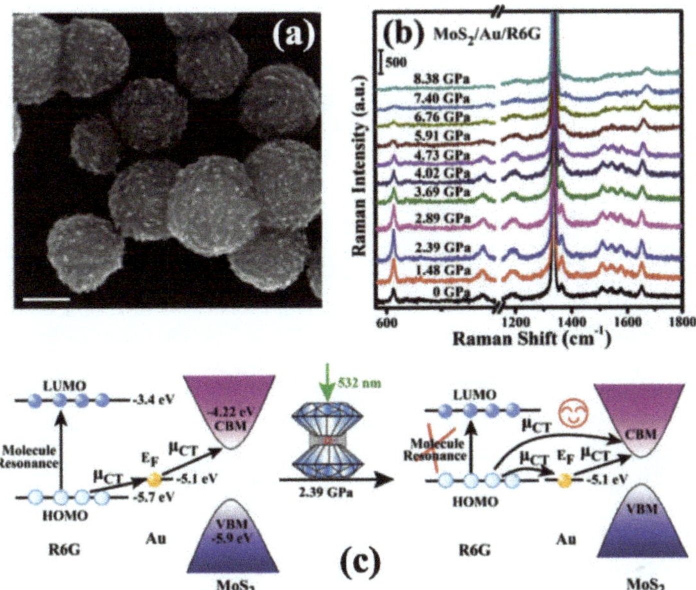

Figure 10. (a) SEM picture of AuNPs/MoS_2 NFs (scale bar = 500 nm). (b) SERS spectra of MoS_2/Au/R6G system at different pressures. (c) Charge transfer mechanism for the SERS enhancement induced pressure. All of the figures are reproduced from [88] with permission from the Royal Society of Chemistry.

The first CT mechanism was an electron transfer from the HOMO level of R6G to the conduction band minimum (CBM) of MoS_2 NFs. Subsequently, the second one was a two-step mechanism: (i) electron transfer from the HOMO level of R6G to the Fermi energy level of Au and (ii) electron transfer from the Fermi energy level of Au to CBM of MoS_2 NFs. In this study, the band gap energy of MoS_2 NFs was determined by using the following expression [98]:

$$E_g = 1.68 - 0.07P + 0.00113P^2 \tag{4}$$

where E_g is the band gap energy of MoS_2 NFs (in eV), and P is the pressure (in GPa). Thus, the E_g value for MoS_2 NFs was calculated at the pressure of 2.39 GPa, and this latter was equal to 1.51 eV, which was weaker than its value at ambient pressure (1.68 eV) proposing that CBM and valence band maximum (VBM) levels of MoS_2 NFs became smaller. Furthermore, the SERS mechanism at ambient pressure was a two-step mechanism: (i) electron transfer from the HOMO level of R6G to the Fermi energy level of gold enabled by the excitation laser wavelength (532 nm) and (ii) transfer of hot electrons produced by the plasmon resonance of AuNPs to the CBM of MoS_2 NFs (see Figure 10c). Finally, for the pressures superior to 2.39 GPa, the SERS intensity decreased due to the fact that the HOMO level of R6G molecules has exceeded the Fermi energy level of gold [88].

3. Future Directions

The nanoplasmonics in high pressure environment has generally been studied for gold. It would be interesting to apply this type of studies to other plasmonic materials that are well-known, such as silver, copper, palladium, and aluminum [39,40,43] in order to have the influence of the nature of the plasmonic material on the LSPR shift in high pressure environment. We can extend this investigation type to other alternative plasmonic materials, such as transition-metal nitride nanoparticles [44], transparent conductive oxides [46], and iron carbide nanoparticles encapsulated by graphene [99], which are materials at lower costs having a better temperature stability. Moreover, the domain of the nanoplasmonics in high pressure environment can be applied to sensing of analytes, pollutants in high pressure media as the marine medium, for instance. Another future direction of this domain is the SERS effect induced by high pressure, which represents a novel frontier in the SERS field.

4. Conclusions

In this review, we discussed the emerging topic of the nanoplasmonics in high pressure environment. First, we spoke about the effect of high pressure on the localized surface plasmon resonance and the optical density of gold nanoparticles with different shapes and sizes. LSPR shifts were observed for gold nanoparticles and were dependent on the shape, size, pressure regime (hydrostatic and non-hydrostatic), and surrounding medium. Subsequently, we presented studies on the supercrystal formation in a high pressure environment. The supercrystal formation was dependent on the shape of gold nanoparticles influencing the form of the crystal superlattice. The phenomenon of the supercrystal formation was reversible and very fast (a few seconds). To finish, we reported on the detection of phase transitions and refractive index variations for different liquids as water and urea, by measuring the LSPR shift of gold nanoparticles in these liquids. Moreover, we also reported on the detection of chemical molecules by using the SERS effect that was induced by high pressure. In summary, the nanoplasmonics in high pressure environment can be very useful for obtaining structural information on solvents or studying optical, thermodynamic properties of several liquids (organic and inorganic) or solids.

Funding: This research received no external funding.

Conflicts of Interest: The author declares no conflict of interest.

References

1. Moakhar, R.S.; Gholipour, S.; Masudy-Panah, S.; Seza, A.; Mehdikhani, A.; Riahi-Noori, N.; Tafazoli, S.; Timasi, N.; Lim, Y.F.; Saliba, M. Recent Advances in Plasmonic Perovskite Solar Cells. *Adv. Sci.* **2020**, *30*, 1908408.
2. Qin, P.; Wu, T.; Wang, Z.; Xiao, L.; Ma, L.; Ye, F.; Xiong, L.; Chen, X.; Li, H.; Yu, X.; et al. Grain Boundary and Interface Passivation with Core–Shell Au@CdS Nanospheres for High-Efficiency Perovskite Solar Cells. *Adv. Funct. Mater.* **2020**, *30*, 1908408.

3. Bi, W.B.; Wu, Y.J.; Chen, C.; Zhou, D.L.; Song, Z.L.; Li, D.Y.; Chen, G.Y.; Dai, Q.L.; Zhu, Y.S.; Song, H.W. Dye Sensitization and Local Surface Plasmon Resonance-Enhanced Upconversion Luminescence for Efficient Perovskite Solar Cells. *ACS App. Mater. Interfaces* **2020**, *12*, 24737–24746.

4. Li, H.; Zhou, S.J.; Yin, L.W. Surface Plasmon Resonance Effect Enhanced $CsPbBr_3$ Inverse Opals for High-Performance Inorganic Perovskite Solar Cells. *Adv. Mater. Interfaces* **2020**, *7*, 1901885.

5. Kaur, N.; Bhullar, V.; Singh, D.P. Bimetallic Implanted Plasmonic Photoanodes for TiO_2 Sensitized Third Generation Solar Cells. *Sci. Rep.* **2020**, *10*, 7657.

6. Ho, W.J.; Yang, H.Y.; Liu, J.J.; Lin, P.J.; Ho, C.H. Plasmonic effects of two-dimensional indium-nanoparticles embedded within SiO_2 anti-reflective coating on the performance of silicon solar cells. *Appl. Surf. Sci.* **2020**, *508*, 145275.

7. Ono, M.; Hata, M.; Tsunekawa, M.; Nozaki, K.; Sumikura, H.; Chiba, H.; Notomi, M. Ultrafast and energy-efficient all-optical switching with graphene-loaded deep-subwavelength plasmonic waveguides. *Nat. Photonics* **2020**, *14*, 37–43.

8. Winkler, J.M.; Ruckriegel, M.J.; Rojo, H.; Keitel, R.C.; De Leo, E.; Rabouw, F.T.; Norris, D.J. Dual-Wavelength Lasing in Quantum-Dot Plasmonic Lattice Lasers. *ACS Nano* **2020**, *14*, 5223–5232.

9. Sergent, S.; Takiguchi, M.; Tsuchizawa, T.; Taniyama, H.; Notomi, M. Low-Threshold Lasing up to 360 K in All-Dielectric Subwavelength-Nanowire Nanocavities. *ACS Photonics* **2020**, *7*, 1104–1110.

10. Guo, X.D.; Liu, R.N.; Hu, D.B.; Hu, H.; Wei, Z.; Wang, R.; Dai, Y.Y.; Cheng, Y.; Chen, K.; Liu, K.H.; et al. Efficient All-Optical Plasmonic Modulators with Atomically Thin Van Der Waals Heterostructures. *Adv. Mater.* **2020**, *32*, 1907105.

11. Zhou, X.B.; Jiang, M.M.; Wu, Y.T.; Ma, K.J.; Liu, Y.; Wan, P.; Kan, C.X.; Shi, D.N. Hybrid quadrupole plasmon induced spectrally pure ultraviolet emission from a single AgNPs@ZnO:Ga microwire based heterojunction diode. *Nanoscale Adv.* **2020**, *2*, 1340–1351.

12. Barbillon, G.; Ivanov, A.; Sarychev, A.K. Applications of Symmetry Breaking in Plasmonics. *Symmetry* **2020**, *12*, 896.

13. Li, Z.; Jin, J.; Yang, F.; Song, N.; Yin, Y. Coupling magnetic and plasmonic anisotropy in hybrid nanorods for mechanochromic responses. *Nat. Commun.* **2020**, *11*, 2883.

14. Ahmadivand, A.; Gerislioglu, B.; Ramezani, Z. Generation of magnetoelectric photocurrents using toroidal resonances: A new class of infrared plasmonic photodetectors. *Nanoscale* **2019**, *11*, 13108–13116.

15. Tomitaka, A.; Arami, H.; Ahmadivand, A.; Pala, N.; McGoron, A.J.; Takemura, Y.; Febo, M.; Nair, M. Magneto-plasmonic nanostars for image-guided and NIR-triggered drug delivery. *Sci. Rep.* **2020**, *10*, 10115.

16. Qiu, G.G.; Gai, Z.B.; Tao, Y.L.; Schmitt, J.; Kullak-Ublick, G.A.; Wang, J. Dual-Functional Plasmonic Photothermal Biosensors for Highly Accurate Severe Acute Respiratory Syndrome Coronavirus 2 Detection. *ACS Nano* **2020**, *14*, 5268–5277.

17. Portella, A.; Calvo-Lozano, O.; Estevez, M.C.; Escuela, A.M.; Lechuga, L.M. Optical nanogap antennas as plasmonic biosensors for the detection of miRNA biomarkers. *J. Mater. Chem. B* **2020**, *8*, 4310–4317.

18. Moakhar, R.S.; AbdelFatah, T.; Sanati, A.; Jalali, M.; Flynn, S.E.; Mahshid, S.S.; Mahshid, S. A Nanostructured Gold/Graphene Microfluidic Device for Direct and Plasmonic-Assisted Impedimetric Detection of Bacteria. *ACS App. Mater. Interfaces* **2020**, *12*, 23298–23310.

19. Huang, J.A.; Mousavi, M.Z.; Zhao, Y.Q.; Hubarevich, A.; Omeis, F.; Giovannini, G.; Schutte, M.; Garoli, D.; De Angelis, F. SERS discrimination of single DNA bases in single oligonucleotides by electro-plasmonic trapping. *Nat. Commun.* **2019**, *10*, 5321.

20. Dolci, M.; Bryche, J.-F.; Moreau, J.; Leuvrey, C.; Begin-Colin, S.; Barbillon, G.; Pichon, B.P. Investigation of the structure of iron oxide nanoparticle assemblies in order to optimize the sensitivity of surface plasmon resonance-based sensors. *Appl. Surf. Sci.* **2020**, *527*, 146773.

21. Gutierrez, Y.; Giangregorio, M.M.; Palumbo, F.; Gonzalez, F.; Brown, A.S.; Moreno, F.; Losurdo, M. Sustainable and Tunable Mg/MgO Plasmon-Catalytic Platform for the Grand Challenge of SF_6 Environmental Remediation. *Nano Lett.* **2020**, *20*, 3352–3360.

22. Wang, Y.L.; Fang, L.L.; Gong, M.; Deng, Z.X. Chemically modified nanofoci unifying plasmonics and catalysis. *Chem. Sci.* **2019**, *10*, 5929–5934.

23. Alekseeva, S.; Nedrygailov, I.I.; Langhammer, C. Single Particle Plasmonics for Materials Science and Single Particle Catalysis. *ACS Photonics* **2019**, *6*, 1319–1330.

24. Saboktakin, M.; Ye, X.; Chettiar, U.K.; Engheta, N.; Murray, C.B.; Kagan, C.R. Plasmonic Enhancement of Nanophosphor Upconversion Luminescence in Au Nanohole Arrays. *ACS Nano* **2013**, *7*, 7186–7192.

25. Park, W.; Lu, D.; Ahn, S. Plasmon enhancement of luminescence upconversion. *Chem. Soc. Rev.* **2015**, *44*, 2940–2962.

26. Wang, Y.; Ding, T. Optical tuning of plasmon-enhanced photoluminescence. *Nanoscale* **2019**, *11*, 10589–10594.

27. Runowski, M.; Stopikowska, N.; Goderski, S.; Lis, S. Luminescent-plasmonic, lanthanide-doped core/shell nanomaterials modified with Au nanorods—Up-conversion luminescence tuning and morphology transformation after NIR laser irradiation. *J. Alloys Compd.* **2018**, *762*, 621–630.

28. Zhong, J.H.; Vogelsang, J.; Yi, J.M.; Wang, D.; Wittenbecher, L.; Mikaelsson, S.; Korte, A.; Chimeh, A.; Arnold, C.L.; Schaaf, P.; et al. Nonlinear plasmon-exciton coupling enhances sum-frequency generation from a hybrid metal/semiconductor nanostructure. *Nat. Commun.* **2020**, *11*, 1464.

29. Shen, Q.X.; Jin, W.L.; Yang, G.; Rodriguez, A.W.; Mikkelsen, M.H. Active Control of Multiple, Simultaneous Nonlinear Optical Processes in Plasmonic Nanogap Cavities. *ACS Photonics* **2020**, *7*, 901–907.

30. Gao, M.; He, Y.H.; Chen, Y.; Shih, T.M.; Yang, W.M.; Chen, H.Y.; Yang, Z.L.; Wang, Z.H. Enhanced sum frequency generation for ultrasensitive characterization of plasmonic modes. *Nanophotonics* **2020**, *9*, 815–822.

31. Dalstein, L.; Humbert, C.; Ben Haddada, M.; Boujday, S.; Barbillon, G.; Busson, B. The Prevailing Role of Hotspots in Plasmon-Enhanced Sum-Frequency Generation Spectroscopy. *J. Phys. Chem. Lett.* **2019**, *10*, 7706–7711.

32. Barbillon, G.; Noblet, T.; Busson, B.; Tadjeddine, A.; Humbert, C. Localised detection of thiophenol with gold nanotriangles highly structured as honeycombs by nonlinear sum frequency generation spectroscopy. *J. Mater. Sci.* **2018**, *53*, 4554–4562.

33. Baibakov, M.; Patra, S.; Claude, J.-B.; Moreau, A.; Lumeau, J.; Wenger, J. Extending Single-Molecule Förster Resonance Energy Transfer (FRET) Range beyond 10 Nanometers in Zero-Mode Waveguides. *ACS Nano* **2019**, *13*, 8469–8480.

34. Petreto, A.; Dos Santos, M.C.; Lefebvre, O.; Dos Santos, G.R.; Ponzellini, P.; Garoli, D.; De Angelis, F.; Ammar, M.; Hildebrandt, N. Optimizing FRET on Aluminum Surfaces via Controlled Attachment of Fluorescent Dyes. *ACS Omega* **2018**, *3*, 18867–18876.

35. Rotho, D.J.; Nasir, M.E.; Ginzburg, P.; Wang, P.; Le Marois, A.; Suhling, K.; Richards, D.; Zayats, A.V. Forster Resonance Energy Transfer inside Hyperbolic Metamaterials. *ACS Photonics* **2018**, *5*, 4594–4603.

36. de Torres, J.; Mivelle, M.; Moparthi, S.B.; Rigneault, H.; Van Hulst, N.F.; Garcia-Parajo, M.F.; Margeat, E.; Wenger, J. Plasmonic Nanoantennas Enable Forbidden Forster Dipole-Dipole Energy Transfer and Enhance the FRET Efficiency. *Nano Lett.* **2016**, *163*, 6222–6230.

37. Bidault, S.; Devilez, A.; Ghenuche, P.; Stout, B.; Bonod, N.; Wenger, J. Competition between Forster Resonance Energy Transfer and Donor Photodynamics in Plasmonic Dimer Nanoantennas. *ACS Photonics* **2016**, *3*, 895–903.

38. Castro-Lopez, M.; Brinks, D.; Sapienza, R.; van Hulst, N.F. Aluminum for Nonlinear Plasmonics: Resonance-Driven Polarized Luminescence of Al, Ag, and Au Nanoantennas. *Nano Lett.* **2011**, *11*, 4674–4678.

39. Ostovar, B.; Su, M.-N.; Renard, D.; Clark, B.D.; Dongare, P.D.; Dutta, C.; Gross, N.; Sader, J.E.; Landes, C.F.; Chang, W.-S.; et al. Acoustic Vibrations of Al Nanocrystals: Size, Shape, and Crystallinity Revealed by Single-Particle Transient Extinction Spectroscopy. *J. Phys. Chem. A* **2020**, *124*, 3924–3934.

40. Ma, R.; Wu, D.; Liu, Y.M.; Ye, H.; Sutherland, D. Copper plasmonic metamaterial glazing for directional thermal energy management. *Mater. Des.* **2020**, *188*, 108407.

41. Bohme, A.; Sterl, F.; Kath, E.; Ubl, M.; Manninen, V.; Giessen, H. Electrochemistry on Inverse Copper Nanoantennas: Active Plasmonic Devices with Extraordinarily Large Resonance Shift. *ACS Photonics* **2019**, *6*, 1863–1868.

42. Sugawa, K.; Tahara, H.; Yamashita, A.; Otsuki, J.; Sagara, T.; Harumoto, T.; Yanagida, S. Refractive Index Susceptibility of the Plasmonic Palladium Nanoparticle: Potential as the Third Plasmonic Sensing Material. *ACS Nano* **2015**, *9*, 1895–1904.

43. Sterl, F.; Strohfeldt, N.; Both, S.; Herkert, E.; Weiss, T.; Giessen, H. Design Principles for Sensitivity Optimization in Plasmonic Hydrogen Sensors. *ACS Sens.* **2020**, *5*, 917–927.

44. Karaballi, R.A.; Monfared, Y.E.; Dasog, M. Overview of Synthetic Methods to Prepare Plasmonic Transition-Metal Nitride Nanoparticles. *Chem. Eur. J.* **2020**, *26*, 8499–8505.

45. Askes, S.H.C.; Schilder, N.J.; Zoethout, E.; Polman, A.; Garnett, E.C. Tunable plasmonic HfN nanoparticles and arrays. *Nanoscale* **2019**, *11*, 20252–20260.

46. Noginov, M.A.; Gu, L.; Livenere, J.; Zhu, G.; Pradhan, A.K.; Mundle, R.; Bahoura, M.; Barnakov, Y.A.; Podolskiy, V.A. Transparent conductive oxides: Plasmonic materials for telecom wavelengths. *Appl. Phys. Lett.* **2011**, *99*, 021101.

47. Boltasseva, A.; Atwater, H.A. Low-Loss Plasmonic Metamaterials. *Science* **2011**, *301*, 290–291.

48. Ding, S.-Y.; You, E.-M.; Tian, Z.-Q.; Moskovits, M. Electromagnetic theories of surface-enhanced Raman spectroscopy. *Chem. Soc. Rev.* **2017**, *46*, 4042–4076.

49. Reguera, J.; Langer, J.; Jimenez de Aberasturi, D.; Liz-Marzan, L.M. Anistropic metal nanoparticles for surface enhanced Raman scattering. *Chem. Soc. Rev.* **2017**, *46*, 3866–3885.

50. Bryche, J.-F.; Bélier, B.; Bartenlian, B.; Barbillon, G. Low-cost SERS substrates composed of hybrid nanoskittles for a highly sensitive sensing of chemical molecules. *Sens. Actuators B* **2017**, *239*, 795–799.

51. Barbillon, G. Latest Novelties on Plasmonic and Non-Plasmonic Nanomaterials for SERS Sensing. *Nanomaterials* **2020**, *10*, 1200.

52. Sheena, T.S.; Devaraj, V.; Lee, J.-M.; Balaji, P.; Gnanasekar, P.; Oh, J.-W.; Akbarsha, M.A.; Jeganathan, K. Sensitive and label-free shell isolated Ag NPs@Si architecture based SERS active substrate: FDTD analysis and in-situ cellular DNA detection. *Appl. Surf. Sci.* **2020**, *515*, 145955.

53. Zhang, D.J.; Peng, L.Q.; Shang, X.L.; Zheng, W.X.; You, H.J.; Xu, T.; Ma, B.; Ren, B.; Fang, J.X. Buoyant particulate strategy for few-to-single particle-based plasmonic enhanced nanosensors. *Nat. Commun.* **2020**, *11*, 2603.

54. Zhu, S.; Li, H.; Yang, M.; Pang, S.W. Highly sensitive detection of exosomes by 3D plasmonic photonic crystal biosensor. *Nanoscale* **2018**, *10*, 19927–19936.

55. Cai, J.X.; Zhang, C.P.; Liang, C.W.; Min, S.Y.; Cheng, X.; Li, W.D. Solution-Processed Large-Area Gold Nanocheckerboard Metasurfaces on Flexible Plastics for Plasmonic Biomolecular Sensing. *Adv. Opt. Mater.* **2019**, *7*, 1900516.

56. Kawasaki, D.; Yamada, H.; Maeno, K.; Sueyoshi, K.; Hisamoto, H.; Endo, T. Core-Shell-Structured Gold Nanocone Array for Label-Free DNA Sensing. *ACS Appl. Nano Mater.* **2019**, *2*, 4983–4990.

57. Zhang, C.; Paria, D.; Semancik, S.; Barman, I. Composite-Scattering Plasmonic Nanoprobes for Label-Free, Quantitative Biomolecular Sensing. *Small* **2019**, *15*, 1901165.

58. Park, J.; Ndao, A.; Cai, W.; Hsu, L.; Kodigala, A.; Lepetit, T.; Lo, Y.; Kanté, B. Symmetry-breaking-induced plasmonic exceptional points and nanoscale sensing. *Nat. Phys.* **2020**, *16*, 462–468.

59. Dhawan, A.; Duval, A.; Nakkach, M.; Barbillon, G.; Moreau, J.; Canva, M.; Vo-Dinh, T. Deep UV nano-microstructuring of substrates for surface plasmon resonance imaging. *Nanotechnology* **2011**, *22*, 165301.

60. Barbillon, G.; Ivanov, A.; Sarychev, A.K. Hybrid Au/Si Disk-Shaped Nanoresonators on Gold Film for Amplified SERS Chemical Sensing. *Nanomaterials* **2019**, *9*, 1588.

61. Manfrinato, V.R.; Camino, F.E.; Stein, A.; Zhang, L.H.; Lu, M.; Stach, E.A.; Black, C.T. Patterning Si at the 1 nm Length Scale with Aberration-Corrected Electron-Beam Lithography: Tuning of Plasmonic Properties by Design. *Adv. Funct. Mater.* **2019**, *29*, 1903429.

62. Quilis, N.G.; Hageneder, S.; Fossati, S.; Auer, S.K.; Venugopalan, P.; Bozdogan, A.; Petri, C.; Moreno-Cencerrado, A.; Toca-Herrera, J.L.; Jonas, U.; et al. UV-Laser Interference Lithography for Local Functionalization of Plasmonic Nanostructures with Responsive Hydrogel. *J. Phys. Chem. C* **2020**, *124*, 3297–3305.

63. Yang, L.T.; Lee, J.H.; Rathnam, C.; Hou, Y.N.; Choi, J.W.; Lee, K.B. Dual-Enhanced Raman Scattering-Based Characterization of Stem Cell Differentiation Using Graphene-Plasmonic Hybrid Nanoarray. *Nano Lett.* **2019**, *19*, 8138–8148.

64. Chau, Y.F.C.; Chen, K.H.; Chiang, H.P.; Lim, C.M.; Huang, H.J.; Lai, C.H.; Kumara, N.T.R.N. Fabrication and Characterization of a Metallic-Dielectric Nanorod Array by Nanosphere Lithography for Plasmonic Sensing Applications. *Nanomaterials* **2019**, *9*, 1691.

65. Goetz, S.; Bauch, M.; Dimopoulos, T.; Trassi, S. Ultrathin sputter-deposited plasmonic silver nanostructures. *Nanoscale Adv.* **2020**, *2*, 869–877.

66. Driencourt, L.; Federspiel, F.; Kazazis, D.; Tseng, L.T.; Frantz, R.; Ekinci, Y.; Ferrini, R.; Gallinet, B. Electrically Tunable Multicolored Filter Using Birefringent Plasmonic Resonators and Liquid Crystals. *ACS Photonics* **2020**, *7*, 444–453.

67. Maier, S.A. *Plasmonics: Fundamentals and Applications*; Springer: New York, NY, USA, 2007; pp. 3–220.

68. Enoch, S.; Bonod, N. *Plasmonics: From Basics to Advanced Topics*; Springer: Berlin/Heidelberg, Germany, 2012; pp. 3–317.

69. Li, Z.; Butun, S.; Aydin, K. Ultranarrow Band Absorbers Based on Surface Lattice Resonances in Nanostructured Metal Surfaces. *ACS Nano* **2014**, *8*, 8242–8248.

70. Kravets, V.G.; Kabashin, A.V.; Barnes, W.L.; Grigorenko, A.N. Plasmonic Surface Lattice Resonances: A Review of Properties and Applications. *Chem. Rev.* **2018**, *118*, 5912–5951.

71. Sobhani, A.; Manjavacas, A.; Cao, Y.; McClain, M.J.; Javier Garcia de Abajo, F.; Nordlander, P.; Halas, N.J. Pronounced Linewidth Narrowing of an Aluminum Nanoparticle Plasmon Resonance by Interaction with an Aluminum Metallic Film. *Nano Lett.* **2015**, *15*, 6946–6951.

72. Yue, W.; Wang, Z.; Whittaker, J.; Lopez-Royo, F.; Yang, Y.; Zayats, A.V. Amplification of surface-enhanced Raman scattering due to substrate-mediated localized surface plasmons in gold nanodimers. *J. Mater. Chem. C* **2017**, *5*, 4075–4084.

73. Bao, Y.J.; Zhao, B.; Hou, D.J.; Liu, J.S.; Wang, F.; Wang, X.; Cui, T. The redshift of surface plasmon resonance of colloidal gold nanoparticles induced by pressure with diamond anvil cell. *J. Appl. Phys.* **2014**, *115*, 223503.

74. Li, B.S.; Wen, X.D.; Li, R.P.; Wang, Z.W.; Clem, P.G.; Fan, H.Y. Stress-induced phase transformation and optical coupling of silver nanoparticle superlattices into mechanically stable nanowires. *Nat. Commun.* **2014**, *5*, 4179.

75. Bao, Y.J.; Zhao, B.; Tang, X.Y.; Hou, D.J.; Cai, J.; Tang, S.; Liu, J.S.; Wang, F.; Cui, T. Tuning surface plasmon resonance by the plastic deformation of Au nanoparticles within a diamond anvil cell. *Appl. Phys. Lett.* **2015**, *107*, 201909.

76. Gu, X.W.; Hanson, L.A.; Eisler, C.N.; Koc, M.A.; Alivisatos, A.P. Pseudoelasticity at Large Strains in Au Nanocrystals. *Phys. Rev. Lett.* **2018**, *121*, 056102.

77. Martin-Sanchez, C.; Barreda-Argüeso, J.A.; Seibt, S.; Mulvaney, P.; Rodriguez, F. Effects of Hydrostatic Pressure on the Surface Plasmon Resonance of Gold Nanocrystals. *ACS Nano* **2019**, *13*, 498–504.

78. Jacobs, K.; Wickham, J.; Alivisatos, A.P. Threshold Size for Ambient Metastability of Rocksalt CdSe Nanocrystals. *J. Phys. Chem. B* **2002**, *106*, 3759–3762.

79. Choi, C.L.; Koski, K.J.; Sivasankar, S.; Alivisatos, A.P. Strain-Dependent Photoluminescence Behavior of CdSe/CdS Nanocrystals with Spherical, Linear, and Branched Topologies. *Nano Lett.* **2009**, *9*, 3544–3549.

80. Heinz, D.L.; Jeanloz, R. The Equation of State of the Gold Calibration Standard. *J. Appl. Phys.* **1984**, *55*, 885–893.

81. Takemura, K.; Dewaele, A. Isothermal equation of state for gold with a He-pressure medium. *Phys. Rev. B* **2008**, *78*, 104119.

82. Schroer, M.A.; Lehmkühler, F.; Möller, J.; Lange, H.; Grübel, G.; Schulz, F. Pressure-Stimulated Supercrystal Formation in Nanoparticle Suspensions. *J. Phys. Chem. Lett.* **2018**, *9*, 4720–4724.

83. Lehmkühler, F.; Schroer, M.A.; Markmann, V.; Frenzel, L.; Möller, J.; Lange, H.; Grübel, G.; Schulz, F. Kinetics of pressure-induced nanocrystal superlattice formation. *Phys. Chem. Chem. Phys.* **2019**, *21*, 21349–21354.

84. Schroer, M.A.; Lehmkühler, F.; Markmann, V.; Frenzel, L.; Möller, J.; Lange, H.; Grübel, G.; Schulz, F. Supercrystal Formation of Gold Nanorods by High Pressure Stimulation. *J. Phys. Chem. C* **2019**, *123*, 29994–30000.

85. Martin-Sanchez, C.; Sanchez-Iglesias, A.; Mulvaney, P.; Liz-Marzan, L.M.; Rodriguez, F. Plasmonic Sensing of Refractive Index and Density in Methanol–Ethanol Mixtures at High Pressure. *J. Phys. Chem. C* **2020**, *124*, 8978–8983.

86. Martin-Sanchez, C.; Gonzalez-Rubio, G.; Mulvaney, P.; Guerrero-Martinez, A.; Liz-Marzan, L.M.; Rodriguez, F. Monodisperse Gold Nanorods for High-Pressure Refractive Index Sensing. *J. Phys. Chem. Lett.* **2019**, *10*, 1587–1593.

87. Runowski, M.; Sobczak, S.; Marciniak, J.; Bukalska, I.; Lis, S.; Katrusiak, A. Gold nanorods as a high-pressure sensor of phase transitions and refractive-index gauge. *Nanoscale* **2019**, *11*, 8718–8726.

88. Sun, H.H.; Yao, M.G.; Song, Y.P.; Zhu, L.Y.; Dong, J.J.; Liu, R.; Li, P.; Zhao, B.; Liu, B.B. Pressure-induced SERS enhancement in a MoS_2/Au/R6G system by a two-step charge transfer process. *Nanoscale* **2019**, *11*, 21493–21501.

89. Eggert, J.H.; Xu, L.; Che, R.; Chen, L.; Wang, J. High pressure refractive index measurements of 4:1 methanol:ethanol. *J. Appl. Phys.* **1992**, *72*, 2453–2461.

90. Ahrens, T.J.; Ruderman, M.H. Immersed-Foil Method for Measuring Shock Wave Profiles in Solids. *J. Appl. Phys.* **1966**, *37*, 4758–4765.

91. Petersen, C.F.; Rosenberg, J.T. Index of Refraction of Ethanol and Glycerol under Shock. *J. Appl. Phys.* **1969**, *40*, 3044–3046.

92. Vedam, K.; Limsuwan, P. Piezo- and elasto-optic properties of liquids under high pressure. II. Refractive index vs density. *J. Chem. Phys.* **1978**, *69*, 4772–4778.

93. Chen, C.C.; Vedam, K. Piezo- and elasto-optic properties of liquids under high pressure. III. Results on twelve more liquids. *J. Chem. Phys.* **1980**, *73*, 4577–4584.

94. Polian, A.; Grimsditch, M. Brillouin scattering from H_2O: Liquid, ice VI, and ice VII. *Phys. Rev. B* **1983**, *27*, 6409–6412.

95. Dewaele, A.; Eggert, J.H.; Loubeyre, P.; Le Toullec, R. Measurement of refractive index and equation of state in dense He, H_2, H_2O, and Ne under high pressure in a diamond anvil cell. *Phys. Rev. B* **2003**, *67*, 094112.

96. Zha, C.; Hemley, R.J.; Gramsch, S.A.; Mao, H.; Basset, W.A. Optical study of H_2O ice to 120 GPa: Dielectric function, molecular polarizability, and equation of state. *J. Chem. Phys.* **2007**, *126*, 074506.

97. Roszak, K.; Katrusiak, A. Giant Anomalous Strain between High-Pressure Phases and the Mesomers of Urea. *J. Phys. Chem. C* **2017**, *121*, 778–784.

98. Nayak, A.P.; Bhattacharyya, S.; Zhu, J.; Liu, J.; Wu, X.; Pandey, T.; Jin, C.; Singh, A.K.; Akinwande, D.; Lin, J.F. Pressure-induced semiconducting to metallic transition in multilayered molybdenum disulphide. *Nat. Commun.* **2014**, *5*, 3731.

99. Alahmadi, M.; Siaj, M. Graphene-Assisted Magnetic Iron Carbide Nanoparticles Growth. *ACS Appl. Nano Mater.* **2018**, *1*, 7000–7005.

MDPI

St. Alban-Anlage 66

4052 Basel

Switzerland

Tel. +41 61 683 77 34

Fax +41 61 302 89 18

www.mdpi.com

Photonics Editorial Office

E-mail: photonics@mdpi.com

www.mdpi.com/journal/photonics